Advanced Modeling and Research in Hybrid Microgrid Control and Optimization

Advanced Modeling and Research in Hybrid Microgrid Control and Optimization

Editor

Nicu Bizon

MDPI • Basel • Beijing • Wuhan • Barcelona • Belgrade • Manchester • Tokyo • Cluj • Tianjin

Editor
Nicu Bizon
Faculty of Electronics,
Communication and Computers
University of Pitesti
Pitesti
Romania

Editorial Office
MDPI
St. Alban-Anlage 66
4052 Basel, Switzerland

This is a reprint of articles from the Special Issue published online in the open access journal *Mathematics* (ISSN 2227-7390) (available at: www.mdpi.com/journal/mathematics/special_issues/Hybrid_Microgrid_Control_Optimization).

For citation purposes, cite each article independently as indicated on the article page online and as indicated below:

LastName, A.A.; LastName, B.B.; LastName, C.C. Article Title. *Journal Name* **Year**, *Volume Number*, Page Range.

ISBN 978-3-0365-1886-2 (Hbk)
ISBN 978-3-0365-1885-5 (PDF)

© 2021 by the authors. Articles in this book are Open Access and distributed under the Creative Commons Attribution (CC BY) license, which allows users to download, copy and build upon published articles, as long as the author and publisher are properly credited, which ensures maximum dissemination and a wider impact of our publications.

The book as a whole is distributed by MDPI under the terms and conditions of the Creative Commons license CC BY-NC-ND.

Contents

About the Editor vii

Nicu Bizon and Phatiphat Thounthong
A Simple and Safe Strategy for Improving the Fuel Economy of a Fuel Cell Vehicle
Reprinted from: *Mathematics* **2021**, *9*, 604, doi:10.3390/math9060604 1

Krishan Arora, Ashok Kumar, Vikram Kumar Kamboj, Deepak Prashar, Bhanu Shrestha and Gyanendra Prasad Joshi
Impact of Renewable Energy Sources into Multi Area Multi-Source Load Frequency Control of Interrelated Power System
Reprinted from: *Mathematics* **2021**, *9*, 186, doi:10.3390/math9020186 31

José Antonio Cortajarena, Oscar Barambones, Patxi Alkorta and Jon Cortajarena
Grid Frequency and Amplitude Control Using DFIG Wind Turbines in a Smart Grid
Reprinted from: *Mathematics* **2021**, *9*, 143, doi:10.3390/math9020143 51

Milad Bahrami, Jean-Philippe Martin, Gaël Maranzana, Serge Pierfederici, Mathieu Weber, Farid Meibody-Tabar and Majid Zandi
Multi-Stack Lifetime Improvement through Adapted Power Electronic Architecture in a Fuel Cell Hybrid System
Reprinted from: *Mathematics* **2020**, *8*, 739, doi:10.3390/math8050739 69

Hossein Khoun Jahan, Naser Vosoughi Kurdkandi, Mehdi Abapour, Kazem Zare, Seyed Hossein Hosseini, Yongheng Yang and Frede Blaabjerg
Common-Ground-Type Single-Source High Step-Up Cascaded Multilevel Inverter for Transformerless PV Applications
Reprinted from: *Mathematics* **2020**, *8*, 1716, doi:10.3390/math8101716 97

Deepak Kumar Gupta, Ankit Kumar Soni, Amitkumar V. Jha, Sunil Kumar Mishra, Bhargav Appasani, Avireni Srinivasulu, Nicu Bizon and Phatiphat Thounthong
Hybrid Gravitational–Firefly Algorithm-Based Load Frequency Control for Hydrothermal Two-Area System
Reprinted from: *Mathematics* **2021**, *9*, 712, doi:10.3390/math9070712 115

Gerardo Humberto Valencia-Rivera, Luis Ramon Merchan-Villalba, Guillermo Tapia-Tinoco, Jose Merced Lozano-Garcia, Mario Alberto Ibarra-Manzano and Juan Gabriel Avina-Cervantes
Hybrid LQR-PI Control for Microgrids under Unbalanced Linear and Nonlinear Loads
Reprinted from: *Mathematics* **2020**, *8*, 1096, doi:10.3390/math8071096 131

Burin Yodwong, Phatiphat Thounthong, Damien Guilbert and Nicu Bizon
Differential Flatness-Based Cascade Energy/Current Control of Battery/Supercapacitor Hybrid Source for Modern e–Vehicle Applications
Reprinted from: *Mathematics* **2020**, *8*, 704, doi:10.3390/math8050704 157

Seung-Ju Lee and Yourim Yoon
Electricity Cost Optimization in Energy Storage Systems by Combining a Genetic Algorithm with Dynamic Programming
Reprinted from: *Mathematics* **2020**, *8*, 1526, doi:10.3390/math8091526 175

Laurentiu-Mihai Ionescu, Nicu Bizon, Alin-Gheorghita Mazare and Nadia Belu
Reducing the Cost of Electricity by Optimizing Real-Time Consumer Planning Using a New Genetic Algorithm-Based Strategy
Reprinted from: *Mathematics* **2020**, *8*, 1144, doi:10.3390/math8071144 . . . . **195**

Song-Kyoo (Amang) Kim
Enhanced IoV Security Network by Using Blockchain Governance Game
Reprinted from: *Mathematics* **2021**, *9*, 109, doi:10.3390/math9020109 . . . . **221**

Maria-Simona Răboacă, Irina Băncescu, Vasile Preda and Nicu Bizon
An Optimization Model for the Temporary Locations of Mobile Charging Stations
Reprinted from: *Mathematics* **2020**, *8*, 453, doi:10.3390/math8030453 . . . . **235**

About the Editor

Nicu Bizon

Nicu Bizon (senior member, IEEE) was born in Albesti de Muscel, Arges county, Romania, in 1961. He received his 5-year B.S. degree in electronic engineering from the University "Polytechnic" of Bucharest, Romania, in 1986 and his Ph.D. degree in automatic systems and control from the same university in 1996. From 1996 to 1989, he was involved in hardware design with Dacia Renault SA, Romania. Since 2000, he has served as a professor with the University of Pitesti, Romania, and has received two awards from the Romanian Academy, in 2013 and 2016. He is the editor and author of 5 books published in Springer and the author of 206 scientific papers (including 7 and 84 papers in IEEE transactions and conferences, respectively) published in Scopus, which have been cited 1678 times, corresponding to an h-index = 29. His current research interests include power electronic converters, fuel cell and electric vehicles, renewable energy, energy storage system, microgrids, and control and optimization.

Article

A Simple and Safe Strategy for Improving the Fuel Economy of a Fuel Cell Vehicle

Nicu Bizon [1,2,3,*] and Phatiphat Thounthong [4,5]

[1] Faculty of Electronics, Communication and Computers, University of Pitesti, 1 Targul din Vale, 110040 Pitesti, Romania
[2] ICSI Energy Department, National Research and Development Institute for Cryogenic and Isotopic Technologies, 1 Uzinei, 240050 Ramnicu Valcea, Romania
[3] Doctoral School, Polytechnic University of Bucharest, 313 Splaiul Independentei, 060042 Bucharest, Romania
[4] Renewable Energy Research Centre (RERC), Department of Teacher Training in Electrical Engineering, Faculty of Technical Education, King Mongkut's University of Technology North Bangkok, 1518 Pracharat 1 Road, Wongsawang, Bangsue, Bangkok 10800, Thailand; phatiphat.t@fte.kmutnb.ac.th
[5] Group of Research in Electrical Engineering of Nancy (GREEN), University of Lorraine, 2 Avenue de la Forêt de Haye, 54518 Vandeuvre lès Nancy CEDEX, F-54000 Nancy, France
* Correspondence: nicu.bizon@upit.ro

Citation: Bizon, N.; Thounthong, P. A Simple and Safe Strategy for Improving the Fuel Economy of a Fuel Cell Vehicle. *Mathematics* **2021**, *9*, 604. https://doi.org/10.3390/math9060604

Academic Editor: Anatoliy Swishchuk

Received: 10 February 2021
Accepted: 5 March 2021
Published: 11 March 2021

Publisher's Note: MDPI stays neutral with regard to jurisdictional claims in published maps and institutional affiliations.

Copyright: © 2021 by the authors. Licensee MDPI, Basel, Switzerland. This article is an open access article distributed under the terms and conditions of the Creative Commons Attribution (CC BY) license (https://creativecommons.org/licenses/by/4.0/).

Abstract: A new real-time strategy is proposed in this article to optimize the hydrogen utilization of a fuel cell vehicle, by switching the control references of fueling regulators, based on load-following. The advantages of this strategy are discussed and compared, with advanced strategies that also use the aforementioned load-following mode regulator of fueling controllers, but in the entire loading range, respectively, with a benchmark strategy utilizing the static feed-forward control of fueling controllers. Additionally, the advantages of energy-storage function in a charge-sustained mode, such as a longer service life and reduced size due to the implementation of the proposed switching strategy, are presented for the dynamic profiles across the entire load range. The optimization function was designed to improve the fuel economy by adding to the total power of the fuel utilization efficiency (in a weighted way). The proposed optimization loop will seek the reference value to control the fueling regulator in real-time, which is not regulated by a load-following approach. The best switching threshold between the high and low loading scales were obtained using a sensitivity analysis carried out for both fixed and dynamic loads. The results obtained were promising—(1) the fuel economy was two-times higher than the advanced strategies mentioned above; and (2) the total fuel consumption was 13% lower than the static feed-forward strategy. This study opens new research directions for fuel cell vehicles, such as for obtaining the best fuel economy or estimating fuel consumption up to the first refueling station on the planned road.

Keywords: fuel economy; load-following; switching strategy; real-time optimization; fuel cell vehicle; fuel cell system

1. Introduction

In the coming decades, energy and environmental issues will become the most important challenges for researchers working in the sustainable development of energy sector [1,2]. The proton exchange membrane fuel cell (PEMFC) system is usually utilized as green secondary energy generator for hybrid power systems (HPS), based on renewable energy for balancing power flow stability on the DC grid, due to the variability of energy flows from the load and from renewable sources [3,4]. Thanks to its advantages as compared to other fuel cell (FC) technologies, such as very low pollutant emissions, high specific energy, low operating temperature, and a fast start, the PEMFC system is now the most widely used FC type in portable [5,6] and space applications [7,8].

The design of effective control approaches [8,9] and optimization methods [10,11] needs accurate PEMFC dynamic models [12,13]. It was demonstrated that high ripples on

FC power (especially in the low frequency band up to hundreds of Hz) or load pulses [14], produce mechanical stress on the proton exchange membrane (PEM), and as a consequence, cause a rapid degradation of the FC lifetime [15,16]. The main stresses were analyzed in [15], highlighting how the mechanical damage of the PEM membrane was effectively produced. The causes and consequences of gas starvation, together with potential mitigation methods, are presented in [16]. The issue of cold start under subfreezing temperatures is experimentally analyzed in [17]. The consequences of frequent start–stop operations are addressed in [18]. Additionally, some recommendations to boost the FC durability and lifetime are given in [16,17]. Thus, fault-tolerant strategies and advanced control methods are proposed to operate the PEMFC power source in safe conditions [19,20].

This paper proposes a load-following (LFW) switching management of fueling flow rates that might optimally operate the PEMFC system. The safety measures for smooth switching of the fueling controllers were considered as well.

The load-following management was studied to handle the load dynamics [21,22] and then extended to an unknown load profile [23]. Impacts of load profiles on the PEMFC system efficiency, but also on safe functioning of the PEMFC power source, are analyzed in [24,25].

The energy supplied by the FC power source can be controlled by using the air regulator [26–36] or the fuel regulator [37–40], or by switching the fueling regulators, utilizing the strategy proposed in this study.

The air regulator is set to ensure the needed oxygen parameters for the FC stack cathode (the flow and pressure depending on the system load demand) [26]. The air compressor delivers the needed air (with about 21% oxygen) for the electrochemical reaction, ensuring an oxygen excess ratio higher than 1, to avoid the oxygen starvation phenomenon (which might appear during the load pulses) [6,27,28]. The oxygen excess ratio is controlled using different techniques, as follows—(1) feed-forward control [26,29]; (2) PID-based control [30] and its variants (such as optimal PID plus fuzzy controller [31]), feed-forward PID controller [32], and robust PI control [33]; (3) control techniques based on artificial intelligence concepts like fuzzy logic [34,35], neural networks [36], and genetic operators [37]; (4) model predictive control (MPC) based on constrained model [38], linearization method of the model [39,40], and multivariable nonlinear MPC [41,42]; (5) sliding control using the adaptive sliding mode (Lyapunov-based) [43], high-order sliding mode [44,45], sliding mode (nonlinear multivariable) [46], cascade adaptive sliding mode [47], or a combination of sliding mode with flatness control [48] and super-twisting algorithms [49,50]; (6) robust control based on reduced order model [51] or a load governor [52]; and (7) model reference adaptive control (MRAC) [53].

The MRAC technique [53] can prevent compressor surge better than the MPC techniques mentioned above [38–42]. The robustness analysis performed in [33,51,52] for robust control techniques highlights the advantages of using robust control as compared to classic control techniques, such as feed-forward control [26,29] or PID-based control [30–32] techniques. However, the feed-forward control [26] is simple, involves low computation, as compared to control techniques based on artificial intelligence concepts [34–37], and is already implemented in commercial solutions. Consequently, the feed-forward regulator is utilized as a reference in this study.

Nonlinear control techniques based on different sliding modes that combine the nonlinear terms of the super-twisting approach have the advantage of high robustness and good response to disturbances (achieving a short convergence time) [49,50]. For example, the nonlinear multivariable sliding mode control proposed in [46] reduces the dynamics of oxygen excess percentage and also the power fed by the air compressor.

It is known that the power fed by the air compressor is equal to 15% of the FC power. Therefore, new air compressors must be developed, with lower energy consumption and better dynamics [54]. The dynamics of the air compressor makes the air-feed subsystem slower, as compared to the hydrogen subsystem [55,56]. Thus, the authors of this paper focused on oxygen excess percentage control techniques to improve the response of the air-feed subnetwork. This study uses the nonlinear dynamic model proposed in [57] for the air compressor.

The fuel-feed and air-feed subsystems give great importance to safety, when generating the FC power requested by the dynamic load [9,29,58,59], in order to avoid fuel starvation [60]. The following techniques are proposed to control the fuel-feed subsystem—multi-input-multi-output (MIMO) nonlinear control [60], linear quadratic Gaussian (LQG) control [61], nonlinear control [62,63], linear and nonlinear control [64,65], nonlinear MPC [66,67], and sliding mode control [67,68]. An important task is to maintain the stoichiometric ratio for gases under variable load demand profiles [69], due to nitrogen addition [70] and fuel recirculation, using both ejector and blower [65,71]. Thus, in order to optimize the anodic purge techniques, different control methods were proposed to find the optimal purge interval [72], based on a state observer [73], adaptive strategies [74,75], bleeding strategies [75,76], and intelligent control [77]. The anode bleeding strategies manage the nitrogen diffusion through a continuous and controlled leak of gases, by combining the discontinuous classic purge mode with nonlinear control of the fuel-feed subsystem [78,79].

Degradation analysis of the PEMFC system in dead-end operating modes were analyzed in [80,81] and [82] as an effect of variable thermal and pressure conditions, and low-quality hydrogen, respectively. The methods to increase the lifetime of the FC system and avoid potential carbon monoxide poisoning are also presented. To progress the lifetime of the FC generator [29,56,59], the research was focused on thermal [83,84] and water [85,86] subsystems, but also on other FC subsystems, because these are all interconnected and interdependent [87]. Therefore, a systemic methodology based on an integrated approach is recommended to optimize the FC system [88,89]. Thus, new optimization strategies were studied for the FC system [90], FC cars [28,91], and FC hybrid power sources [92], based on nonlinear control [93], state diagram [94,95], MPC [96,97], droop control [98], fuzzy control [99,100], data fusion approach [101], global optimization using the extremum seeking (GES) approach [102], dynamic programming method [103], or Pontryagin's minimum algorithm [104]. Except for the global optimization strategies, the other strategies mentioned above showed a less than 100% strike ratio in searching for the optimal level of the considered optimization function or could not find it [102]. The optimization function might be designed for different objectives, such as fuel economy [105,106], lifetime extension for FC system [107], or battery stack [99,108], and safe operation of all subsystems [109].

Thus, the real-time strategy analyzed here (and mentioned below as the SW–LFW strategy) minimizes the hydrogen consumption, using a new load-following switching control for the oxygen and hydrogen controllers of the FC generator, which in comparison with the real-time strategy proposed in [110] is simpler (because it uses only one optimization loop instead of two), safer (because the second optimization loop is not involved in setting the set-points of the fueling controllers, and thus some perturbation (pulses) in the desired points of the fueling regulators due to the switching control are avoided), and almost the same fuel economy is obtained. Therefore, the same load profiles used in [110] are also used here to evaluate the results obtained.

Another aim of this work was to compare the performance of the SW–LFW strategy with other strategies, such as the Air–LFW technique [111] and the Fuel–LFW approach [112], which use the load-following set for the hydrogen regulator or oxygen regulator, respectively, in the entire loading scale. The main benefit of the load-following regulator (employed for one of the fueling regulators [111,112] or for both regulators [110]) is the battery operation in a charge-sustained approach [113], which substantially improved the battery's lifespan [114]. The FC power source generates the required power on the DC grid, according to the load demand. Thus, the hybrid ultracapacitor/battery energy storage system (ESS) only compensate minor differences between the load demand and the FC power. Therefore, the size of the battery might be reduced, as compared to rule-based techniques (where the battery is charged and discharged to sustain the power flow stability on the DC grid due to load dynamics and renewable energy variability, if the renewable energy is available to sustain a part of the load demand [115]).

Therefore, as the results obtained show, the advantages of the load-following control proposed in this paper are the following—(1) increases the battery lifetime; (2) reduces the battery size; and (3) removes need for monitoring the state-of-charge (SOC).

For a fair comparison of the techniques analyzed here, the same optimization function and test conditions are used, and a sensitivity analysis is functioned to obtain the best threshold between the high and low loading ranges (in a manner similar to [116], to obtain the best weighting parameters of the optimization function).

The optimization function is designed (using appropriate weighting parameters [116]) to minimize the hydrogen consumption by adding fuel efficiency to the FC net power (which is usually used as the optimization function of the FC energy efficiency [34–36]). The optimization loop solves the optimum value of the FC current, in real-time, which is used to regulate the fuel controller and the air regulator under the Air–LFW and Fuel–LFW approaches, respectively. In case of the SW–LFW strategy, the fueling regulator, which is not controlled on the basis of the load-following technique is used to improve the fuel economy, using this real-time optimization loop.

The results obtained in the case of a variable load underline a hydrogen economy for the proposed strategy, which is double in comparison to the basic strategies mentioned above (about 2.23-times and 2.47-times higher for the Air–LFW and Fuel–LFW techniques, respectively).

The hydrogen economy using the proposed strategy was compared to that of the benchmark strategy using the static feed-forward (sFF) control for the fueling regulators [59]. It is worth mentioning that the reduction of the total fuel consumption was more than 13% using the SW–LFW method, as compared to the sFF approach in the case of a variable outline of the system load demand.

Additionally, the better fuel saving method that uses the SW–LFW technique, as compared to basic strategies of the Air–LFW and Fuel–LFW, was obtained by choosing the best strategy from these two for a certain load level. It was proven that the optimum fuel saving was obtained using the Air–LFW for high-load levels and the Fuel–LFW strategy for low-load levels. Therefore, the best load threshold (between the high and low loading ranges) would be obtained on the basis of the sensitivity analysis functioned for both constant and variable loads.

Therefore, the key findings and novelty of this work are as follows—(1) a scheme for a new optimization approach (the SW–LFW strategy) for better fuel economy of the FC system; (2) the fuel efficiency was compared using the SW–LFW strategy, and the advanced Air–LFW and Fuel–LFW strategies; (3) the fuel economy obtained with the SW–LFW strategy and the sFF control-based strategy was also compared; (4) the advantages related to battery lifetime and size were advanced by switching the system from load-following control to fueling regulators; (5) the optimum switching threshold was advanced on the basis of a sensitivity investigation; (6) the optimization function was designed to achieve the best fuel economy; (7) almost the same fuel economy was advanced with the SW–LFW strategy and switching strategy proposed in [110], but the one proposed here was simpler and safer; and finally, a method for estimating fuel consumption up to the first refueling station on the planned road was proposed.

The innovative solutions proposed here to develop the fuel economy could help increase the system performance of FC vehicles.

2. Fuel Cell Hybrid Power System

The fuel cell hybrid power system (FCHPS) diagram using an equivalent DC load for the powertrain of fuel cell electric vehicle is illustrated in Figure 1. The 6 kW/45 V FC system can supply the load demand in a scale up to the power of 8 kW. The FC rated power of 6 kW was obtained for the nominal values of the fueling flow rates, Air Flow rate (*AirFr*) and Fuel Flow rate (*FuelFr*), of 300 and 50 L per minute [lpm], respectively. The FC power could be controlled by changing the fueling flow rates using the energy management and optimization unit (EMOU), via the fueling flow rate regulators (see

Figure 2). The EMOU would produce the control set-points $I_{ref(Air)}$ and $I_{ref(Fuel)}$ based on the load-following strategy that would be detailed in the following section, which compared to the strategy proposed in [110] was simpler (because we used only one optimization loop instead of two), safer (because the second optimization loop was not involved in setting the references of the fueling controllers), and almost the same fuel consumption was obtained on the same load cycle.

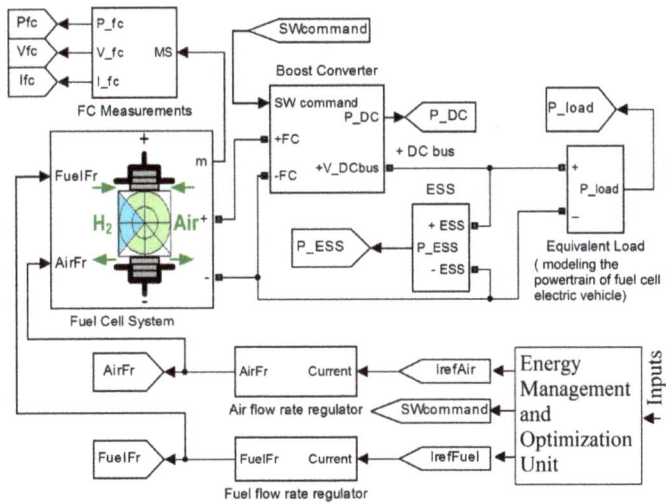

Figure 1. Fuel cell hybrid power system (FCHPS) diagram using an equivalent DC load for the powertrain of fuel cell electric vehicle.

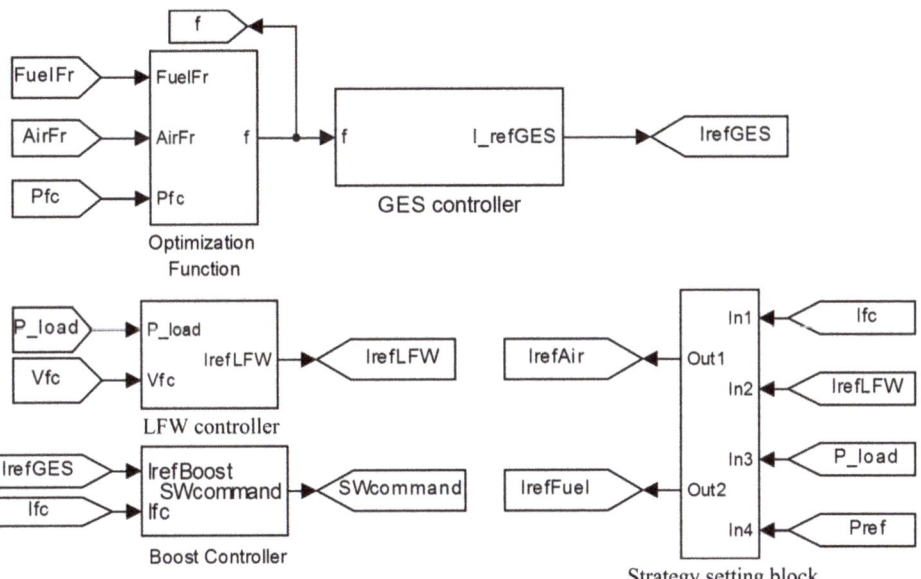

Figure 2. Energy management and optimization unit (EMOU) diagram.

The *AirFr* controller and *FuelFr* controller were regulated by the control set-points $I_{ref(Air)}$ and $I_{ref(Fuel)}$ based on (1) and (2) [59]:

$$AirFr = \frac{60000 \cdot R \cdot (273 + \theta) \cdot N_C \cdot I_{ref(Air)}}{4F \cdot \left(101325 \cdot P_{f(O2)}\right) \cdot \left(U_{f(O2)}/100\right) \cdot (y_{O2}/100)} \quad (1)$$

$$FuelFr = \frac{60000 \cdot R \cdot (273 + \theta) \cdot N_C \cdot I_{ref(Fuel)}}{2F \cdot \left(101325 \cdot P_{f(H2)}\right) \cdot \left(U_{f(H2)}/100\right) \cdot (x_{H2}/100)} \quad (2)$$

The FC parameters ($N_C, \theta, U_{f(H2)}, U_{f(O2)}, P_{f(H2)}, P_{f(O2)}, x_{H2}, y_{O2}$) were fixed to default values [117], the FC time constant was fixed to 0.2 s, the slope limiters of the fueling controllers were set at 100 A/s, R = 8.3145 J/(mol K), and F = 96485 As/mol.

Another method to regulate the FC power was via the DC/DC boost power circuit that links the FC generator with the DC bus. The EMOU generates the switching (SW) command, based on the real-time optimization technique, in order to advance the fuel economy; this is explained in the next section.

Therefore, in summary, the optimization loop controls the power generated by the FC system to be as much as necessary for the equivalent load on the DC bus, and the LFW control loop ensures a fuel and oxygen flow corresponding to this power. Thus, the battery from the ESS operates in a charge-sustained mode, as explained below and in Section 3, where the optimization loop and the LFW control setting are detailed in Figures 2 and 3.

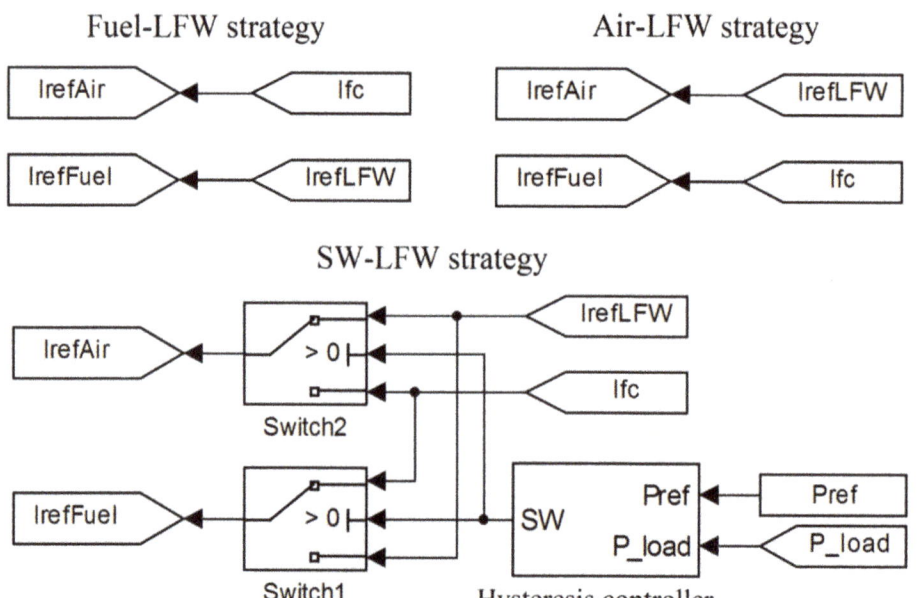

Figure 3. The load-following control setting.

The output power of the DC/DC boost circuit depends on the circuit efficiency coefficient (η_{boost}) and the FC generated power (P_{FCgen}):

$$P_{DC} = \eta_{boost} P_{FCgen} \quad (3)$$

The FC generated power is the FC net power (P_{FCnet}):

$$P_{FCgen} = P_{FCnet} \cong P_{FC} - P_{cm} \quad (4)$$

where the power losses are considered to be largely due to the air compressor (P_{cm}) [57]:

$$P_{cm} = I_{cm} \cdot V_{cm} = \left(a_2 \cdot AirFr^2 + a_1 \cdot AirFr + a_0\right) \cdot (b_1 \cdot I_{FC} + b_0) \quad (5)$$

The air compressor model uses the following coefficients [57]. $a_0 = 0.6$, $a_1 = 0.04$, $a_2 = -0.00003231$, $b_0 = 0.9987$, and $b_1 = 46.02$, and the dynamics part is modeled by a second order system with 0.7 damping ratio and 100 Hz natural frequency [57,110–112].

The power flow balance (6):

$$C_{DC} u_{dc} du_{dc}/dt = p_{DC} + p_{ESS} - p_{load} \quad (6)$$

could be ensured under a variable load (p_{load}) by controlling (the FC power generated via the DC/DC boost circuit) or p_{ESS} (the ESS power transferred with the DC distribution network).

If the aim is to minimize the ESS size, then the first control mode is recommended. In this case, the FC generator supplies the DC bus with the smooth part ($P_{load(MV)}$) of the variable load during a load cycle (LC), therefore:

$$P_{FCgen} = V_{FC} \cdot I_{FC} \cong P_{load(MV)}/\eta_{boost} \quad (7)$$

$$P_{ESS(MV)} \cong 0 \quad (8)$$

Besides the filtering technique based on the mean-value (MV), other low-pass filtering circuits might also be obtained [118].

The load-following method is realized based on (7) and the switching technique of the load-following control method for the fueling flow rates are detailed in the following section.

Considering (8), the power transferred by the hybrid battery/ultracapacitor ESS with the DC grid is close to zero, during a load cycle. The DC bus voltage control implemented on the ESS side stabilizes the DC bus voltage level to a reference of 200 V ($u_{DC} \cong V_{DC(ref)}$ = 200 V). The 100 µF capacitor (C_{DC}) filters the DC voltage ripple. The power and energy transients on the DC grid is balanced by the battery and ultracapacitors stack, using appropriate control [7,14]. For example, a power pulse can appear during a sharp load variation (such as a stair profile), because the power supplied by the FC system follows the smooth part of the system load cycle profile, but with a delay due to the 100 A/s slope limiters, 0.2 s FC time constant, 0.1 s time constant of the air compressor, and the response time of the optimization loop (which is discussed in the next section). Therefore, the lack of power (during a step-up in load) or excess of power (during a step-down in load) is stabilized by the 100 F ultracapacitor module, via a two-quadrant bi-directional DC/DC power converter, regulated by the DC bus voltage controller. The energy storage device (100 Ah/200 V battery with 10 s constant time) is directly connected to the DC bus and is used to compensate minor energy differences that can appear on the DC bus due to the optimization loop (which would change the FC operating point set by the load-following control toward the optimum point found in the searching range of the optimization function).

3. Energy Management and Optimization Unit

The optimization function (9) is proposed to improve the fuel economy:

$$f(x, AirFr, FuelFr, P_{Load}) = 0.5 \cdot P_{FCnet} + k_{fuel} \cdot Fuel_{eff} \quad (9)$$

where x is the state vector and $Fuel_{eff} \cong P_{FCnet}/FuelFr$ is the fuel consumption efficiency.

The weighting parameter k_{fuel} [lpm/W] is considered in the sensitivity analysis to progress the fuel economy. The optimization function (9) is implemented in the EMOU diagram presented in Figure 2.

The regulation of the DC/DC boost circuit is of hysteretic technique (with 0.1 A hysteresis band), therefore, the FC current would track the set-point supplied by the global extremum seeking (GES) technique:

$$I_{FC} \cong I_{ref(boost)} = I_{ref(GES)} \qquad (10)$$

The GES technique (see the top of Figure 2) finds the optimum based on relationships (11)–(19):

$$y = f(v_1, v_2), \; y_N = k_{Ny} \cdot y \qquad (11)$$

$$\dot{y}_f = -\omega_h \cdot y_f + \omega_h \cdot y_N, \; y_{HPF} = y_N - y_f, \; \dot{y}_{BPF} = -\omega_l \cdot y_{BPF} + \omega_l \cdot y_{HPF} \qquad (12)$$

$$y_{DM} = y_{BPF} \cdot s_d, \; s_d = \sin(\omega t) \qquad (13)$$

$$\dot{y}_{Int} = y_{DM} \qquad (14)$$

$$G_d = |y_{MV}|, \; y_{MV} = \frac{1}{T_d} \cdot \int y_{BPF} dt \qquad (15)$$

$$y_M = G_d \qquad (16)$$

$$p_1 = k_1 \cdot y_{Int}, \; k_1 = \gamma_{sd} \cdot \omega \qquad (17)$$

$$p_2 = k_2 \cdot y_M \cdot s_d \qquad (18)$$

$$I_{ref(GES)} = k_{Np} \cdot (p_1 + p_2) \qquad (19)$$

where f_d is the dither's frequency. The cut-off frequencies of the high-pass filter (HPF) and low-pass filter (LPF) are tuned by the parameters b_h = 0.1 and b_l = 1.5 [119,120]. The setting control parameters and the normalization gains are designed using [11]: k_1 = 1, k_2 = 2, k_{Np} = 20, and k_{Ny} = 1/1000.

The searching time is less than 0.1 s (which means less than 10 periods of 100 Hz dither used in this study) [102], therefore, this GES technique would find in real-time the optimum of the optimization function. As mentioned before, the response time of the FC generator due to an interruption in load was higher than a 0.1 s search time.

The response time of the optimization loop is set by the dynamic model (20):

$$\dot{x} = g(x, AirFr, FuelFr, P_{Load}), x \in X \qquad (20)$$

where g is a smooth function [48].

For example, 100 A/s slope limiters of the fueling controllers and dynamic model of the air compressor would limit the speed variation of the *AirFr* and *FuelFr* values, due to a disturbance in load. Considering (7), the FC current is set by the load demand to (21):

$$I_{FC} \cong P_{load(MV)} / (\eta_{boost} \cdot V_{FC}) \qquad (21)$$

Therefore, the reference $I_{ref(LFW)}$ is set by (22):

$$I_{ref(LFW)} \cong P_{load(MV)} / (\eta_{boost} \cdot V_{FC}) \qquad (22)$$

Considering Equations (10) and (21), the reference $I_{ref(LFW)}$ is equal to the FC current (I_{FC}), but during the optimization cycle, these signals are a bit different between them, and are obviously different as compared to reference $I_{ref(GES)}$. Therefore,

$$I_{FC} \cong I_{ref(LFW)} \neq I_{ref(GES)} \qquad (23)$$

The reference $I_{ref(GES)}$ is used in the optimization loop for all three strategies—the two reference strategies based on the load-following of the *AirFr* and *FuelFr* (called the Fuel–LFW strategy and the Air–LFW strategy) and the switching algorithm studied in this work (named "SW–LFW strategy"), which switches the load-following management mode for the *AirFr* or for the *FuelFr*, if $P_{load} > P_{ref}$ or $P_{load} < P_{ref}$, respectively.

The reference $I_{ref(LFW)}$ is utilized in the load-following loop for all three strategies and is selected as mentioned before, by the strategy setting block displayed in Figure 2 and explained in Figure 3.

Thus, the Air–LFW strategy uses the settings $I_{ref(Fuel)} = I_{FC}$, $I_{ref(Air)} = I_{ref(LFW)}$, and $I_{ref(boost)} = I_{ref(GES)}$, whereas the Fuel–LFW strategy uses $I_{ref(Air)} = I_{FC}$, $I_{ref(Fuel)} = I_{ref(LFW)}$, and $I_{ref(boost)} = I_{ref(GES)}$.

The SW–LFW strategy uses the settings (24)–(26):

$$I_{ref(Fuel)} = \begin{cases} I_{ref(LFW)}, & if\, P_{load} \leq P_{ref} \\ I_{FC}, & if\, P_{load} > P_{ref} \end{cases} \quad (24)$$

$$I_{ref(Air)} = \begin{cases} I_{FC}, & if\, P_{load} \leq P_{ref} \\ I_{ref(LFW)}, & if\, P_{load} > P_{ref} \end{cases} \quad (25)$$

$$I_{ref(boost)} = I_{ref(GES)} \quad (26)$$

The settings for the fueling controllers involves only the desired-point $I_{ref(LFW)}$, instead of the settings used in switching strategy proposed in [110], which, in addition to this reference, uses the reference $I_{ref(GES2)}$ generated by the second optimization loop. The reference $I_{FC} + I_{ref(GES2)}$ is obviously different to the FC current (I_{FC}) and reference $I_{ref(LFW)}$ given by Equation (22), therefore, the smooth and safe operation of the fueling regulators might be perturbed using the switching strategy proposed in [110].

The threshold P_{ref} is tuned after a sensitivity analysis of the whole fuel consumption ($Fuel_T = \int FuelFr(t)dt$) for a constant load, and then this threshold is validated for the variable load cycles. The improvements in sum fuel consumption using the SW–LFW technique is compared to those given by utilizing the strategies Fuel–LFW and Air–LFW, considering the Static Feed-Forward (sFF) technique as presented in [59].

The sFF technique uses the settings $I_{ref(Fuel)} = I_{FC}$, $I_{ref(Air)} = I_{FC}$, and $I_{ref(boost)} = I_{ref(LFW)}$ [59]. The sFF technique was chosen as a reference because it is the most known technique and is usually used as a reference for new strategies. Furthermore, the sFF strategy is successfully implemented in FC systems for research or commercial use.

The fuel economy is given by (27)–(29):

$$\Delta Fuel_{T(SW)} = Fuel_{T(SW)} - Fuel_{T(sFF)} \quad (27)$$

$$\Delta Fuel_{T(Air)} = Fuel_{T(Air)} - Fuel_{T(sFF)} \quad (28)$$

$$\Delta Fuel_{T(Fuel)} = Fuel_{T(Fuel)} - Fuel_{T(sFF)} \quad (29)$$

4. Performance Validation

4.1. Constant Load Cycle

The total fuel consumption was evaluated for the sFF strategy and the SW–LFW strategy (with $k_{fuel} = 0$ and $P_{ref} = 5$ kW) using the FCHPS with the appropriate settings mentioned above. The results for different loading levels (mentioned in the 1st column of Table 1) are shown in the 2nd and 3rd column of Table 1.

Table 1. Fuel economy for $k_{fuel} = 0$.

P_{load}	$Fuel_{T(sFF)}$	$Fuel_{T(SW)}$	$\Delta Fuel_{T(SW)}$	$\Delta Fuel_{T(Air)}$	$\Delta Fuel_{T(Fuel)}$
[kW]	[L]	[L]	[L]	[L]	[L]
2	34.02	33.56	−0.46	11.26	−0.46
3	56.3	55.08	−1.22	4.14	−1.22
4	74.88	72.6	−2.28	2.08	−2.28
5	98.6	93	−5.6	−0.08	−5.6
6	125.58	123.3	−2.28	−2.28	−7.66
7	158.34	146.18	−12.16	−12.16	−13.56
8	176	147.52	−28.48	−28.48	−22.92

The fuel economy was computed using (27) and is shown in the 4th column of Table 1. The total fuel consumption for the Fuel–LFW strategy and Air–LFW strategy was evaluated using the FCHPS, with the appropriate settings mentioned above and $k_{fuel} = 0$. The fuel economy computed using (28) and (29) was recorded in the last two columns in Table 1.

In the same manner, the fuel economy was recorded, as compared to the sFF strategy in Tables 2 and 3, for $k_{fuel} = 25$ and $k_{fuel} = 50$, in case of the strategies SW–LFW (with $P_{ref} = 5$ kW), Air–LFW, and Fuel–LFW, respectively. This fuel economy is represented in Figure 4c, for $k_{fuel} = 25$ and $k_{fuel} = 50$.

Table 2. Fuel economy for $k_{fuel} = 25$.

P_{load}	$Fuel_{T(sFF)}$	$Fuel_{T(SW)}$	$\Delta Fuel_{T(SW)}$	$\Delta Fuel_{T(Air)}$	$\Delta Fuel_{T(Fuel)}$
[kW]	[L]	[L]	[L]	[L]	[L]
2	34.02	33.376	−0.644	12.14	−0.644
3	56.3	52.424	−3.876	5.548	−3.876
4	74.88	69.704	−5.176	1.2	−5.176
5	98.6	89.84	−8.76	−6.44	−8.76
6	125.58	111.44	−14.14	−14.14	−12.54
7	158.34	129.92	−28.42	−28.42	−24.26
8	176	144.92	−31.08	−31.08	−26

Table 3. Fuel economy for $k_{fuel} = 50$.

P_{load}	$Fuel_{T(sFF)}$	$Fuel_{T(SW)}$	$\Delta Fuel_{T(SW)}$	$\Delta Fuel_{T(Air)}$	$\Delta Fuel_{T(Fuel)}$
[kW]	[L]	[L]	[L]	[L]	[L]
2	34.02	33.92	−0.1	7.628	−0.1
3	56.3	52.6	−3.7	2.764	−3.7
4	74.88	69.616	−5.264	0.288	−5.264
5	98.6	89.84	−8.76	−5.8	−8.76
6	125.58	112.56	−13.02	−13.02	−13.98
7	158.34	133.52	−24.82	−24.82	−20.74
8	176	146.2	−29.8	−29.8	−25

The fuel economy for strategies SW–LFW, Fuel–LFW, and Air–LFW using $k_{fuel} = 0$ is represented in Figure 4a in comparison to the sFF technique.

The threshold $P_{ref} = 5$ kW was chosen by considering the best values of the fuel economy for the Fuel–LFW and Air–LFW strategies that were obtained in case $k_{fuel} = 25$ (see Table 2 and Figure 4b).

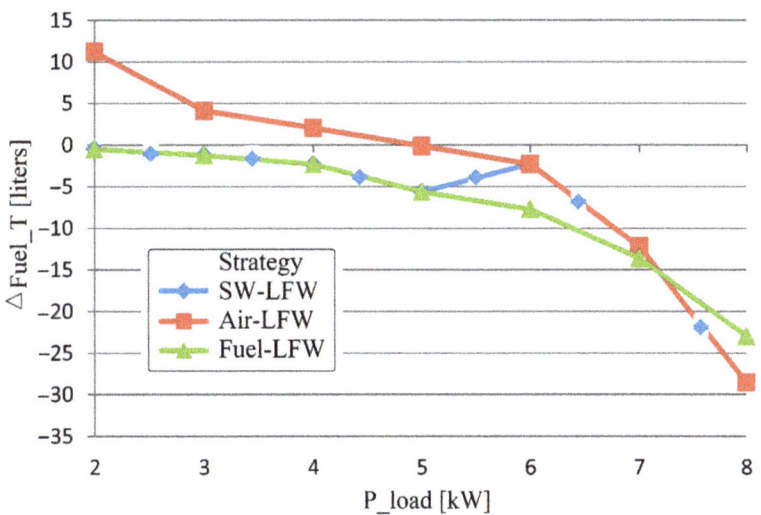

(**a**) $k_{fuel} = 0$

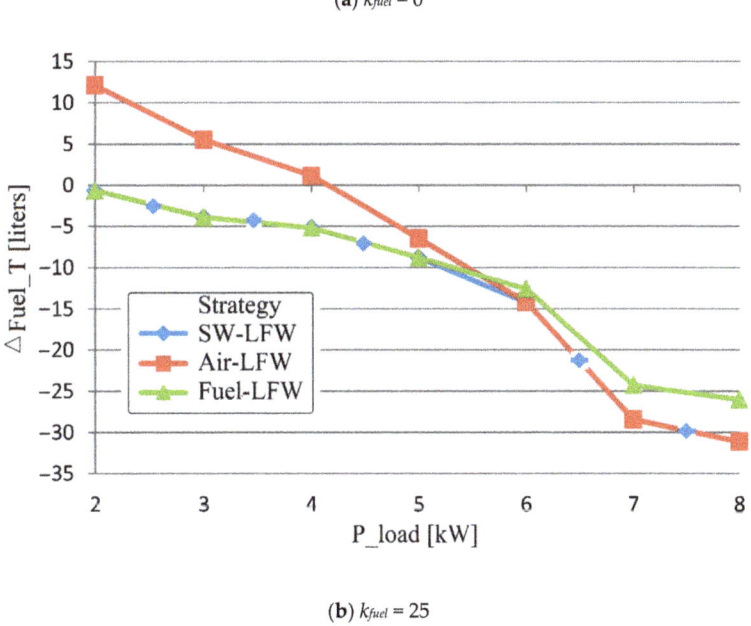

(**b**) $k_{fuel} = 25$

Figure 4. *Cont.*

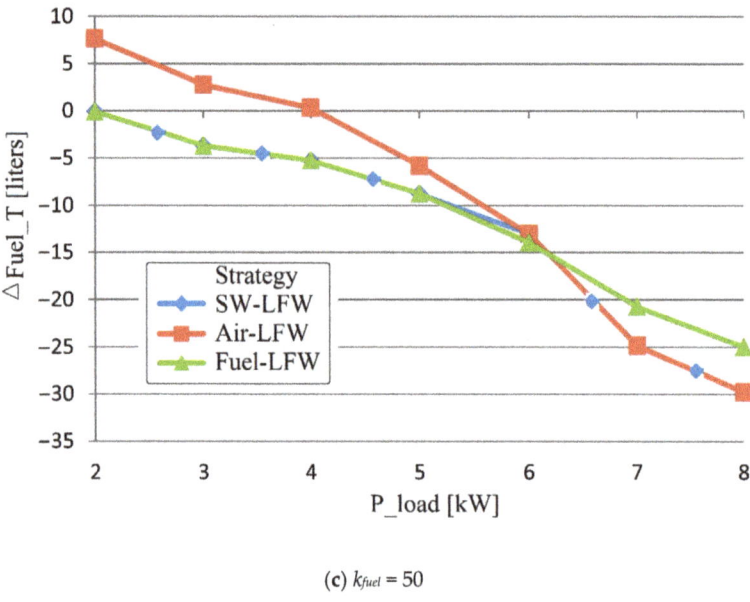

(c) k_{fuel} = 50

Figure 4. Fuel economy for constant load.

In fact, the threshold P_{ref} could be located between 5 kW and 6 kW (5 kW $\leq P_{ref} <$ 6 kW) for the 1 kW resolution used in the sensitivity analysis performed for constant load.

Figure 4c suggests that the threshold P_{ref} could be situated between 6 kW and 7 kW (6 kW $\leq P_{ref} <$ 7 kW), but the fuel economy in case k_{fuel} = 25 was smaller than that in case k_{fuel} = 50. The smallest fuel efficiency was gained in case k_{fuel} = 0, because the system optimization function is given only by the fuel cell total power. Consequently, the maximum of the optimization function was named Maximum Efficiency Point (MEP). If $k_{fuel} \neq 0$, then the optimization function was fuel economy oriented by adding the fuel utilization efficiency ($Fuel_{eff} \cong P_{FCnet}/FuelFr$) to the FC total power, by the appropriate weighting factors (to make both terms comparable in order of magnitude). The best fuel efficiency was gained for k_{fuel} = 25, therefore this value was used for the next simulations, unless otherwise mentioned.

4.2. Load Profile: Variable Load Cycle

4.2.1. The First Variable Load Cycle with Different Power $P_{load(AV)}$ Levels

The first variable load cycle with different $P_{load(AV)}$ values was used to test the Air–LFW and Fuel–LFW strategies [110,111]. Therefore, this 12 s load cycle was also utilized in this work to estimate the fuel economy point of the SW–LFW technique. The power levels were $0.75 \cdot P_{load(AV)}$, $1.25 \cdot P_{load(AV)}$, and $1.00 \cdot P_{load(AV)}$, during 4 s for each level. Therefore, the average value (AV) of this 12 s load cycle was power $P_{load(AV)}$. The values employed for $P_{load(AV)}$ were 2, 3, 4, 5, and 6 kW (see Table 4) in order to operate the FC system within the admissible limit for FC power (up to 8 kW). For example, the power levels for the load cycle with power $P_{load(AV)}$ = 6 kW were 4.5, 7.5, and 6 kW, the levels for the 4 kW load cycle were 3, 5, and 4 kW, and the levels for the 2 kW load cycle were 1.5, 2.5, and 2 kW.

Table 4. Fuel economy for the first load profile using different $P_{load(AV)}$ values.

$P_{load(AV)}$ [kW]	$Fuel_{T(sFF)}$ [L]	$Fuel_{T(SW)}$ [L]	$\Delta Fuel_{T(SW)}$ [L]	$\Delta Fuel_{T(Air)}$ [L]	$\Delta Fuel_{T(Fuel)}$ [L]
2	34.14	36.06	1.92	7.18	1.92
3	53.92	54.74	0.82	6.24	0.82
4	75.8	75.49	−0.31	3.32	−0.64
5	100.62	96.8	−3.82	−3.16	−4.16
6	130.2	116.92	−13.28	−13.28	−10.08

If the threshold P_{ref} is selected in the middle of the range of these levels, 4 kW ≤ P_{ref} < 4.5 kW, then the SW–LFW technique performs as the Fuel–LFW technique for the load cycles with power $P_{load(AV)}$ of 2 kW and $P_{load(AV)}$ of 3 kW, as the Air–LFW technique for the load drive cycle with power $P_{load(AV)}$ of 2 kW, and specifically for the load drive cycles with power $P_{load(AV)}$ of 4 kW and power $P_{load(AV)}$ of 5 kW, using the switching rules (24)–(26). This kind of operation for the SW–LFW strategy could be observed in Table 4, by analyzing the fuel economy recorded, and was easier observed in fuel economy in Figure 5.

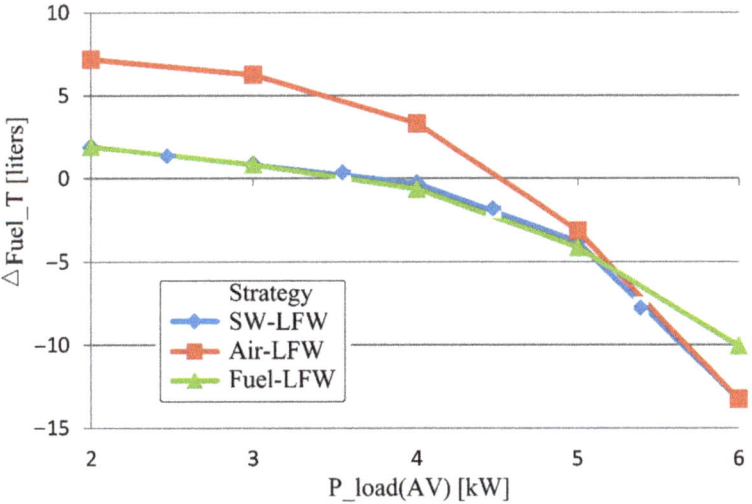

Figure 5. Fuel economy using different $P_{load(AV)}$ values for the first load profile.

4.2.2. The Second Variable Load Cycle with 3/7 kW Load Pulses

The second load profile used power values of 7 kW and 3 kW during 3 s for each level, resulting in a pulsed load profile (see the first curve in Figure 6). Figure 6 presents the characteristics of the FCHPS under the second load profile for strategies Air–LFW (Figure 6a), Fuel–LFW (Figure 6b), and SW–LFW (Figure 6c), with the plots structured as follows—the 1st plot illustrates the 3/7 kW pulsed load profile; the 2nd and 3rd curves portray the ESS power (P_{ESS}) and fuel cell net power (P_{FCnet}) and; the 4th and 5th curves display the fueling flow rates ($FuelFr$ and $AirFr$); the 6th plot in Figure 6c represents the airflow rate ($AirFr$) in case of the strategy proposed in [110]; the total fuel utilization ($Fuel_T$), the fuel consumption efficiency ($Fuel_{eff}$), and the fuel cell electrical efficiency ($\eta_{sys} = P_{FCnet}/P_{FC}$) are shown in the last three curves.

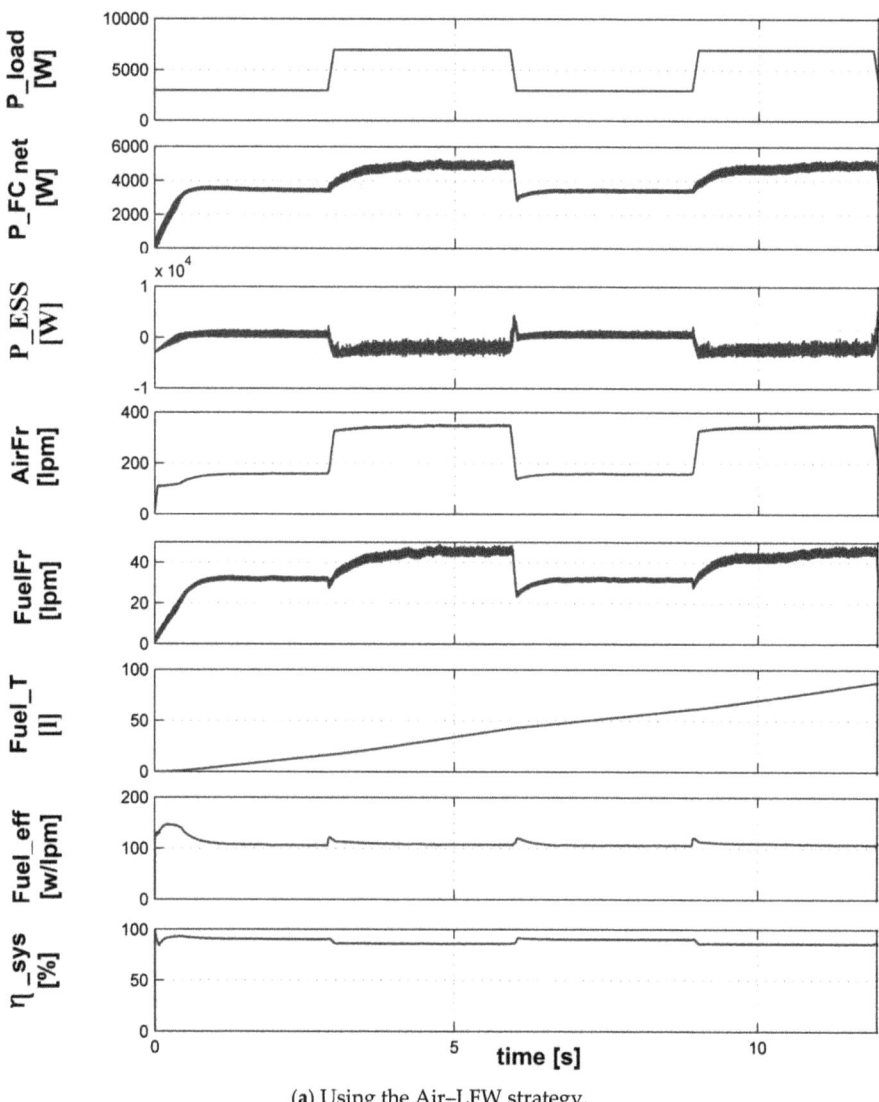

(**a**) Using the Air–LFW strategy.

Figure 6. *Cont.*

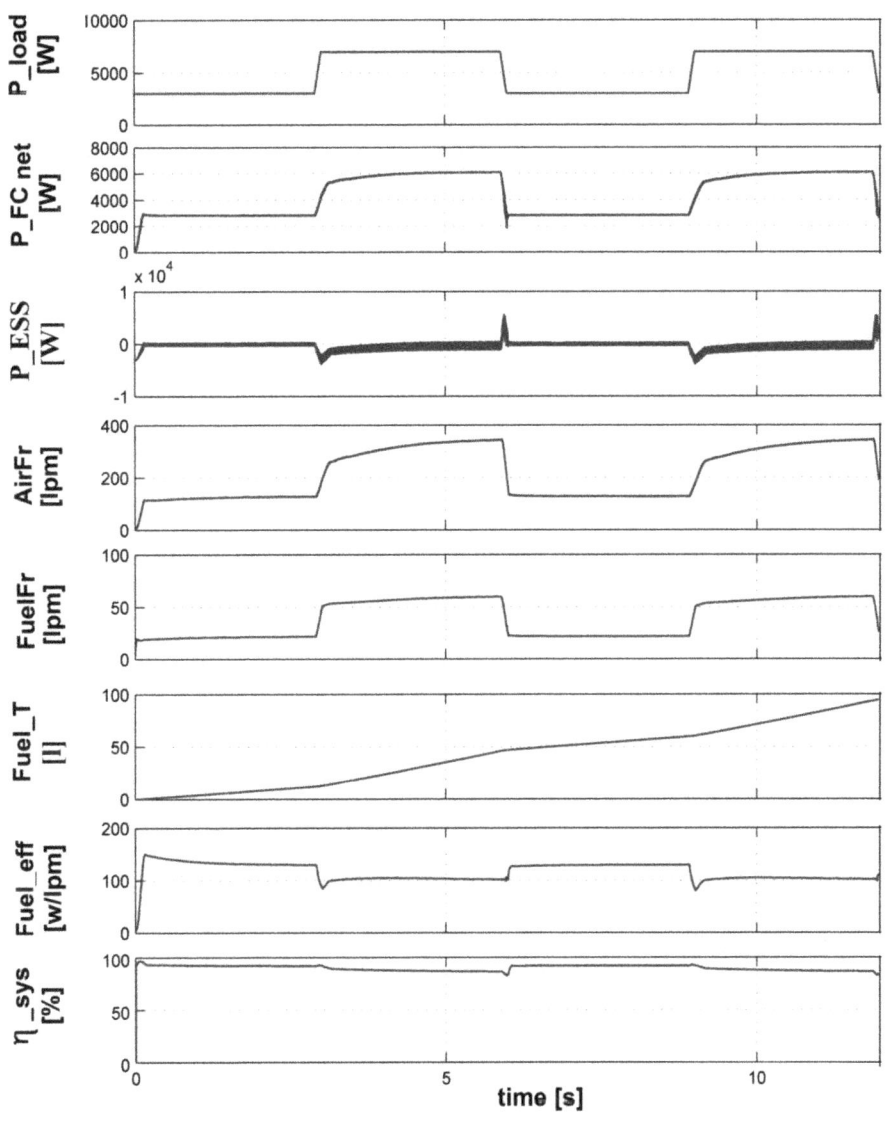

(**b**) Using the Fuel–LFW strategy.

Figure 6. *Cont.*

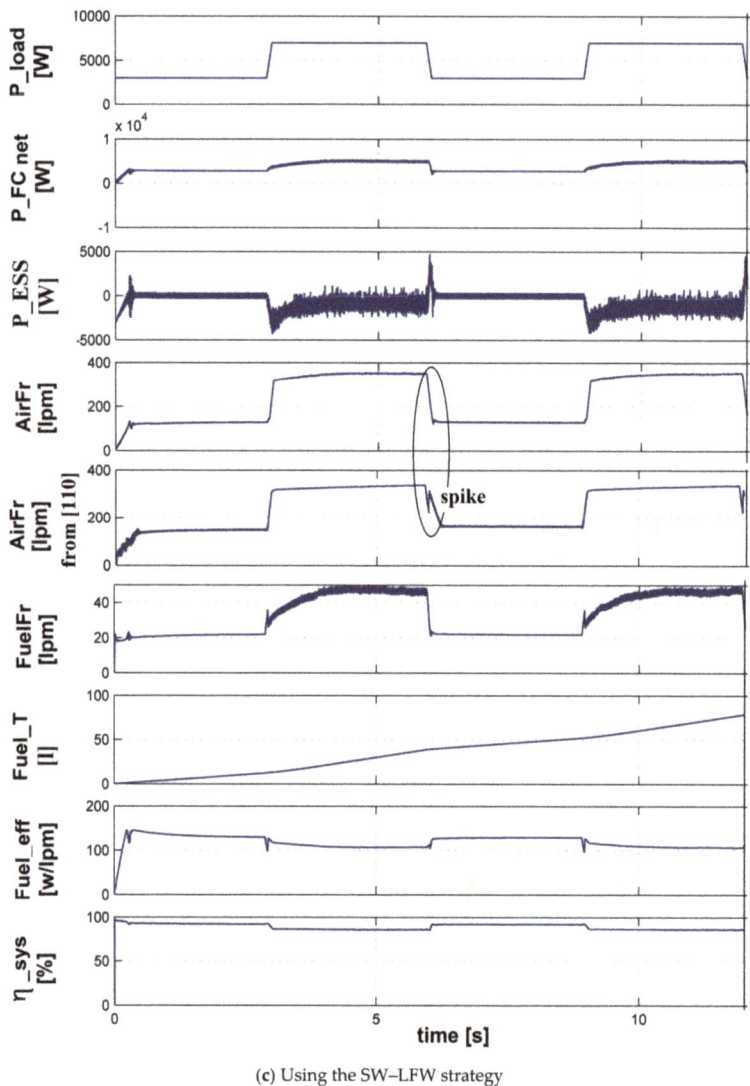

(c) Using the SW–LFW strategy

Figure 6. Behavior of the FCHPS under the second load profile (3/7 kW load pulses).

It is worth mentioning the observations. (1) The load-following management mode operated the *AirFr* regulator (refer to the 4th curve in Figure 6a) or the *FuelFr* regulator (see the 5th plot in Figure 6b) if the Air–LFW technique or Fuel–LFW technique was used. (2) The optimization loop set $I_{FC} \cong I_{ref(GES)}$ for the *FuelFr* regulator (see the search of the optimum of the *FuelFr* in the 5th plot in Figure 6a) or the *AirFr* regulator (see the search of the optimum of the *AirFr* in the 4th plot in Figure 6b) if the Air–LFW technique or the Fuel–LFW technique was used. (3) The FC system supplied the requested load demand based on the load-following regulation mode designed using Equation (6), $P_{FCgen} \cong P_{load(MV)}/\eta_{boost}$, therefore, the battery functions in the charge–sustained mode (refer to the 3rd curve in Figure 6). (4) The ultracapacitor bank transiently stabilizes the power equilibrium during the dynamic load (see the 3rd plot in Figure 6). (5) The fuel consumption efficiency ($Fuel_{eff}$) had values in the range of 95 to 140 W/lpm (refer to the 7th curve in Figure 6a,b, and 8th plot in Figure 6c). Therefore, the optimization terms $k_{fuel} \cdot Fuel_{eff}$ and $0.5 \cdot P_{FCnet}$ have comparable values for $k_{fuel} = 25$. (6) The FC electrical efficiency (η_{sys}) had values in the range of 85 to 94% (see the 8th plot in Figure 6a,b, and 9th plot in Figure 6c), therefore, a new optimization function $f = \eta_{sys} + k_{fuel} \cdot Fuel_{eff}$ might be defined for excellent fuel economy using k_{fuel} in the scale of 0.5 to 2 lpm/W. (7) The proposed strategy is safer than that analyzed in [110] (where the second optimization loop is not involved in setting the desired-points of the fueling regulators and thus some perturbation (spikes) might appear in the desired-points of the fueling regulators due to the switching control (see the spike in the 5th plot of Figure 6c, as compared to the 4th plot of Figure 6c, which appears when the set-points $I_{ref(LFW)}$ and $I_{FC} + I_{ref(GES2)}$ are switched [110], and the reference $I_{FC} + I_{ref(GES2)}$ is clearly different from the reference $I_{ref(LFW)}$).

It is important mentioning that when using the Air–LFW technique or the Fuel–LFW technique, such spikes do not appear in the fueling flow rated, because the fueling references ($I_{ref(Fuel)} = I_{FC}$ and $I_{ref(Air)} = I_{ref(LFW)}$ for the Air–LFW technique, and $I_{ref(Air)} = I_{FC}$ and $I_{ref(Fuel)} = I_{ref(LFW)}$ for the Fuel–LFW technique) are based on the references $I_{ref(LFW)}$ and I_{FC}. The proposed strategy switches these references, but $I_{FC} \cong I_{ref(LFW)}$ considering Equation (14).

The total fuel consumption is shown in the 6th curve of Figure 6a,b, and in the 7th plot of Figure 6c, and the fuel efficiency is shown in Table 5. Fuel economy for the SW–LFW technique, as compared to the sFF technique, represented around 25% of $Fuel_{T(sFF)}$. The fuel efficiency for the SW–LFW technique as compared to the Air–LFW and Fuel–LFW techniques was 1.49-times and 2.38-times higher, respectively.

Table 5. Fuel economy for the second load profile (3/7 kW load pulses).

$P_{load(pulse)}$	$Fuel_{T(sFF)}$	$Fuel_{T(SW)}$	$\Delta Fuel_{T(SW)}$	$\Delta Fuel_{T(Air)}$	$\Delta Fuel_{T(Fuel)}$
[kW]	[L]	[L]	[L]	[L]	[L]
3/7 kW	105.9	78.91	−26.99	−18.15	−11.32

In the case of pulsed load, the threshold P_{ref} could be chosen between the levels, 3 kW $< P_{ref} <$ 7 kW, but, considering the results for a constant load, it was obvious that the best value must be established for a variable load, based on a sensitivity analysis. For this, the third load profile (symmetrical stair up and down) was designed.

4.2.3. The Third Load Profile (Symmetrical Stair Up and Down)

The levels for the symmetrical stair were 3, 4, 5, 6, and 7 kW, with 2 s for each level in the stair up and other 2 s for each level in the stair down (see the first plot in Figure 8). The threshold P_{ref} was chosen between the levels, in order to analyze the fuel efficiency obtained in each case (see Table 6).

Table 6. Fuel economy for the third load profile (symmetrical stair up and down).

P_{ref}	$Fuel_{T(sFF)}$	$Fuel_{T(SW)}$	$\Delta Fuel_{T(SW)}$	$\Delta Fuel_{T(Air)}$	$\Delta Fuel_{T(Fuel)}$	$\Delta Fuel_{T(SW)}$ [110]
[kW]	[L]	[L]	[L]	[L]	[L]	[l]
2.5	286.5	268.6	−17.9	−17.4	−15.7	−19.4
3.5	286.5	258.8	−27.7	−17.4	−15.7	−28.7
4.5	286.5	250.4	−36.1	−17.4	−15.7	−35.2
5.5	286.5	247.6	−38.9	−17.4	−15.7	−39.3
6.5	286.5	252	−34.5	−17.4	−15.7	−33.7

The best fuel efficiency for the SW–LFW technique compared to the sFF technique was observed for $P_{ref} = 5.5$ kW, at about 13.6% of $Fuel_{T(sFF)}$ ($100 \times 38.9/286.5 \cong 13.58\%$). The fuel efficiency for the SW–LFW technique compared to the Air–LFW and Fuel–LFW techniques was 2.23- times and 2.47-times higher, respectively. Fuel economy using different values for the threshold P_{ref} is shown in Figure 7. The fuel efficiency utilizing the SW-FLW technique with power $P_{ref} = 2.5$ kW was equal to that gained utilizing the Air-FLW strategy, with small differences appearing due to the use of the Fuel-FLW strategy during the starting phase (until the load became higher than 2.5 kW).

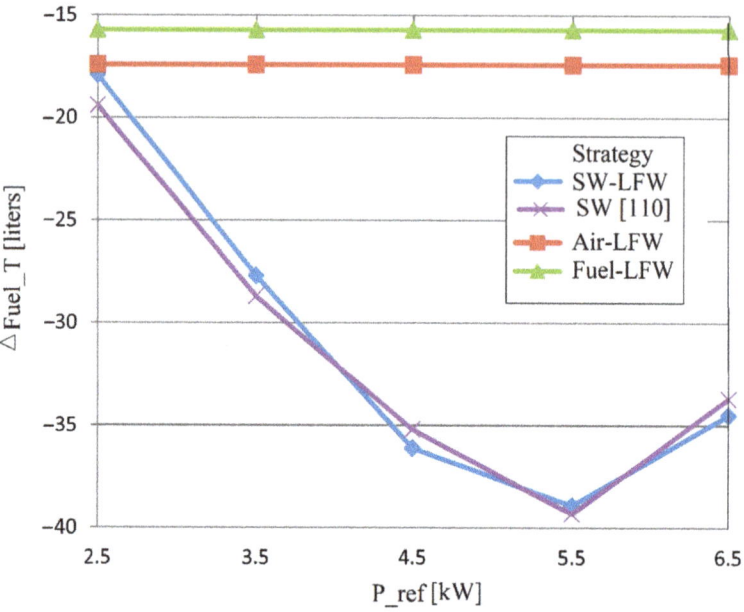

Figure 7. Fuel economy for the third load profile using different values of the threshold P_{ref}.

It is worth mentioning that only minor differences appeared in the fuel economy obtained for different levels of the threshold P_{ref} using the SW–LFW technique, as compared to the switching technique proposed in [110] (see the last column of Table 6 and Figure 7). Thus, at the same fuel economy that might be obtained during a load cycle, the proposed strategy had the advantages of being simpler and safer than the switching strategy proposed in [110].

Figure 8 presents the behavior of the FCHPS under the second load profile for the strategies Air–LFW (Figure 8a), Fuel–LFW (Figure 8b), and SW–LFW (Figure 8c), with the plots structured as in Figure 6. Beside the above-mentioned findings for transient load, these further observations were of significance. (1) The search for the optimum using the Air–LFW or the SW–LFW strategies for $P_{load} > P_{ref} = 4.5$ kW is performed via the $FuelFr$ regulator because $I_{FC} \cong I_{ref(GES)}$; because the optimization function was defined to minimize the fuel economy, the minimum values of the $FuelFr$ is tracked (refer to the 5th plot in Figure 8a,c) and the FC power is different from the requested load. (2) This power difference (given by the power transfer balance on the DC distributed network) is sustained by the storage device (here a battery), which is charged or discharged during the load cycle, but the SOC remains the same at the end (see the 3th plot in Figure 8a). (3) The search for the optimum using the Fuel–LFW or the SW–LFW strategies for $P_{load} < P_{ref} = 4.5$ kW operates via the $AirFr$ regulator, so the $FuelFr$ controller is controlled using the load-following technique (see the 5th plot in Figure 8b); thus, the FC power is almost close to the requested load. (4) the battery storage device functions in a charge-sustained mode (refer to the 3rd curve in Figure 8b). (5) In the case of a dynamic load in the full range, the fuel efficiency for the Air–LFW and Fuel–LFW techniques seems to be the same (see Table 6); the difference of 1.7 L means about 0.63% of the total fuel utilization.

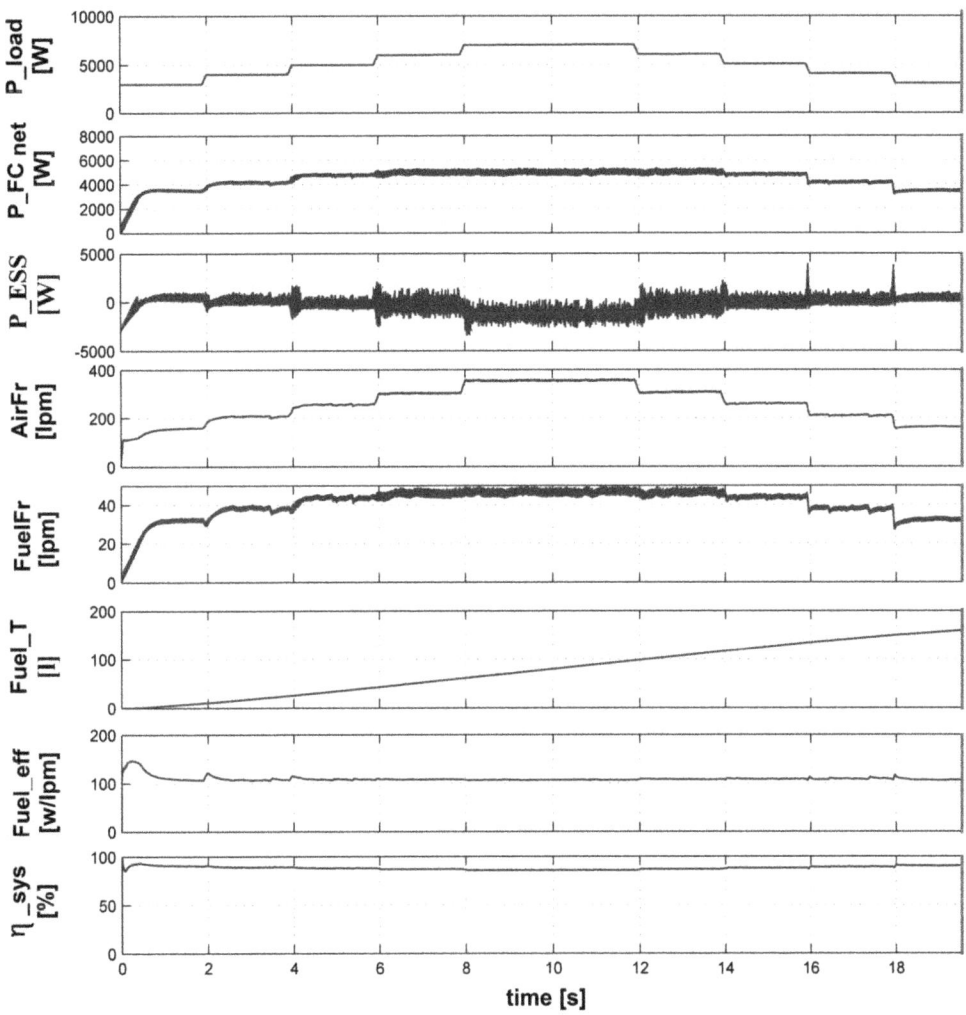

(a) Using the Air–LFW strategy.

Figure 8. *Cont.*

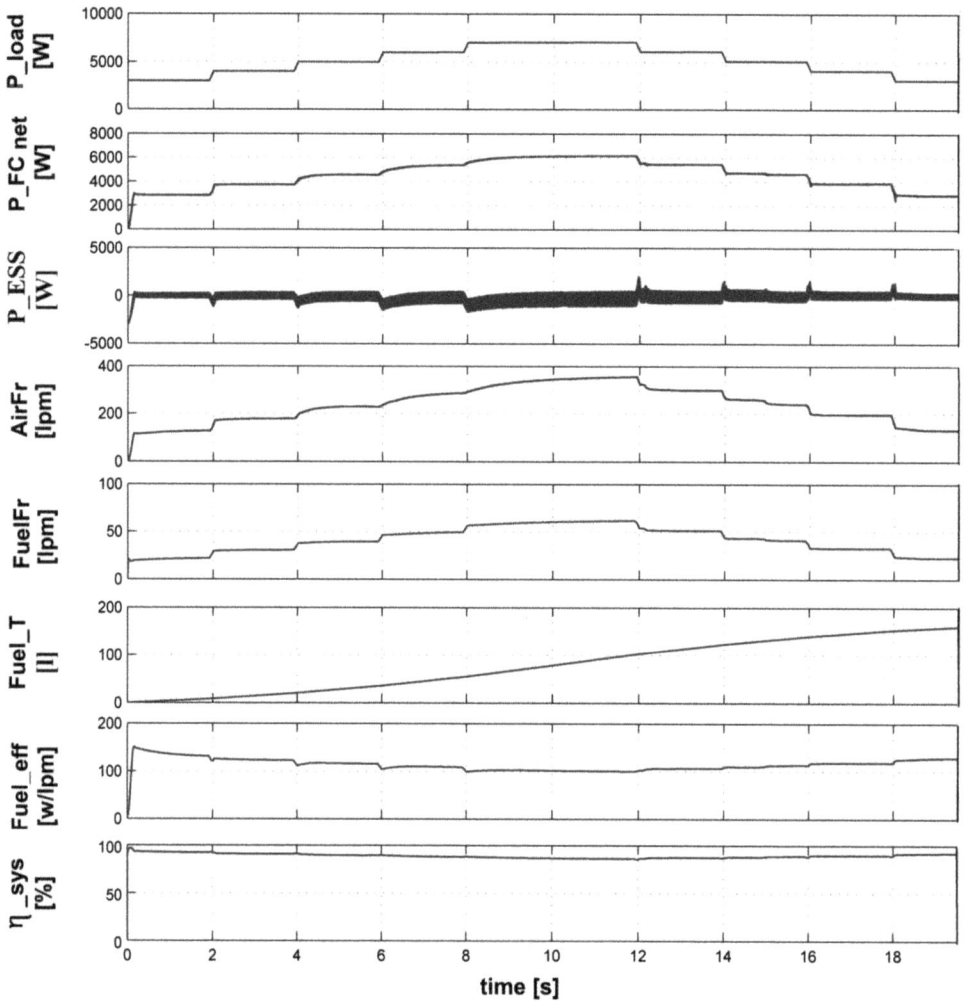

(b) Using the Fuel–LFW strategy.

Figure 8. *Cont.*

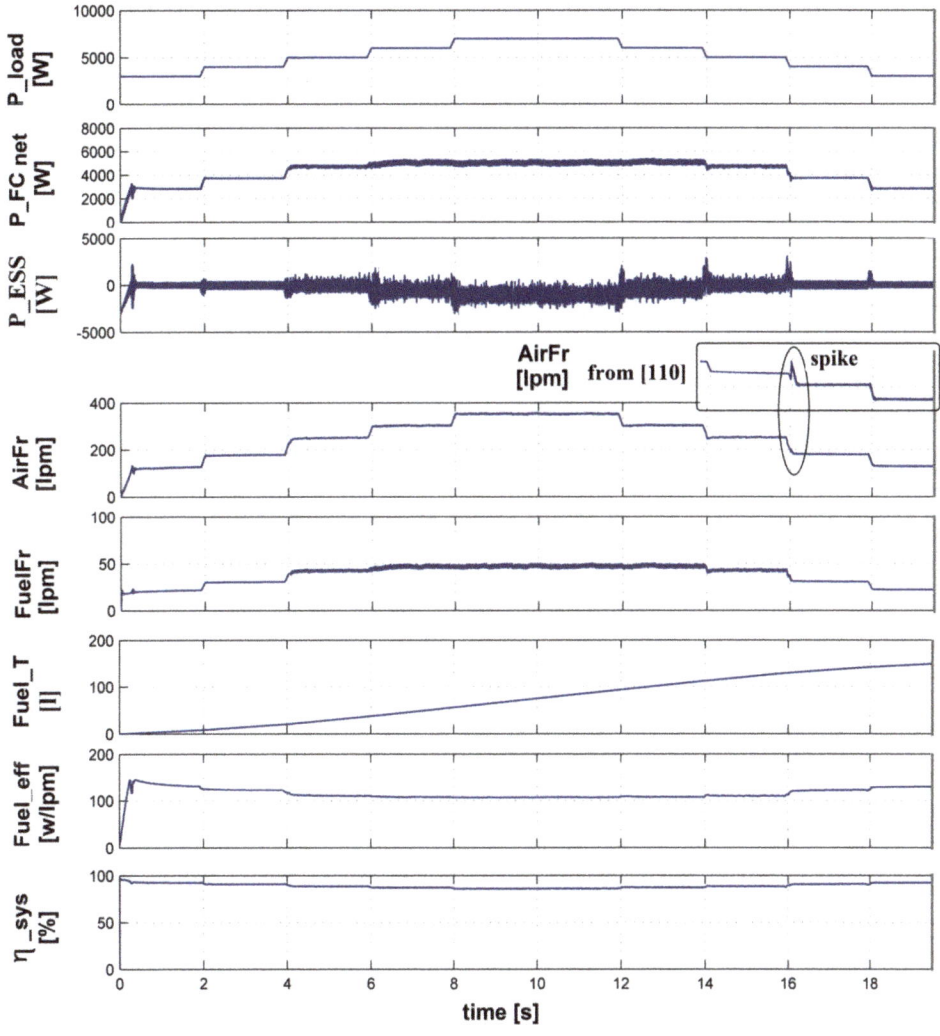

(c) Using the SW–LFW strategy with P_{ref} = 4.5 kW.

Figure 8. Behavior of the FCHPS under the third load profile (symmetrical stair up and down).

Additionally, it is worth mentioning the aforementioned spike when the references $I_{ref(LFW)}$ and $I_{FC} + I_{ref(GES2)}$ are switched [110] (refer to the 4th curve of Figure 8c).

5. Discussion and Next Works

The results obtained for the FCHPS in previous sections and [110] are summarized in Tables 7–10, for the case of k_{eff} = 25 and P_{ref} = 4.5 kW, where the fuel economy indicator was calculated for a given load profile using the relationship $\%Fuel^{load}_{T(strategy)} = 100 \cdot \left(\frac{Fuel^{load}_{T(reference)} - Fuel^{load}_{T(strategy)}}{Fuel^{load}_{T(reference)}} \right)$.

Table 7. Percent of fuel economy for different P_{load} values in case of $k_{eff} = 25$ and $P_{ref} = 4.5$ kW.

Parameter [unit] \ Strategy	Fuel-LFW	Air-LFW	SW-LFW	SW [110]	P_{load} [kW]
$\%Fuel^{P\_load}_{T(strategy)}$ [%]	1.89	−35.68	1.89	1.65	2
	6.88	−9.85	6.88	3.55	3
	6.91	−1.60	6.91	5.02	4
	8.88	6.53	8.88	11.58	5
	9.99	11.26	11.26	14.19	6
	15.32	17.95	17.95	19.10	7
	14.77	17.66	17.66	27.11	8
$\frac{1}{8}\Sigma_{P\_load} \%Fuel^{P\_load}_{T(strategy)}$ [%]	8.08	0.78	8.93	10.28	

Table 8. Percent of fuel economy for the first load profile in case of $k_{eff} = 25$ and $P_{ref} = 4.5$ kW.

Parameter [unit] \ Strategy	Fuel-LFW	Air-LFW	SW-LFW	SW [110]	$P_{load(AV)}$ [kW]
$\%Fuel^{P\_load(AV)}_{T(strategy)}$ [%]	−5.62	−21.03	−5.62	0.29	2
	−1.52	−11.57	−1.52	1.93	3
	0.84	−4.38	0.41	4.09	4
	4.13	3.14	3.80	11.33	5
	7.74	10.20	10.20	32.67	6
$\frac{1}{6}\Sigma_{P\_load(AV)} \%Fuel^{P\_load(AV)}_{T(strategy)}$ [%]	0.93	−3.94	1.21	8.39	

Table 9. Percent of fuel economy for the second load profile in case of $k_{eff} = 25$ and $P_{ref} = 4.5$ kW.

Parameter [unit] \ Strategy	Fuel-LFW	Air-LFW	SW-LFW	SW [110]	$P_{load(AV)}$ [kW]
$\%Fuel^{P\_load}_{T(strategy)}$ [%]	10.69	17.14	25.49	25.87	5

Table 10. Percent of fuel economy for the third load profile in case of $k_{eff} = 25$ and $P_{ref} = 4.5$ kW

Parameter [unit] \ Strategy	Fuel-LFW	Air-LFW	SW-LFW	SW [110]	$P_{load(AV)}$ [kW]
$\%Fuel^{P\_load}_{T(strategy)}$ [%]	5.48	6.07	12.60	12.29	5

Table 7 presents the percent of fuel economy for different constant P_{load} values (mentioned in the last column), highlighting that the SW–LFW strategy achieved the best fuel economy through the switching technique of the Air–LFW and Fuel–LFW strategies. The average value of the fuel economy percentages for the eight load levels was calculated using $\frac{1}{8}\Sigma_{P\_load} \%Fuel^{P\_load}_{T(strategy)}$. The value obtained showed that the SW–LFW strategy and the strategy proposed in [110] were of different classes (the first used a single GHG controller, the second used two GES controllers, which led to the expansion of the search field by using two variables and therefore better performance).

Better performance for the strategy proposed in [110] compared to the SW–LFW strategy proposed in this study was also obtained for a variable load profile but with low dynamics (note that the power levels used were $0.75 \cdot P_{load(AV)}$, $1.25 \cdot P_{load(AV)}$, and $1.00 \cdot P_{load(AV)}$ for the first load profile). Table 8 presents the percent of fuel economy for different first load profiles, mentioning the $P_{load(AV)}$ values in the last column.

For a variable load profile with a dynamic across the full range of loads allowed, the performance indicator had similar values for the SW–LFW and SW strategies [110] (see percentage of fuel economy for the second and third load profile presented in Tables 9 and 10).

A possible explanation would be that the load dynamics could no longer be tracked due to the limited search speed, to avoid fuel starvation, and the advantage of searching with two variables was lost. Although the discussion on the performance of the SW–LFW strategy remained open, as compared to the strategy proposed in [110], it was obvious that the current limitations imposed by the safe operation of the FC system limited the search speed. Therefore, the search slopes of maximum 100 A/s imposed in the air and fuel regulators limits the search speed and, in the end, similar performances are obtained (regardless of the search speed of the algorithm used). However, the advantage of the simplicity of the SW–LFW strategy still holds, as compared to the strategy proposed in [110].

The fuel consumption for the SW–LFW strategy, mentioned in Tables 2 and 4, is shown in Figure 9 by load (a) and average load (b).

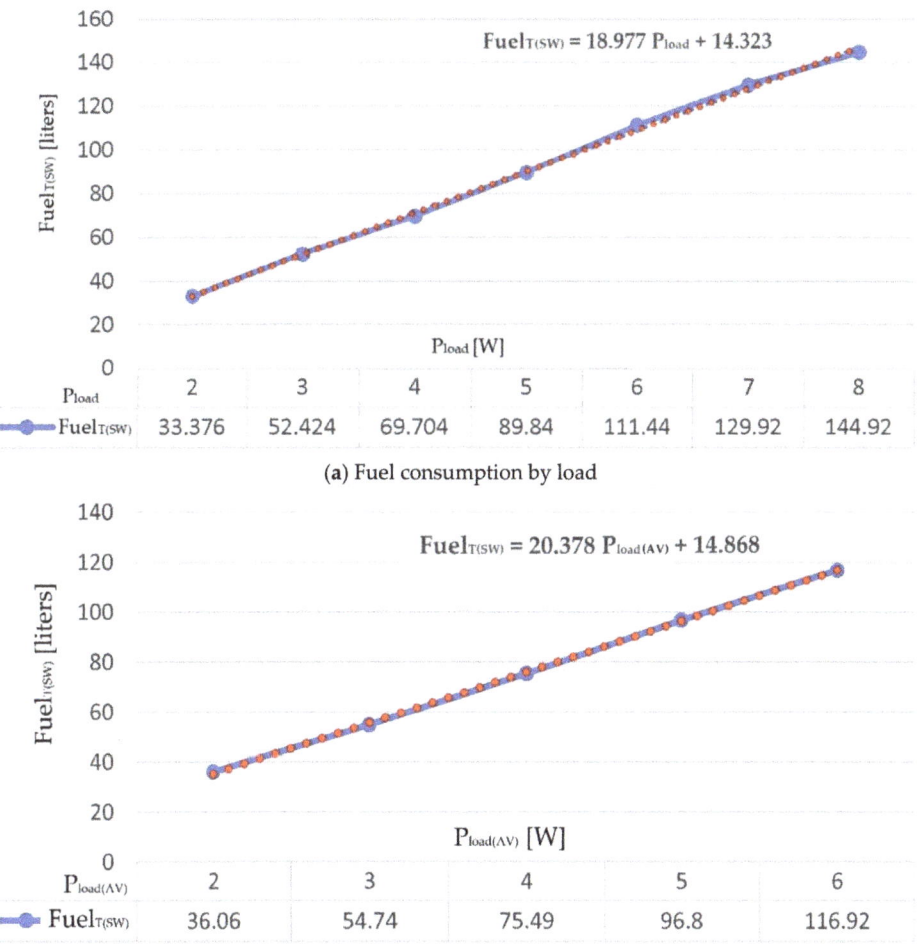

(a) Fuel consumption by load

(b) Fuel consumption by average load

Figure 9. Fuel consumption for the SW–LFW strategy.

It is worth mentioning that the fuel consumption for the SW–LFW strategy was almost linear in both cases, the trend line depending on the load and the average load, according to the relations $Fuel_{T(SW)} = 20.378\ P_{load} + 14.868$ and $Fuel_{T(SW)} = 18.977\ P_{load(AV)} + 14.323$, respectively. Therefore, the fuel consumption could be estimated on the basis of the average load, up to the first refueling station on the planned road.

Consequently, the next works is focused on the following actions:

1. Testing the relationship between fuel consumption and average load for different load demand profiles.
2. Finding the best threshold between the high and low loading ranges. This value of P_{ref} = 4.5 kW was obtained on the basis of the sensitivity investigation functions for both variable and constant loads. The load-following management switches to the fuel and air regulators, if the load is higher or lower than this threshold. This value and other values close to this are to be considered for further tests under a different load profile.
3. Inclusion of a monitoring and energy management system for the battery. In this study, the system load demand was clearly sustained by the power supplied by the FC system and the battery storage device functions in a charge-sustained mode. Thus, the size of the battery might be reduced. Additionally, the battery's life time increases by avoiding the charge–discharge cycles that appear in most strategies, based on SOC monitoring. An advanced energy management system [121,122] and a battery aging modeling [123,124] is to be included to evaluate the advantage of the LFW control over the battery life.
4. Reducing the number of switches of the fueling regulators' references in the event of a high dynamic load. The SW–LFW strategy used a 200 W hysteresis controller (see Figure 3) instead of a simple comparator, to switch the reference $I_{ref(LFW)}$ to the inputs of the fueling regulators ($I_{ref(Fuel)}$ and $I_{ref(Air)}$). Different values of the hysteresis band for the hysteresis controller is to be tested, in combination with an appropriate filtering of the load power.

In summary, besides the practical recommendations and next works mentioned above, the implementation of the SW–LFW strategy could be approached through the following step-by-step research methodology. (1) First, the strategies Air–LFW and Fuel–LFW must be tested for high and low load levels, using a simple searching algorithm and a Fuel Cell simulator based on a Hardware in the Loop Simulation (HILS) approach. (2) The air and fuel flow rates obtained with the techniques Air–LFW and Fuel–LFW is to be recorded in a look-up table to use for the SW–LFW strategy (a look-up table approach of the load-following control presented in this work). (3) The obtained fuel economy is to be validated on a Fuel Cell system using the SW–LFW strategy for variable load profiles. (4) The fuel efficiency should be improved by testing other thresholds, using recently developed searching algorithms (such as the Global Extremum Seeking algorithm used in this paper), the load-following management in place of the look-up table approach, faster searching algorithms than GES algorithm used here, etc.

6. Conclusions

This work analyzed the performance of the SW–LFW technique in comparison to strategies based on Air–LFW, Fuel–LFW, sFF control, and the switching strategy proposed in [110]. The fuel efficiency was better for the SW–LFW strategy in case of a dynamic load across the full range as it used the best strategy for a certain load level—Air–LFW strategy for high-load values and Fuel–LFW strategy for low-load values.

In addition, the fueling regulator, which was not controlled based on the load-following technique, was utilized to improve the fuel economy employing a real-time optimization loop.

In consequence, the key observations of this work were as follows. (1) A new energy management scheme based on the SW–LFW technique to further improve fuel economy compared to basic techniques (the Air–LFW and Fuel–LFW techniques) and sFF reference strategy. (2) SW–LFW strategy has the advantages of improved battery lifetime and reduced size, by switching the load-following control to the fueling controllers. (3) The fuel economy in case of a dynamic load across the full range was doubled compared to basic strategies (about 2.23-times and 2.47-times higher compared to Air–LFW and Fuel–LFW techniques, respectively). (4) The total fuel utilization was reduced by more than 13%, as compared to commercial strategy based on the sFF technique. (5) At the same fuel economy that could be obtained during a dynamic load cycle, the proposed strategy had the advantages of simplicity and safety in operation, as compared to the switching strategy proposed in [110]. (6) In addition, fuel consumption could be estimated based on the average load to the first refueling station on the planned road, using the linear relationship obtained in this study.

In conclusion, this study would help improve the performance of the FC vehicles, by utilizing the innovative solutions proposed in this paper.

Author Contributions: Research methodology, N.B.; writing—original draft preparation, N.B. and P.T.; supervision, P.T. and N.B.; Investigation, N.B. and P.T.; writing—review and editing, N.B. and P.T. All authors have read and agreed to the published version of the manuscript.

Funding: This work was partially supported by the International Research Partnerships: Electrical Engineering Thai-French Research Center (EE-TFRC) between King Mongkut's University of Technology North Bangkok and University of Lorraine under Grant KMUTNB-64-KNOW-31.

Institutional Review Board Statement: Not applicable.

Informed Consent Statement: Not applicable.

Data Availability Statement: Not applicable.

Conflicts of Interest: The authors declare no conflict of interest.

References

1. Gielen, D.; Boshell, F.; Saygin, D.; Bazilian, M.D.; Wagner, N.; Gorini, R. The role of renewable energy in the global energy transformation. *Energy Strategy Rev.* **2019**, *24*, 38–50. [CrossRef]
2. Hesselink, L.X.W.; Chappin, E.J.L. Adoption of energy efficient technologies by households—Barriers, policies and agent-based modelling studies. *Renew Sustain. Energy Rev.* **2019**, *99*, 29–41. [CrossRef]
3. Wang, F.-C.; Hsiao, Y.-S.; Yang, Y.-Z. The Optimization of Hybrid Power Systems with Renewable Energy and Hydrogen Generation. *Energies* **2018**, *11*, 1948. [CrossRef]
4. Bukar, A.L.; Tan, C.W. A review on stand-alone photovoltaic-wind energy system with fuel cell: System optimization and energy management strategy. *J. Clean. Prod.* **2019**, *221*, 73–88. [CrossRef]
5. Sulaiman, N.; Hannan, M.; Mohamed, A.; Ker, P.; Majlan, E.; Daud, W.W. Optimization of energy management system for fuel-cell hybrid electric vehicles: Issues and recommendations. *Appl. Energy* **2018**, *228*, 2061–2079. [CrossRef]
6. Sorrentino, M.; Cirillo, V.; Nappi, L. Development of flexible procedures for co-optimizing design and control of fuel cell hybrid vehicles. *Energy Convers. Manag.* **2019**, *185*, 537–551. [CrossRef]
7. Bizon, N. Hybrid power sources (HPSs) for space applications: Analysis of PEMFC/Battery/SMES HPS under unknown load containing pulses. *Renew. Sustain. Energy Rev.* **2019**, *105*, 14–37. [CrossRef]
8. Pan, Z.; An, L.; Wen, C. Recent advances in fuel cells based propulsion systems for unmanned aerial vehicles. *Appl. Energy* **2019**, *240*, 473–485. [CrossRef]
9. Bizon, N. Real-time optimization strategy for fuel cell hybrid power sources with load-following control of the fuel or air flow. *Energy Convers. Manag.* **2018**, *157*, 13–27. [CrossRef]
10. Olatomiwa, L.; Mekhilef, S.; Ismail, M.; Moghavvemi, M. Energy management strategies in hybrid renewable energy systems: A review. *Renew. Sustain. Energy Rev.* **2016**, *62*, 821–835. [CrossRef]
11. Bizon, N. Optimal Operation of Fuel Cell/Wind Turbine Hybrid Power System under Turbulent Wind and Variable Load. *Appl. Energy* **2018**, *212*, 196–209. [CrossRef]
12. Priya, K.; Sathishkumar, K.; Rajasekar, N. A comprehensive review on parameter estimation techniques for Proton Exchange Membrane fuel cell modelling. *Renew. Sustain. Energy Rev.* **2018**, *93*, 121–144. [CrossRef]
13. Yue, M.; Jemei, S.; Gouriveau, R.; Zerhouni, N. Review on health-conscious energy management strategies for fuel cell hybrid electric vehicles: Degradation models and strategies. *Int. J. Hydrogen Energy* **2019**, *44*, 6844–6861. [CrossRef]
14. Bizon, N. Effective mitigation of the load pulses by controlling the battery/SMES hybrid energy storage system. *Appl. Energy* **2018**, *229*, 459–473. [CrossRef]
15. Dafalla, A.M.; Jiang, J. Stresses and their impacts on proton exchange membrane fuel cells: A review. *Int. J. Hydrogen Energy* **2018**, *43*, 2327–2348. [CrossRef]
16. Chen, H.; Zhao, X.; Zhang, T.; Pei, P. The reactant starvation of the proton exchange membrane fuel cells for vehicular applications: A review. *Energy Convers. Manag.* **2019**, *182*, 282–298. [CrossRef]

17. Luo, Y.; Jiao, K. Cold start of proton exchange membrane fuel cell. *Prog. Energy Combust. Sci.* **2018**, *64*, 29–61. [CrossRef]
18. Zhang, T.; Wang, P.; Chen, H.; Pei, P. A review of automotive proton exchange membrane fuel cell degradation under start-stop operating condition. *Appl. Energy* **2018**, *223*, 249–262. [CrossRef]
19. Dijoux, E.; Steiner, N.Y.; Benne, M.; Péra, M.-C.; Pérez, B.G. A review of fault tolerant control strategies applied to proton exchange membrane fuel cell systems. *J. Power Sources* **2017**, *359*, 119–133. [CrossRef]
20. Das, V.; Padmanaban, S.; Venkitusamy, K.; Selvamuthukumaran, R.; Blaabjerg, F.; Siano, P. Recent advances and challenges of fuel cell based power system architectures and control—A review. *Renew. Sustain. Energy Rev.* **2017**, *73*, 10–18. [CrossRef]
21. Bizon, N. Load-following mode control of a standalone renewable/fuel cell hybrid power source. *Energy Convers. Manag.* **2014**, *77*, 763–772. [CrossRef]
22. Daud, W.; Rosli, R.; Majlan, E.; Hamid, S.; Mohamed, R.; Husaini, T. PEM fuel cell system control: A review. *Renew. Energy* **2017**, *113*, 620–638. [CrossRef]
23. Bizon, N.; Radut, M.; Oproescu, M. Energy control strategies for the Fuel Cell Hybrid Power Source under unknown load profile. *Energy* **2015**, *86*, 31–41. [CrossRef]
24. Ahmadi, P.; Torabi, S.H.; Afsaneh, H.; Sadegheih, Y.; Ganjehsarabi, H.; Ashjaee, M. The effects of driving patterns and PEM fuel cell degradation on the lifecycle assessment of hydrogen fuel cell vehicles. *Int. J. Hydrogen Energy* **2020**, *45*, 3595–3608. [CrossRef]
25. Wang, F.-C.; Lin, K.-M. Impacts of Load Profiles on the Optimization of Power Management of a Green Building Employing Fuel Cells. *Energies* **2018**, *12*, 57. [CrossRef]
26. Zhao, D.; Xu, L.; Huangfu, Y.; Dou, M.; Liu, J. Semi-physical modeling and control of a centrifugal compressor for the air feeding of a PEM fuel cell. *Energy Convers. Manag.* **2017**, *154*, 380–386. [CrossRef]
27. Han, J.; Yu, S. Ram air compensation analysis of fuel cell vehicle cooling system under driving modes. *Appl. Therm. Eng.* **2018**, *142*, 530–542. [CrossRef]
28. Zhang, H.; Li, X.; Liu, X.; Yan, J. Enhancing fuel cell durability for fuel cell plug-in hybrid electric vehicles through strategic power management. *Appl. Energy* **2019**, *241*, 483–490. [CrossRef]
29. Pukrushpan, J.T.; Stefanopoulou, A.G.; Peng, H. Control of fuel cell breathing. *IEEE Control Syst. Mag.* **2004**, *24*, 30–46.
30. Ahn, J.-W.; Choe, S.-Y. Coolant controls of a PEM fuel cell system. *J. Power Sources* **2008**, *179*, 252–264. [CrossRef]
31. Beirami, H.; Shabestari, A.Z.; Zerafat, M.M. Optimal PID plus fuzzy controller design for a PEM fuel cell air feed system using the self-adaptive differential evolution algorithm. *Int. J. Hydrogen Energy* **2015**, *40*, 9422–9434. [CrossRef]
32. Liu, Z.; Li, L.; Ding, Y.; Deng, H.; Chen, W. Modeling and control of an air supply system for a heavy duty PEMFC engine. *Int. J. Hydrogen Energy* **2016**, *41*, 16230–16239. [CrossRef]
33. Talj, R.; Ortega, R.; Astolfi, A. Passivity and robust PI control of the air supply system of a PEM fuel cell model. *Automatica* **2011**, *47*, 2554–2561. [CrossRef]
34. Cano, M.H.; Mousli, M.I.A.; Kelouwani, S.; Agbossou, K.; Hammoudi, M.; Dubéc, Y. Improving a free air breathing proton exchange membrane fuel cell through the Maximum Efficiency Point Tracking method. *J. Power Sources* **2017**, *345*, 264–274. [CrossRef]
35. Baroud, Z.; Benmiloud, M.; Benalia, A.; Ocampo-Martinez, C. Novel hybrid fuzzy-PID control scheme for air supply in PEM fuel-cell-based systems. *Int. J. Hydrogen Energy* **2017**, *42*, 10435–10447. [CrossRef]
36. Hasikos, J.; Sarimveis, H.; Zervas, P.; Markatos, N. Operational optimization and real-time control of fuel-cell systems. *J. Power Sources* **2009**, *193*, 258–268. [CrossRef]
37. Nejad, H.C.; Farshad, M.; Gholamalizadeh, E.; Askarian, B.; Akbarimajd, A. A novel intelligent-based method to control the output voltage of Proton Exchange Membrane Fuel Cell. *Energy Convers. Manag.* **2019**, *185*, 455–464. [CrossRef]
38. Arce, A.; Alejandro, J.; Bordons, C.; Daniel, R. Real-time implementation of a constrained MPC for efficient airflow control in a PEM fuel cell. *IEEE Trans. Ind. Electron.* **2010**, *57*, 1892–1905. [CrossRef]
39. Ziogou, C.; Papadopoulou, S.; Pistikopoulos, E.; Georgiadis, M.; Voutetakis, S. Model-Based Predictive Control of Integrated Fuel Cell Systems—From Design to Implementation. *Adv. Energy Syst. Eng.* **2016**, *2017*, 387–430. [CrossRef]
40. Ziogou, C.; Papadopoulou, S.; Georgiadis, M.C.; Voutetakis, S. On-line nonlinear model predictive control of a PEM fuel cell system. *J. Process. Control.* **2013**, *23*, 483–492. [CrossRef]
41. Barzegari, M.M.; Alizadeh, E.; Pahnabi, A.H. Grey-box modeling and model predictive control for cascade-type PEMFC. *Energy* **2017**, *127*, 611–622. [CrossRef]
42. Ziogou, C.; Voutetakis, S.; Georgiadis, M.C.; Papadopoulou, S. Model predictive control (MPC) strategies for PEM fuel cell systems—A comparative experimental demonstration. *Chem. Eng. Res. Des.* **2018**, *131*, 656–670. [CrossRef]
43. Laghrouche, S.; Harmouche, M.; Ahmed, F.S.; Chitour, Y. Control of PEMFC Air-Feed System Using Lyapunov-Based Robust and Adaptive Higher Order Sliding Mode Control. *IEEE Trans. Control. Syst. Technol.* **2014**, *23*, 1. [CrossRef]
44. Pilloni, A.; Pisano, A.; Usai, E. Observer-Based Air Excess Ratio Control of a PEM Fuel Cell System via High-Order Sliding Mode. *IEEE Trans. Ind. Electron.* **2015**, *62*, 5236–5246. [CrossRef]
45. Deng, H.; Li, Q.; Chen, W.; Zhang, G. High-Order Sliding Mode Observer Based OER Control for PEM Fuel Cell Air-Feed System. *IEEE Trans. Energy Convers.* **2017**, *33*, 232–244. [CrossRef]
46. Sankar, K.; Jana, A.K. Nonlinear multivariable sliding mode control of a reversible PEM fuel cell integrated system. *Energy Convers. Manag.* **2018**, *171*, 541–565. [CrossRef]

47. Deng, H.; Li, Q.; Cui, Y.; Zhu, Y.; Chen, W. Nonlinear controller design based on cascade adaptive sliding mode control for PEM fuel cell air supply systems. *Int. J. Hydrogen Energy* **2019**, *44*, 19357–19369. [CrossRef]
48. Saadi, R.; Kraa, O.; Ayad, M.; Becherif, M.; Ghodbane, H.; Bahri, M.; Aboubou, A. Dual loop controllers using PI, sliding mode and flatness controls applied to low voltage converters for fuel cell applications. *Int. J. Hydrogen Energy* **2016**, *41*, 19154–19163. [CrossRef]
49. Derbeli, M.; Farhat, M.; Barambones, O.; Sbita, L. Control of PEM fuel cell power system using sliding mode and super-twisting algorithms. *Int. J. Hydrogen Energy* **2017**, *42*, 8833–8844. [CrossRef]
50. Kunusch, C.; Puleston, P.F.; Mayosky, M.A.; Riera, J. Sliding Mode Strategy for PEM Fuel Cells Stacks Breathing Control Using a Super-Twisting Algorithm. *IEEE Trans. Control. Syst. Technol.* **2008**, *17*, 167–174. [CrossRef]
51. Hernández-Torres, D.; Riu, D.; Sename, O. Reduced-order Robust Control of a Fuel Cell Air Supply System. *IFAC-PapersOnLine* **2017**, *50*, 96–101. [CrossRef]
52. Sun, J.; Kolmanovsky, I. Load governor for fuel cell oxygen starvation protection: A robust nonlinear reference governor approach. *IEEE Trans. Contr. Syst. Technol.* **2005**, *13*, 911–920.
53. Han, J.; Yu, S.; Yi, S. Adaptive control for robust air flow management in an automotive fuel cell system. *Appl. Energy* **2017**, *190*, 73–83. [CrossRef]
54. He, Y.; Xing, L.; Zhang, Y.; Zhang, J.; Cao, F.; Xing, Z. Development and experimental investigation of an oil-free twin-screw air compressor for fuel cell systems. *Appl. Therm. Eng.* **2018**, *145*, 755–762. [CrossRef]
55. Li, Q.; Chen, W.; Liu, Z.; Guo, A.; Liu, S. Control of proton exchange membrane fuel cell system breathing based on maximum net power control strategy. *J. Power Sources* **2013**, *241*, 212–218. [CrossRef]
56. Mane, S.; Mejari, M.; Kazi, F.; Singh, N. Improving Lifetime of Fuel Cell in Hybrid Energy Management System by Lure–Lyapunov-Based Control Formulation. *IEEE Trans. Ind. Electron.* **2017**, *64*, 6671–6679. [CrossRef]
57. Ramos-Paja, C.A.; Spagnuolo, G.; Petrone, G.; Mamarelis, E. A perturbation strategy for fuel consumption minimization in polymer electrolyte membrane fuel cells: Analysis, Design and FPGA implementation. *Appl. Energy* **2014**, *119*, 21–32. [CrossRef]
58. Bizon, N.; Lopez-Guede, J.M.; Kurt, E.; Thounthong, P.; Mazare, A.G.; Ionescu, L.M.; Iana, G. Hydrogen economy of the fuel cell hybrid power system optimized by air flow control to mitigate the effect of the uncertainty about available renewable power and load dynamics. *Energy Convers. Manag.* **2019**, *179*, 152–165. [CrossRef]
59. Pukrushpan, J.T.; Stefanopoulou, A.G.; Peng, H. *Control of Fuel Cell Power Systems*; Springer International Publishing: New York, NY, USA, 2004.
60. Zhong, D.; Lin, R.; Liu, D.; Cai, X. Structure optimization of anode parallel flow field for local starvation of proton exchange membrane fuel cell. *J. Power Sources* **2018**, *403*, 1–10. [CrossRef]
61. Hong, L.; Chen, J.; Liu, Z.; Huang, L.; Wu, Z. A nonlinear control strategy for fuel delivery in PEM fuel cells considering nitrogen permeation. *Int. J. Hydrogen Energy* **2017**, *42*, 1565–1576. [CrossRef]
62. Bao, C.; Ouyang, M.; Yi, B. Modeling and control of air stream and hydrogen flow with recirculation in a PEM fuel cell system—I. Control-oriented modeling. *Int. J. Hydrogen Energy* **2006**, *31*, 1879–1896. [CrossRef]
63. Bao, C.; Ouyang, M.; Yi, B. Modeling and control of air stream and hydrogen flow with recirculation in a PEM fuel cell system—II. Linear and adaptive nonlinear control. *Int. J. Hydrogen Energy* **2006**, *31*, 1897–1913. [CrossRef]
64. He, J.; Choe, S.-Y.; Hong, C.-O. Analysis and control of a hybrid fuel delivery system for a polymer electrolyte membrane fuel cell. *J. Power Sources* **2008**, *185*, 973–984. [CrossRef]
65. He, J.; Ahn, J.; Choe, S.-Y. Analysis and control of a fuel delivery system considering a two-phase anode model of the polymer electrolyte membrane fuel cell stack. *J. Power Sources* **2011**, *196*, 4655–4670. [CrossRef]
66. Thounthong, P.; Mungporn, P.; Pierfederici, S.; Guilbert, D.; Bizon, N. Adaptive Control of Fuel Cell Converter Based on a New Hamiltonian Energy Function for Stabilizing the DC Bus in DC Microgrid Applications. *Mathematics* **2020**, *8*, 2035. [CrossRef]
67. Luna, J.; Ocampo-Martinez, C.; Serra, M. Nonlinear predictive control for the concentrations profile regulation under unknown reaction disturbances in a fuel cell anode gas channel. *J. Power Sources* **2015**, *282*, 129–139. [CrossRef]
68. Park, G.; Gajic, Z. A Simple Sliding Mode Controller of a Fifth-Order Nonlinear PEM Fuel Cell Model. *IEEE Trans. Energy Convers.* **2013**, *29*, 65–71. [CrossRef]
69. Matraji, I.; Laghrouche, S.; Jemei, S.; Wack, M. Robust control of the PEM fuel cell air-feed system via sub-optimal second order sliding mode. *Appl. Energy* **2013**, *104*, 945–957. [CrossRef]
70. Baik, K.D.; Kim, M.S. Characterization of nitrogen gas crossover through the membrane in proton-exchange membrane fuel cells. *Int. J. Hydrogen Energy* **2011**, *36*, 732–739. [CrossRef]
71. Steinberger, M.; Geiling, J.; Oechsner, R.; Frey, L. Anode recirculation and purge strategies for PEM fuel cell operation with diluted hydrogen feed gas. *Appl. Energy* **2018**, *232*, 572–582. [CrossRef]
72. Chen, Y.-S.; Yang, C.-W.; Lee, J.-Y. Implementation and evaluation for anode purging of a fuel cell based on nitrogen concentration. *Appl. Energy* **2014**, *113*, 1519–1524. [CrossRef]
73. Piffard, M.; Gerard, M.; Bideaux, E.; Da Fonseca, R.; Massioni, P. Control by state observer of PEMFC anodic purges in dead-end operating mode. *IFAC-PapersOnLine* **2015**, *48*, 237–243. [CrossRef]
74. Rabbani, A.; Rokni, M. Effect of nitrogen crossover on purging strategy in PEM fuel cell systems. *Appl. Energy* **2013**, *111*, 1061–1070. [CrossRef]
75. Mahoney, F.M. Reduction-Oxidation Tolerant Electrodes for Solid Oxide Fuel Cells. U.S. Patent 20,100,159,356, 24 June 2010.

76. Ahluwalia, R.; Wang, X. Buildup of nitrogen in direct hydrogen polymer-electrolyte fuel cell stacks. *J. Power Sources* **2007**, *171*, 63–71. [CrossRef]
77. Pan, T.; Shen, J.; Sun, L.; Lee, K.Y. Thermodynamic modelling and intelligent control of fuel cell anode purge. *Appl. Therm. Eng.* **2019**, *154*, 196–207. [CrossRef]
78. Koski, P.; Perez, L.C.; Ihonen, J. Comparing Anode Gas Recirculation with Hydrogen Purge and Bleed in a Novel PEMFC Laboratory Test Cell Configuration. *Fuel Cells* **2015**, *15*, 494–504. [CrossRef]
79. Promislow, K.; St-Pierre, J.; Wetton, B. A simple, analytic model of polymer electrolyte membrane fuel cell anode recirculation at operating power including nitrogen crossover. *J. Power Sources* **2011**, *196*, 10050–10056. [CrossRef]
80. Taghiabadi, M.M.; Zhiani, M. Degradation analysis of dead-ended anode PEM fuel cell at the low and high thermal and pressure conditions. *Int. J. Hydrogen Energy* **2019**, *44*, 4985–4995. [CrossRef]
81. Yang, Y.; Zhang, X.; Guo, L.; Liu, H. Overall and local effects of operating conditions in PEM fuel cells with dead-ended anode. *Int. J. Hydrogen Energy* **2017**, *42*, 4690–4698. [CrossRef]
82. Pérez, L.C.; Rajala, T.; Ihonen, J.; Koski, P.; Sousa, J.M.; Mendes, A. Development of a methodology to optimize the air bleed in PEMFC systems operating with low quality hydrogen. *Int. J. Hydrogen Energy* **2013**, *38*, 16286–16299. [CrossRef]
83. Mahjoubi, C.; Olivier, J.-C.; Skander-Mustapha, S.; Machmoum, M.; Slama-Belkhodja, I. An improved thermal control of open cathode proton exchange membrane fuel cell. *Int. J. Hydrogen Energy* **2019**, *44*, 11332–11345. [CrossRef]
84. Strahl, S.; Costa-Castelló, R. Temperature control of open-cathode PEM fuel cells. *IFAC-PapersOnLine* **2017**, *50*, 11088–11093. [CrossRef]
85. Chang, Y.; Qin, Y.; Yin, Y.; Zhang, J.; Li, X. Humidification strategy for polymer electrolyte membrane fuel cells—A review. *Appl. Energy* **2018**, *230*, 643–662. [CrossRef]
86. Liu, Z.; Chen, J.; Chen, S.; Huang, L.; Shao, Z. Modeling and Control of Cathode Air Humidity for PEM Fuel Cell Systems. *IFAC-PapersOnLine* **2017**, *50*, 4751–4756. [CrossRef]
87. Bizon, N.; Thounthong, P. Real-time strategies to optimize the fueling of the fuel cell hybrid power source: A review of issues, challenges and a new approach. *Renew. Sustain. Energy Rev.* **2018**, *91*, 1089–1102. [CrossRef]
88. Ou, K.; Yuan, W.-W.; Choi, M.; Yang, S.; Kim, Y.-B. Performance increase for an open-cathode PEM fuel cell with humidity and temperature control. *Int. J. Hydrogen Energy* **2017**, *42*, 29852–29862. [CrossRef]
89. Ahn, J.-W.; He, J.; Choe, S.-Y. Design of Air, Water, Temperature and Hydrogen Controls for a PEM Fuel Cell System. In Proceedings of the ASME 2011 9th International Conference on Fuel Cell Science, Engineering and Technology, Washington, DC, USA, 7–10 August 2011; pp. 711–718. Available online: https://asmedigitalcollection.asme.org/FUELCELL/proceedings-abstract/FUELCELL2011/54693/711/357956 (accessed on 9 February 2021).
90. Bizon, N. Energy optimization of fuel cell system by using global extremum seeking algorithm. *Appl. Energy* **2017**, *206*, 458–474. [CrossRef]
91. Sorlei, I.-S.; Bizon, N.; Thounthong, P.; Varlam, M.; Carcadea, E.; Culcer, M.; Iliescu, M.; Raceanu, M. Fuel Cell Electric Vehicles—A Brief Review of Current Topologies and Energy Management Strategies. *Energies* **2021**, *14*, 252. [CrossRef]
92. Bizon, N.; Mazare, A.G.; Ionescu, L.M.; Enescu, F.M. Optimization of the proton exchange membrane fuel cell hybrid power system for residential buildings. *Energy Convers. Manag.* **2018**, *163*, 22–37. [CrossRef]
93. Bizon, N. Nonlinear control of fuel cell hybrid power sources: Part II—Current control. *Appl. Energy* **2011**, *88*, 2574–2591. [CrossRef]
94. Kaya, K.; Hames, Y. Two new control strategies: For hydrogen fuel saving and extend the life cycle in the hydrogen fuel cell vehicles. *Int. J. Hydrogen Energy* **2019**, *44*, 18967–18980. [CrossRef]
95. Wang, Y.; Sun, Z.; Chen, Z. Rule-based energy management strategy of a lithium-ion battery, supercapacitor and PEM fuel cell system. *Energy Procedia* **2019**, *158*, 2555–2560. [CrossRef]
96. Li, G.; Zhang, J.; He, H. Battery SOC constraint comparison for predictive energy management of plug-in hybrid electric bus. *Appl. Energy* **2017**, *194*, 578–587. [CrossRef]
97. Torreglosa, J.P.; Garcia, P.; Fernandez, L.M.; Jurado, F. Predictive Control for the Energy Management of a Fuel-Cell–Battery–Supercapacitor Tramway. *IEEE Trans. Ind. Inform.* **2014**, *10*, 276–285. [CrossRef]
98. Li, Q.; Wang, T.; Dai, C.; Chen, W.; Ma, L. Power Management Strategy Based on Adaptive Droop Control for a Fuel Cell-Battery-Supercapacitor Hybrid Tramway. *IEEE Trans. Veh. Technol.* **2017**, *67*, 5658–5670. [CrossRef]
99. Ameur, K.; Hadjaissa, A.; Cheikh, M.S.A.; Cheknane, A.; Essounbouli, N. Fuzzy energy management of hybrid renewable power system with the aim to extend component lifetime. *Int. J. Energy Res.* **2017**, *41*, 1867–1879. [CrossRef]
100. Ahmadi, S.; Bathaee, S.; Hosseinpour, A.H. Improving fuel economy and performance of a fuel-cell hybrid electric vehicle (fuel-cell, battery, and ultra-capacitor) using optimized energy management strategy. *Energy Convers. Manag.* **2018**, *160*, 74–84. [CrossRef]
101. Zhou, D.; Al-Durra, A.; Gao, F.; Ravey, A.; Matraji, I.; Simões, M.G. Online energy management strategy of fuel cell hybrid electric vehicles based on data fusion approach. *J. Power Sources* **2017**, *366*, 278–291. [CrossRef]
102. Bizon, N.; Thounthong, P. Fuel economy using the global optimization of the Fuel Cell Hybrid Power Systems. *Energy Convers. Manag.* **2018**, *173*, 665–678. [CrossRef]
103. Fares, D.; Chedid, R.; Panik, F.; Karaki, S.; Jabr, R. Dynamic programming technique for optimizing fuel cell hybrid vehicles. *Int. J. Hydrogen Energy* **2015**, *40*, 7777–7790. [CrossRef]

104. Onori, S.; Tribioli, L. Adaptive Pontryagin's Minimum Principle supervisory controller design for the plug-in hybrid GM Chevrolet Volt. *Appl. Energy* **2015**, *147*, 224–234. [CrossRef]
105. Bizon, N.; Hoarcă, I.C. Hydrogen saving through optimized control of both fueling flows of the Fuel Cell Hybrid Power System under a variable load demand and an unknown renewable power profile. *Energy Convers. Manag.* **2019**, *184*, 1–14. [CrossRef]
106. Ramos-Paja, C.A.; Bordons, C.; Romero, A.; Giral, R.; Martinez-Salamero, L. Minimum Fuel Consumption Strategy for PEM Fuel Cells. *IEEE Trans. Ind. Electron.* **2008**, *56*, 685–696. [CrossRef]
107. Ou, K.; Yuan, W.-W.; Choi, M.; Yang, S.; Jung, S.; Kim, Y.-B. Optimized power management based on adaptive-PMP algorithm for a stationary PEM fuel cell/battery hybrid system. *Int. J. Hydrogen Energy* **2018**, *43*, 15433–15444. [CrossRef]
108. Bizon, N. Real-time optimization strategies of Fuel Cell Hybrid Power Systems based on Load-following control: A new strategy, and a comparative study of topologies and fuel economy obtained. *Appl. Energy* **2019**, *241*, 444–460. [CrossRef]
109. Wang, Y.-X.; Kai, O.; Kim, Y.-B. Power source protection method for hybrid polymer electrolyte membrane fuel cell/lithiumion battery system. *Renew. Energy* **2017**, *111*, 381–391. [CrossRef]
110. Bizon, N. Fuel saving strategy using real-time switching of the fueling regulators in the proton exchange membrane fuel cell system. *Appl. Energy* **2019**, *252*, 113449. [CrossRef]
111. Bizon, N.; Culcer, M.; Oproescu, M.; Iana, G.; Laurentiu, I.; Mazare, A.; Iliescu, M. Real-time strategy to optimize the airflow rate of fuel cell hybrid power source under variable load cycle. In Proceedings of the 2017 International Conference on Applied Electronics, Pilsen, Czech Republic, 5–7 September 2017.
112. Bizon, N. Sensitivity analysis of the fuel economy strategy for a fuel cell hybrid power system using fuel optimization and load-following based on air control. *Energy Convers. Manag.* **2019**, *199*, 111946. [CrossRef]
113. Bizon, N.; Stan, V.A.; Cormos, A.C. Stan Optimization of the Fuel Cell Renewable Hybrid Power System using the Control Mode of the Required Load Power on the DC Bus. *Energies* **2019**, *12*, 1889. [CrossRef]
114. Bizon, N.; Mazare, A.G.; Ionescu, L.M.; Thounthong, P.; Kurt, E.; Oproescu, M.; Serban, G.; Lita, I. Better Fuel Economy by Optimizing Airflow of the Fuel Cell Hybrid Power Systems Using Fuel Flow-Based Load-Following Control. *Energies* **2019**, *12*, 2792. [CrossRef]
115. Bizon, N. Efficient fuel economy strategies for the Fuel Cell Hybrid Power Systems under variable renewable/load power profile. *Appl. Energy* **2019**, *251*, 113400. [CrossRef]
116. Bizon, N. Sensitivity analysis of the fuel economy strategy based on load-following control of the fuel cell hybrid power system. *Energy Convers. Manag.* **2019**, *199*, 111946. [CrossRef]
117. *SimPowerSystems*; Hydro-Québec and the MathWorks, Inc.: Natick, MA, USA, 2010. Available online: http://www.hydroquebec.com/innovation/en/pdf/2010G080-04A-SPS.pdf (accessed on 9 February 2021).
118. Ettihir, K.; Boulon, L.; Agbossou, K. Optimization-based energy management strategy for a fuel cell/battery hybrid power system. *Appl. Energy* **2016**, *163*, 142–153. [CrossRef]
119. Bizon, N.; Kurt, E. Performance analysis of the tracking of the global extreme on multimodal patterns using the Asymptotic Perturbed Extremum Seeking Control scheme. *Int. J. Hydrogen Energy* **2017**, *42*, 17645–17654. [CrossRef]
120. Bizon, N.; Thounthong, P.; Raducu, M.; Constantinescu, L.M. Designing and modelling of the asymptotic perturbed extremum seeking control scheme for tracking the global extreme. *Int. J. Hydrogen Energy* **2017**, *42*, 17632–17644. [CrossRef]
121. Wu, J.; Wei, Z.; Li, W.; Wang, Y.; Li, Y.; Sauer, D. Battery Thermal- and Health-Constrained Energy Management for Hybrid Electric Bus based on Soft Actor-Critic DRL Algorithm. *IEEE Trans. Ind. Inform.* **2021**, *1*. [CrossRef]
122. Wu, J.; Wei, Z.; Liu, K.; Quan, Z.; Li, Y. Battery-Involved Energy Management for Hybrid Electric Bus Based on Expert-Assistance Deep Deterministic Policy Gradient Algorithm. *IEEE Trans. Veh. Technol.* **2020**, *69*, 12786–12796. [CrossRef]
123. Wei, Z.; He, H.; Pou, J.; Tsui, K.-L.; Quan, Z.; Li, Y. Signal-Disturbance Interfacing Elimination for Unbiased Model Parameter Identification of Lithium-Ion Battery. *IEEE Trans. Ind. Inform.* **2020**, *1*. [CrossRef]
124. Wei, Z.; Zhao, J.; He, H.; Ding, G.; Cui, H.; Liu, L. Future smart battery and management: Advanced sensing from external to embedded multi-dimensional measurement. *J. Power Sources* **2021**, *489*, 229462. [CrossRef]

Article

Impact of Renewable Energy Sources into Multi Area Multi-Source Load Frequency Control of Interrelated Power System

Krishan Arora [1,2,†], Ashok Kumar [2], Vikram Kumar Kamboj [1,3], Deepak Prashar [4], Bhanu Shrestha [5,*] and Gyanendra Prasad Joshi [6,*,†]

1. School of Electronics and Electrical Engineering, Lovely Professional University, Phagwara, Punjab 144411, India; er.krishanarora@gmail.com (K.A.); kamboj.vikram@gmail.com (V.K.K.)
2. Department of Electrical Engineering, Maharishi Markandeshwar University Mullana, Haryana 133207, India; ashok1234arora@gmail.com
3. Department of Electrical and Computer Engineering, Schulich School of Engineering, University of Calgary, Calgary, AB T2N 1N4, Canada
4. School of Computer Science and Engineering, Lovely Professional University, Phagwara, Punjab 144411, India; deepak.prashar@lpu.co.in
5. Department of Electronic Engineering, Kwangwoon University, Seoul 01897, Korea
6. Department of Computer Science and Engineering, Sejong University, Seoul 05006, Korea
* Correspondence: bnu@kw.ac.kr (B.S.); joshi@sejong.ac.kr (G.P.J.); Tel.: +82-02-811-0969 (B.S.); +82-2-6935-2481 (G.P.J.)
† These authors contributed equally to this work.

Citation: Arora, K.; Kumar, A.; Kamboj, V.K.; Prashar, D.; Shrestha, B.; Joshi, G.P. Impact of Renewable Energy Sources into Multi Area Multi-Source Load Frequency Control of Interrelated Power System. *Mathematics* **2021**, *9*, 186. https://doi.org/10.3390/math9020186

Received: 21 December 2020
Accepted: 14 January 2021
Published: 18 January 2021

Publisher's Note: MDPI stays neutral with regard to jurisdictional claims in published maps and institutional affiliations.

Copyright: © 2021 by the authors. Licensee MDPI, Basel, Switzerland. This article is an open access article distributed under the terms and conditions of the Creative Commons Attribution (CC BY) license (https://creativecommons.org/licenses/by/4.0/).

Abstract: There is an increasing concentration in the influences of nonconventional power sources on power system process and management, as the application of these sources upsurges worldwide. Renewable energy technologies are one of the best technologies for generating electrical power with zero fuel cost, a clean environment, and are available almost throughout the year. Some of the widespread renewable energy sources are tidal energy, geothermal energy, wind energy, and solar energy. Among many renewable energy sources, wind and solar energy sources are more popular because they are easy to install and operate. Due to their high flexibility, wind and solar power generation units are easily integrated with conventional power generation systems. Traditional generating units primarily use synchronous generators that enable them to ensure the process during significant transient errors. If massive wind generation is faltered due to error, it may harm the power system's operation and lead to the load frequency control issue. This work proposes binary moth flame optimizer (MFO) variants to mitigate the frequency constraint issue. Two different binary variants are implemented for improving the performance of MFO for discrete optimization problems. The proposed model was evaluated and compared with existing algorithms in terms of standard testing benchmarks and showed improved results in terms of average and standard deviation.

Keywords: wind technology (WT); load frequency control; optimization issue; moth flame optimizer (MFO); Harris hawks optimizer (HHO)

1. Introduction

Renewable energy sources (RESs) and especially wind technology are treated to be the most effective viable technology due to their environmental blessings, and their value of procedure and servicing have declined considerably within a few years. Hybrid plants secure the stability of supply commixture various sustainable energy resources like photovoltaic, wind generators, and even diesel generator sets (DGs) used for back-up purposes [1]. Therefore, grid connection of those together with traditional plants is adopted because of enhanced behavior with respect to effective load. It is ascertained that variation in frequency is induced because fluctuation in load is low, strengthening the insertion of inexhaustible resources. Load frequency control (LFC), in addition to the Proporional Integral Derivative (PID) controller [2,3] is suggested to overcome frequency inconsistency for a power system involving wind, hydro, and thermal units due to load and generating power variation induced due to the insertion of inexhaustible resources [4,5]. A system comprising thermal plants, hydro plants, and wind power plants will be designed with the help of MATLAB [6].

Our contributions in this paper are as follows: First, we recommend the two alternatives of binary moth flame optimizers to unravel the frequency restriction matter. We put into practice two diverse binary alternates for civilizing moth flame optimizer (MFO) behavior for distinct optimization tribulations [7]. In the primary variant, i.e., binary moth flame optimizer (BMFO1), coin flipping-based assortment probability of binary statistics is applicable. We applied the superior sigmoid transformation in the subsequent variant known as BMFO2. Along with Harris hawks optimizer (HHO) algorithms, these advanced algorithms are veteran and examined for a variety of unilateral, bilateral, and contract violation optimization problems. Secondly, Section 2 explores the impact of renewable energy sources on the load frequency control problem. Section 3 depicts the load frequency control issue's mathematical behavior when integrated with a renewable energy source. Section 4 shows the transfer function model of the multiarea multisource power system integrated with a renewable energy source. Lastly, in Section 5, all these algorithms are estimated and evaluated in conditions of typical testing benchmarks in which the projected HHO model has superior consequences with regards to mean and standard deviation. Finally, Section 6 winds up the paper.

2. Impact of Renewable Energy Sources

Renewable energy sources (RESs) definitely disturb the vibrant performance of the power system in such a manner that may be diverse from predictable generating units. Traditional generating units primarily use synchronous generators that enable them to ensure the process during major transient errors. If, due to error, a massive amount of wind generation is faltered, then the adverse influence of that error on the power system's operation with the LFC issue could also be expanded. High penetration of renewable energy in control arrangement may raise some reservations during the irregular procedure. It familiarizes numerous technical implications and exposes significant questions regarding what happens to LFC requirement [8,9] after the addition of various RESs to the present production system and whether the outdated mechanism methodologies are sufficient to operate in a fresh situation.

The influence on optimal flow of power, voltage, and management of frequency, power quality, and structure economics are increased due to the addition of RESs into control system grids. According to the behavior of RES power deviation, the impression on the LFC concern has involved rising research attention throughout the last era [10–12]. Substantial interrelated frequency fluctuations can cause over- or underfrequency transmitting and remove certain generations and loads. Under opposed situations, this may affect a dropping disappointment and miscarriage of the system. Figure 1 shows the block diagram of a wind energy generation system [13].

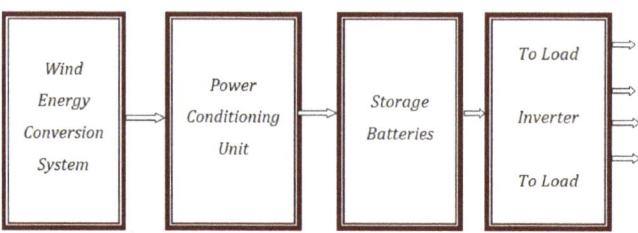

Figure 1. Block diagram of a wind energy generation system.

2.1. Advanced Optimization Methods

Optimization plays a major role in many fields of engineering. It is a path in which a suitable reaction to an exceptional matter is exposed via a search tool [14]. By means of upgradation in expertise, the novel creation of the matter fortitude optimization approach identified as metaheuristic has been used to deliberate mathematical humanity. Metaheuristic algorithms (MA) impersonate a systematic way to obtain the finest consequences for adilemma. MA act as a fantasy search for a good stipulation in an optimization concern [15]. In this paper, three advanced optimization techniques, BMFO1, BMFO2, and HHO, are considered and compared with conventional methods.

2.1.1. Binary Moth Flame Optimizer (BMFO1)

The basic MFO is a nature-inspired heuristic search technique that imitates moths' navigational possessions about artificial lights. BMFO1 is a recently predictable metaheuristics search algorithm proposed by Mirjalili [16], which is re-energized by navigation actions of moth and their meeting close to the beam. It helps to recover the exploitation search of the moths and diminish the quantity of flames. Even if moths have a tough potential to sustain a protected loom with respect to the moon and grip a bearable assembly for traveling in a traditional scratch for broad remoteness. They are also caring in a serious/inoperative bowed pathway over a replicated source of illumination.

2.1.2. Modified SIGMOID Transformation (BMFO2)

The binary calibration of stable quest accommodation and spaces of investigating council, resolution to binary searching domicile could be obligatory for optimizing binary ecological issues such as LFC. A modified sigmoidal transfer function is assumed in the projected work, which has better presentation than any more substitute of it as shown in [17]. Basic moth flame optimizer applied with the modified sigmoidal transformation (BMFO2) is used to carry out the binary chart of actual moth value and flame location for fixing the LFC problem.

2.1.3. Harris Hawks Optimizer

HHO [17,18] is a gradient-free and population-centered algorithm containing unfair and investigative steps for wonder swoop, the fauna of prey assessment, and assorted ploy built on brutal speculation of Harris hawks.

3. LFC Model with Integration of Renewable Energy Source

When renewable energy sources are integrated into the power arrangement, a supplementary cause of the deviation is added to the arrangement's adjustable behavior. To examine the deviations triggered by renewable energy source plants, the entire conclusion is significant, and every modification in renewable energy source output power does not requisite to be accorded with the conversion in other generating units running in the opposite way [19,20]. Sudden variations in load and renewable energy source output power might intensify each other either completely or partially. However, the sluggish renewable energy source power instability and entire average power fluctuation adversely subsidize power inequity and frequency distribution, which could be considered for the LFC arrangement. This energy variation must comprise a predictable LFC arrangement [21].

A general LFC model integrated with renewable energy source [22,23] is shown in Figure 2 in which the conforming blocks for governor dead-band, Generation Rate Constraints (GRC), and time delays are not involved. In this model, altered parametric standards are used for generator regulation and for turbine-governor to shelter the diversity of production categories in the control area. The expressed components and blocks are defined as follows: ΔP_m is mechanical power, Δf is frequency deviation, ΔP_L is load disturbance, ΔP_C is supplementary frequency control action, D_{Sys} is equivalent damping coefficient, H_{Sys} is equivalent inertia constant, β is frequency bias, α_i is participation factor, R_i is drooping characteristic, ΔP_P is primary frequency control action, ΔP_{RES} is renewable energy source power fluctuation, $M_i(s)$ is a governor-turbine model, ACEis area control error [24], and finally, $\Delta P'_L$ and $\Delta P'_{tie}$ are amplified local load alteration and tie-line power vacillation signals, respectively.

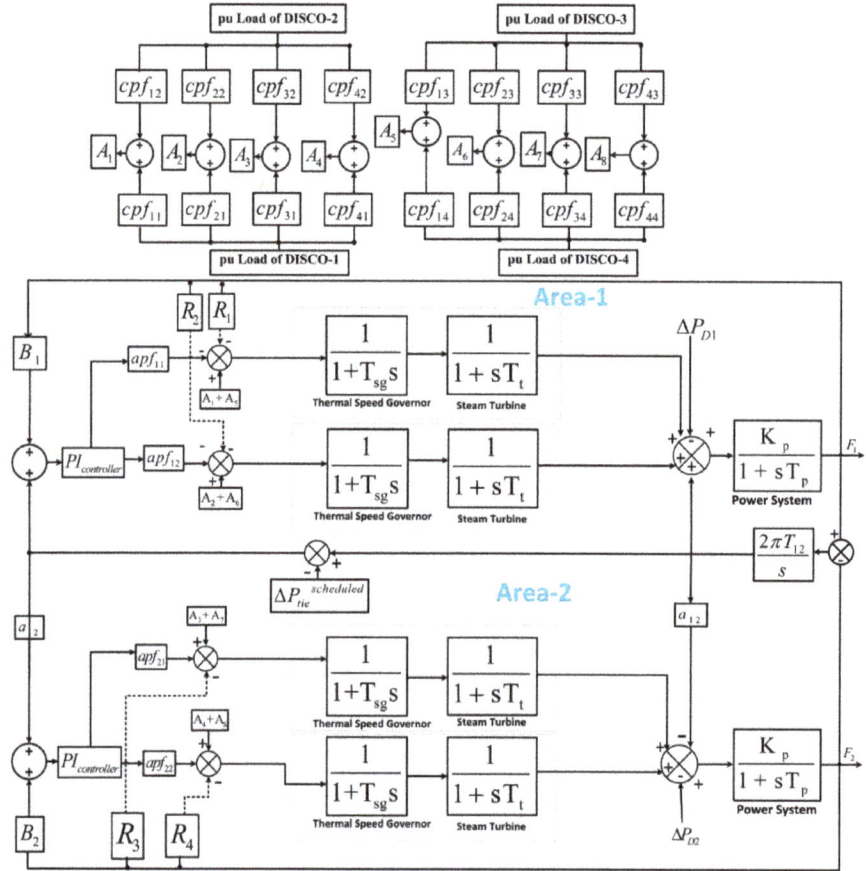

Figure 2. Transfer function model.

In the revised LFC structure, the efficient ACE signal must signify the influences of renewable energy on planned stream over tie-line and local power variation through area frequency. The ACE signal is conventionally described as a linear amalgamation of tie-line power and frequency fluctuations as follows [25,26]:

$$\text{ACE} = B\Delta f + \Delta P_{tie} \tag{1}$$

In the conventional power system, $\Delta P_{tie}(\Delta P_{tie-C})$ is a deviation between actual and scheduled energy flow over the tie-lines.

$$\Delta P_{tie-C} = \sum (P_{tie,actual} - P_{tie,Scheduled}) \tag{2}$$

The difference among the updated LFC model in Figure 2 and the conventional one provides two new signals which represent the transient behavior of RESs on tie-line power and local load variations ($\Delta P'_L$, $\Delta P'_{tie-RES}$):

$$\Delta P'_L(s) = \Delta P_{RES}(s) - \Delta P_L(s) \quad (3)$$

$$\Delta P_{tie-RES} = \sum(\Delta P_{tie-RES,actual} - \Delta P_{tie-RES,Scheduled}) \quad (4)$$

After adding a substantial amount of renewable power to the conventional power flow of tie-lines (ΔP_{tie-C}) in the power system, the updated renewable energy source power via tie-lines ($\Delta P_{tie-RES}$) must be painstaking. Consequently, the restructured tie-line power fluctuation can be articulated as follows:

$$\begin{aligned}\Delta P'_{tie} &= \Delta P_{tie-C} + \Delta P_{tie-RES} \\ &= \sum(\Delta P_{tie-C,actual} - \Delta P_{tie-C,Scheduled}) + \sum(\Delta P_{tie-RES,actual} - \Delta P_{tie-RES,Scheduled})\end{aligned} \quad (5)$$

The entire renewable energy source power flow modification is generally smoother as compared to fluctuation in influences from personal renewable energy source elements. Using Equations (1) and (5), the restructured ACE signal can be accomplished by Equation (6):

$$\begin{aligned}ACE &= B\Delta f + \Delta P'_{tie} \\ &= B\Delta f + \left(\sum(\Delta P_{tie-C,actual} - \Delta P_{tie-C,Scheduled}) + \sum(\Delta P_{tie-RES,actual} - \Delta P_{tie-RES,Scheduled})\right)\end{aligned} \quad (6)$$

where $P_{tie-C,actual}$ is actual conventional tie-line, $P_{tie-C,scheduled}$ is scheduled conventional tie-line, $P_{tie-RES,actual}$ is actual renewable energy source tie-line, and $P_{tie-RES,scheduled}$ is scheduled renewable energy source tie-line powers.

To justify the investigation part of the expected algorithm, various conventional as well as advanced optimization techniques such as Particle Swarm Optimization (PSO), MFO, BMFO1, BMFO2, and HHO were compared in terms of best value, worst value, mean and standard deviation. The different gain values for various algorithms for multiarea modal were also compared. Table 1 presents the performance of various algorithms in the proposed system and Table 2 depicts the Proportional Controller (PI) controller's gain values for various algorithms.

Table 1. Performance of various algorithms in the proposed system.

Parameter	No. of Trials	Mean	Standard Deviation	Best	Worst
PSO	50	1.13807	1.00846	1.02774	1.05084
MFO	50	0.0489	0.00644	0.02835	0.05073
BMFO1	50	0.0284	0.00083	0.02745	0.03216
BMFO2	50	0.0272	0.00011	0.02764	0.02811
HHO	50	7.62×10^{-93}	1.39×10^{-93}	1.4×10^{-112}	4.17×10^{-92}

Table 2. Improved gain values of Proportional Integral (PI) controller for the expected scheme with various algorithms.

	Controller Parameters			
Controller Type	Area1		Area2	
	$K_1{}^P$	$K_1{}^{Int}$	$K_2{}^P$	$K_2{}^{Int}$
PSO	0.53007	-7.12×10^{-6}	0.28516	-0.84672
MFO	0.52333	-0.70889	0.31424	-0.92724
BMFO1	0.53126	-0.0034	0.28435	9.83×10^{-6}
BMFO2	0.55025	-8.07×10^{-7}	0.28545	-1.89×10^{-6}
HHO	0.15025	-9.059×10^{-9}	0.18293	-1.91×10^{-9}

4. Transfer Function Model Multi Area Multi-Source Hydro-Thermal System with Wind Power Plant

The transfer function model of the hydro and thermal generating unit has been derived. The tie-line combines them in two methods, namely two-area, which is conventional, and then multiarea with renewable energy source, a new concept. On analyzing, the thermal system, when exposed to unit step load disturbance of 0.01 per unit in Area1 alone, the area frequency and the tie-line energy oscillates and settles with offset. The offset can be removed by including a secondary controller, which varies the governor's power reference setting. An optimal secondary controller is to be developed for the effective operation of the hydro–thermal power system integrated with renewable energy sources like wind energy.

5. Results and Discussion

Sensitivity analysis with respect to frequency variations, fault in tie-line, actual power flow, and output reaction of dissimilar generators after instant load fluctuation in the planned structure in terms of dispersion of wind under various contracts, such as unilateral, bilateral, and contract-violation case have been publicized in Figure 2.

5.1. Unilateral Transaction

The simulations with penetration of wind have been achieved to check the expected reaction of power scheme with respect to area frequency, power flow among interrelated areas, and the reaction of generating units through an unexpected load modification state in terms of expected agreements of the decontrolled electricity market as shown in Figures 3–8.

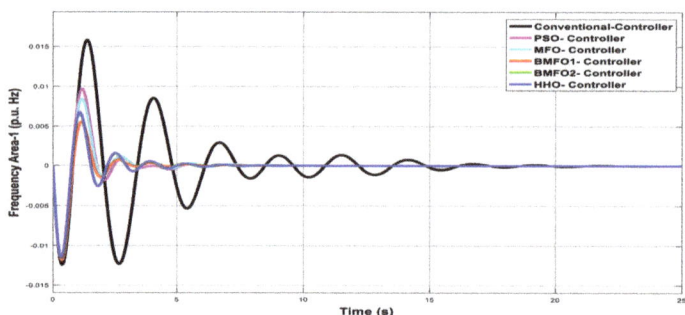

Figure 3. Dynamic response of Area1 frequency with various controllers under unilateral contract.

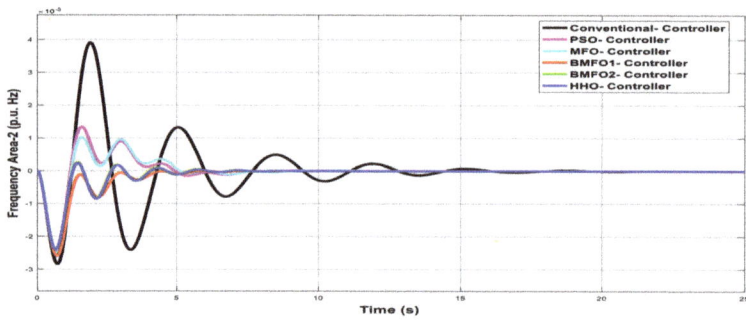

Figure 4. Dynamic response of Area2 frequency with various controllers under unilateral contract.

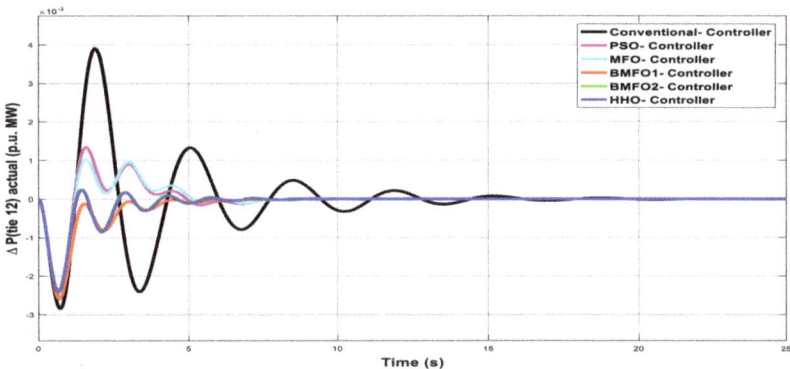

Figure 5. Deviation in actual tie-line power flow with various controllers under unilateral contract.

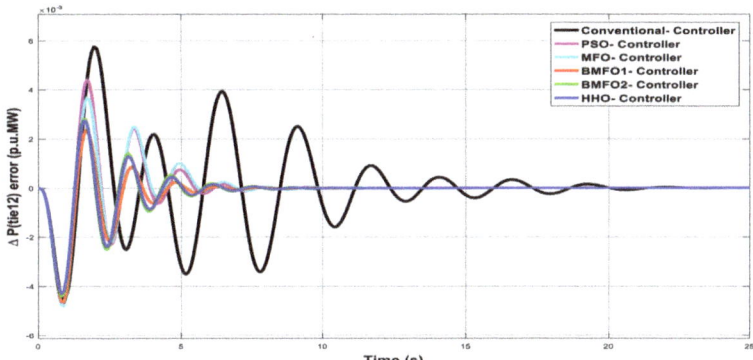

Figure 6. Deviation in tie-line error with various controllers under unilateral contract.

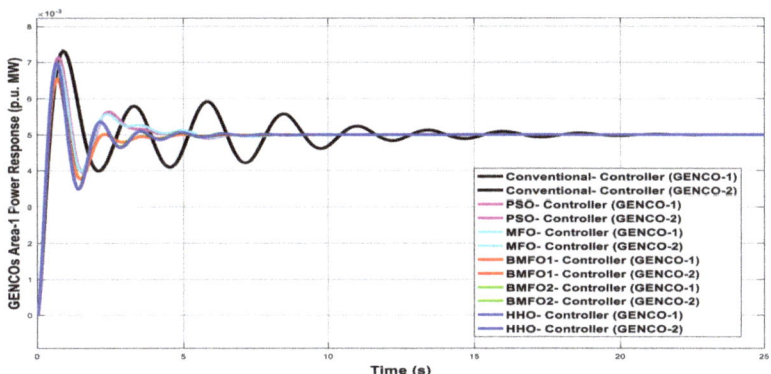

Figure 7. Generation Companies (GENCOs) generation response of Area1 with various controllers under unilateral contract.

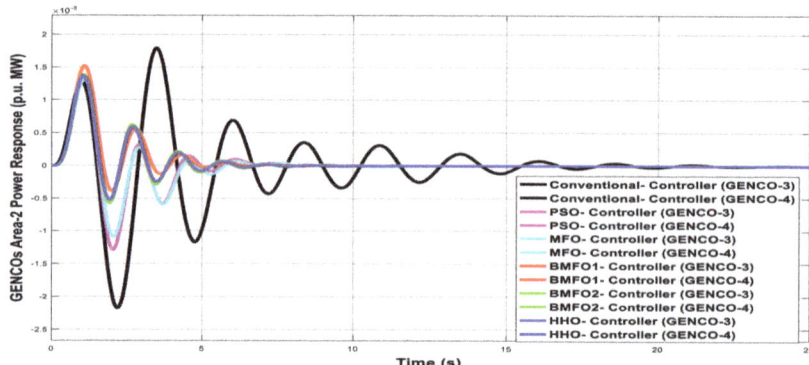

Figure 8. GENCOs generation response of Area2 with various controllers under unilateral contract.

Dynamic response of Area1 frequency with respect to time in seconds with various controllers like conventional, PSO, MFO, BMFO1, BMFO2, and HHO under unilateral contract are compared in Figure 3 and Table 3. The comparative outcomes obtained show improved results of the HHO controller with delay time 0.8633 s and settling time of 19.8 s.

Table 3. Graph analysis of Area1 frequency with various controllers under unilateral contract.

Controller	Delay Time	Rise Time	Peak Overshoot Time	Settling Time
Conventional	1.05	1.186	1.375	22
PSO	0.8833	1.05	1.186	21.430
MFO	0.8833	1.05	1.145	21.3
BMFO1	0.9667	1.05	1.145	21.02
BMFO2	0.8833	1.05	1.087	20.1
HHO	0.8833	0.9667	1.07	19.8

Dynamic response of Area2 frequency with respect to time in seconds with various controllers like conventional, PSO, MFO, BMFO1, BMFO2, and HHO under unilateral contract are compared in Figure 4 and Table 4. The comparative outcomes obtained show improved results of the HHO controller with delay time 1.297 s and settling time of 20.86 s.

Table 4. Graph analysis of Area2 frequency with various controller under unilateral contract.

Controller	Delay Time	Rise Time	Peak Overshoot Time	Settling Time
Conventional	1.474	1.654	1.868	23.39
PSO	1.336	1.414	1.594	21.17
MFO	1.336	1.474	1.594	21.01
BMFO1	−1.201	−1.414	−1.534	21.43
BMFO2	1.297	1.375	1.414	21.43
HHO	1.297	1.375	1.414	20.86

Dynamic response of deviation in actual tie-line power flow with respect to time in seconds with various controllers like conventional, PSO, MFO, BMFO1, BMFO2, and HHO under unilateral contract are compared in Figure 5 and Table 5. The comparative outcomes obtained show improved results of the HHO controller with delay time 0.162 s and settling time of 20.86 s.

Table 5. Graph analysis of deviation in actual tie-line power flow with various controllers under unilateral contract.

Controller	Delay Time	Rise Time	Peak Overshoot Time	Settling Time
Conventional	1.935	3.478	3.865	23.39
PSO	0.664	1.195	1.325	21.17
MFO	0.535	0.975	1.068	21.01
BMFO1	0.435	−0.778	−0.865	21.43
BMFO2	0.535	0.957	1.063	21.43
HHO	0.162	0.292	0.324	20.86

Dynamic response of deviation in tie-line error with respect to time in seconds with various controllers like conventional, PSO, MFO, BMFO1, BMFO2, and HHO under unilateral contract are compared in Figure 6 and Table 6. The comparative outcomes obtained show improved results of the HHO controller with delay time 1.157 s and settling time of 19.8 s [27].

Table 6. Graph analysis of deviation in tie-line error with various controller under unilateral contract.

Controller	Delay Time	Rise Time	Peak Overshoot Time	Settling Time
Conventional	2.935	5.280	5.865	22
PSO	2.118	3.814	4.235	21.430
MFO	1.955	3.498	3.886	21.3
BMFO1	1.115	2.002	2.224	21.02
BMFO2	1.988	3.555	3.950	20.1
HHO	1.157	2.085	2.315	19.8

Dynamic response of Generation Companies (GENCOs) generation of Area1 with respect to time in seconds with various controllers like conventional, PSO, MFO, BMFO1, BMFO2, and HHO under unilateral contract are compared in Figure 7 and Table 7. The comparative outcomes obtained show improved results of the HHO controller with delay time 3.335 s and settling time of 20.86 s.

Table 7. Graph analysis of GENCOs generation response of Area1 with various controllers under unilateral contract.

Controller	Delay Time	Rise Time	Peak Overshoot Time	Settling Time
Conventional	3.665	6.595	7.325	23.39
PSO	3.598	6.478	7.195	21.17
MFO	3.508	6.315	7.015	21.01
BMFO1	3.335	6.002	6.668	21.43
BMFO2	3.503	6.305	7.005	21.43
HHO	3.335	6.002	6.668	20.86

Dynamic response of GENCOs generation of Area2 with respect to time in seconds with various controllers like conventional, PSO, MFO, BMFO1, BMFO2, and HHO under unilateral contract are compared in Figure 8 and Table 8. The comparative outcomes obtained show improved results of the HHO controller with delay time 0.695 s and settling time of 19.8 s.

Table 8. Graph analysis of GENCOs generation response of Area2 with various controllers under unilateral contract.

Controller	Delay Time	Rise Time	Peak Overshoot Time	Settling Time
Conventional	0.888	1.598	1.775	22
PSO	0.725	1.305	1.450	21.430
MFO	0.734	1.333	1.468	21.3
BMFO1	0.755	1.355	1.505	21.02
BMFO2	0.723	1.300	1.445	20.1
HHO	0.695	1.251	1.390	19.8

5.2. Bilateral Based Transaction

The various dynamic responses with penetration of wind in mutual areas and the tie line are shown in Figures 9–14.

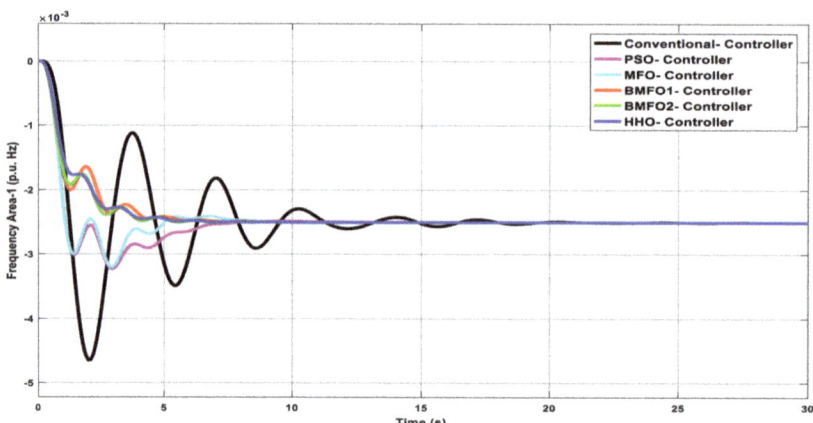

Figure 9. Dynamic response of Area1 frequency with various controllers under bilateral contract.

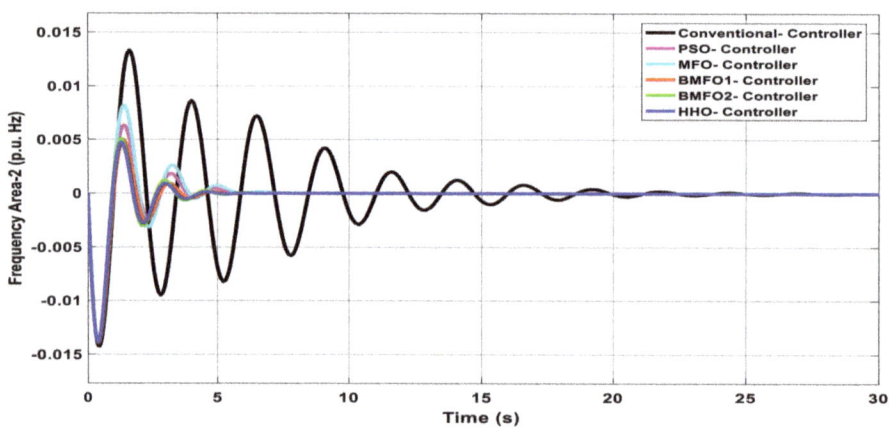

Figure 10. Dynamic response of Area2 frequency with various controllers under bilateral contract.

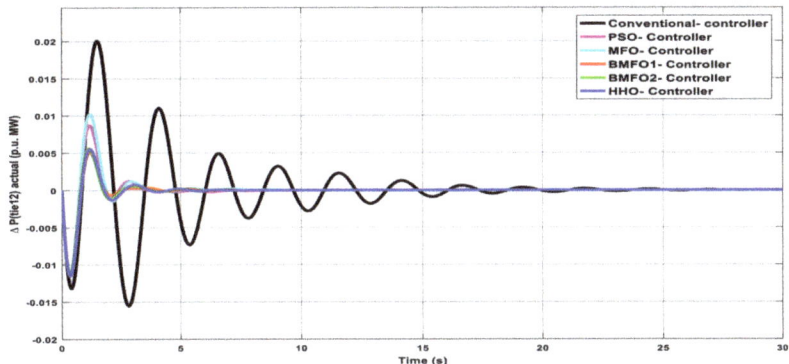

Figure 11. Deviation in actual tie-line power flow with various controllers under bilateral contract.

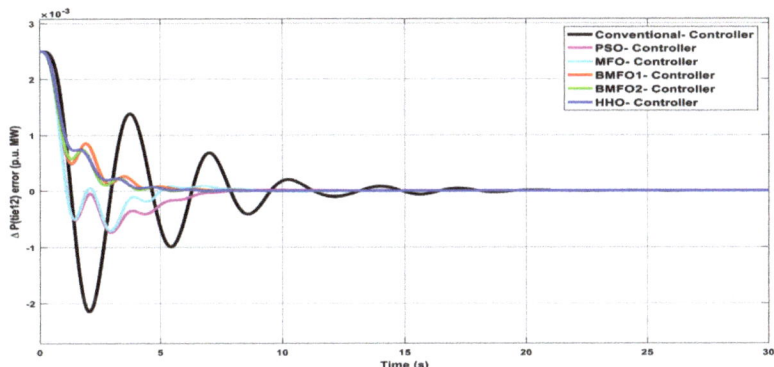

Figure 12. Deviation in tie-line error with various controller under bilateral contract.

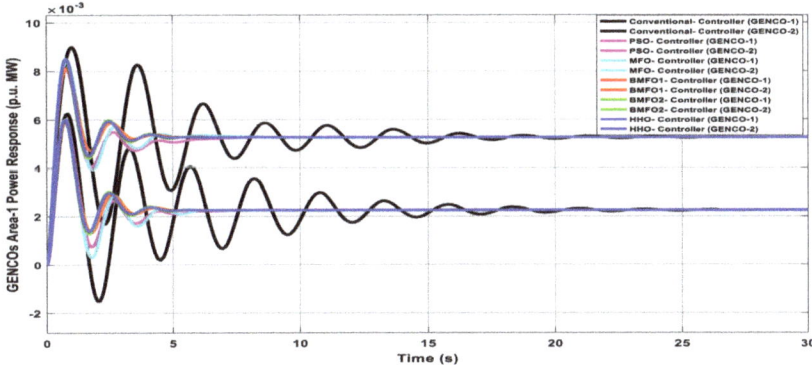

Figure 13. GENCOs generation response of Area1 with various controllers under bilateral contract.

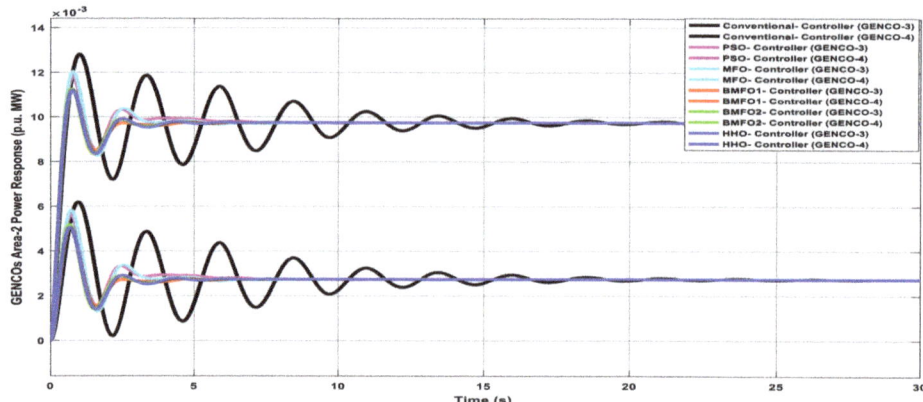

Figure 14. GENCOs generation response of Area2 with various controllers under bilateral contract.

Dynamic response of Area1 frequency with respect to time in seconds with various controllers like conventional, PSO, MFO, BMFO1, BMFO2, and HHO under bilateral contract are compared in Figure 9 and Table 9. The comparative outcomes obtained show improved results of the HHO controller with delay time −0.835 s and settling time of 20.86 s.

Table 9. Graph analysis of Area1 frequency with various controllers under bilateral contract.

Controller	Delay Time	Rise Time	Peak Overshoot Time	Settling Time
Conventional	−2.414	−4.205	−4.668	23.39
PSO	−1.508	−2.715	−3.015	21.17
MFO	−1.503	−2.705	−3.005	21.01
BMFO1	−1.003	−1.805	−2.005	21.43
BMFO2	−0.978	−1.766	−1.955	21.43
HHO	−0.835	−1.499	−1.665	20.86

Dynamic response of Area2 frequency with respect to time in seconds with various controllers like conventional, PSO, MFO, BMFO1, BMFO2, and HHO under Bilateral Contract are compared in Figure 10 and Table 10. The comparative outcomes obtained show improved results of the HHO controller with delay time = 0.0023 s and settling time of 20.86 s.

Table 10. Graph analysis of Area2 frequency with various controllers under bilateral contract.

Controller	Delay Time	Rise Time	Peak Overshoot Time	Settling Time
Conventional	0.0066	0.0118	0.0131	23.39
PSO	0.0031	0.0047	0.0052	21.17
MFO	0.0045	0.0074	0.0082	21.01
BMFO1	0.0026	0.0046	0.0051	21.43
BMFO2	0.0027	0.0048	0.0053	21.43
HHO	0.0023	0.0041	0.0045	20.86

Dynamic response of deviation in actual tie-line power flow with respect to time in seconds with various controllers like Conventional, PSO, MFO, BMFO1, BMFO2, and HHO under Bilateral Contract are compared in Figure 11 and Table 11. The comparative outcomes obtained show improved results of the HHO controller with delay time 0.0029 s and settling time of 20.86 s.

Table 11. Graph analysis of deviation in actual tie-line power flow with various controllers under bilateral contract.

Controller	Delay Time	Rise Time	Peak Overshoot Time	Settling Time
Conventional	0.0106	0.0190	0.0211	23.39
PSO	0.0043	0.0077	0.0085	21.17
MFO	0.0063	0.0114	0.0125	21.01
BMFO1	0.0033	0.0059	0.0065	21.43
BMFO2	0.0031	0.0055	0.0061	21.43
HHO	0.0029	0.0053	0.0058	20.86

Dynamic response of Deviation in tie-line error with respect to time in seconds with various controllers like Conventional, PSO, MFO, BMFO1, BMFO2, and HHO under Bilateral Contract are compared in Figure 12 and Table 12. The comparative outcomes obtained show improved results of the HHO controller with delay time 1.276 s and settling time of 19.8 s.

Table 12. Graph analysis of deviation in tie-line error with various controller under bilateral contract.

Controller	Delay Time	Rise Time	Peak Overshoot Time	Settling Time
Conventional	1.284	2.311	2.568	22
PSO	1.278	2.299	2.555	21.430
MFO	1.277	2.298	2.554	21.3
BMFO1	1.277	2.297	2.553	21.02
BMFO2	1.277	2.296	2.552	20.1
HHO	1.276	2.295	2.551	19.8

Dynamic response of GENCOs to DISCOs generation of Area-1 with respect to time in seconds with various controllers like Conventional, PSO, MFO, BMFO1, BMFO2, and HHO under Bilateral Contract are compared in Figure 13 and Table 13a. The comparative outcomes obtained show improved results of the HHO controller with delay time 3.006 s and settling time of 20.86 s.

Table 13. (a). Graphanalysis of GENCOs to Distribution Companies (DISCOs) generation response of Area1 with various controllers under bilateral contract. (b). Graph analysis of DISCOs to GENCOs generation response of Area1 with various controllers under bilateral contract.

(a)				
Controller	Delay Time	Rise Time	Peak Overshoot Time	Settling Time
Conventional	3.130	5.630	6.255	23.39
PSO	3.116	5.608	6.231	21.17
MFO	3.056	5.514	6.112	21.01
BMFO1	3.056	5.514	6.112	21.43
BMFO2	3.044	5.478	6.088	21.43
HHO	3.006	5.411	6.012	20.86

(b)				
Controller	Delay Time	Rise Time	Peak Overshoot Time	Settling Time
Conventional	4.508	8.114	9.015	22
PSO	4.335	7.799	8.665	21.430
MFO	4.335	7.799	8.665	21.3
BMFO1	4.258	7.663	8.514	21.02
BMFO2	4.258	7.663	8.514	20.1
HHO	4.224	7.600	8.445	19.8

Dynamic response of Distribution Companies (DISCOs) to GENCOs generation of Area1 with respect to time in seconds with various controllers like conventional, PSO,

MFO, BMFO1, BMFO2, and HHO under bilateral contract are compared in Figure 13 and Table 13b. The comparative outcomes obtained show improved results of the HHO controller with delay time 4.224 s and settling time of 19.8 s.

Dynamic response of GENCOs to DISCOs generation of Area2 with respect to time in seconds with various controllers like conventional, PSO, MFO, BMFO1, BMFO2, and HHO under bilateral contract are compared in Figure 14 and Table 14a. The comparative outcomes obtained show improved results of the HHO controller with delay time 2.506 s and settling time of 20.86 s.

Table 14. (a). Graph analysis of GENCOs to DISCOs generation response of Area2 with various controllers under bilateral contract. (b). Graph analysis of DISCOs to GENCOs generation response of Area2 with various controllers under bilateral contract.

(a)				
Controller	Delay Time	Rise Time	Peak Overshoot Time	Settling Time
Conventional	3.114	5.591	6.212	23.39
PSO	2.943	5.297	5.885	21.17
MFO	2.998	5.397	5.996	21.01
BMFO1	2.857	5.143	5.714	21.43
BMFO2	2.663	4.792	5.324	21.43
HHO	2.506	4.511	5.012	20.86
(b)				
Controller	Delay Time	Rise Time	Peak Overshoot Time	Settling Time
Conventional	6.434	11.579	12.865	22
PSO	6.006	10.811	12.012	21.430
MFO	6.066	10.901	12.112	21.3
BMFO1	6.066	10.901	12.112	21.02
BMFO2	5.885	10.595	11.770	20.1
HHO	5.564	10.001	11.112	19.8

Dynamic response of DISCOs to GENCOs generation of Area2 with respect to time in seconds with various controllers like conventional, PSO, MFO, BMFO1, BMFO2, and HHO under bilateral contract are compared in Figure 14 and Table 14b. The comparative outcomes obtained show improved results of the HHO controller with delay time 5.564 s and settling time of 19.8 s.

5.3. Contract Violation Case

The various dynamic responses with penetration of wind energy in two areas and the tie-line are represented in Figures 15–20.

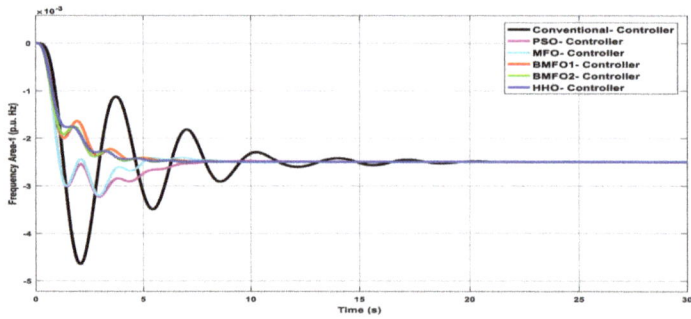

Figure 15. Dynamic response of Area1 frequency with various controllers under contract violation case.

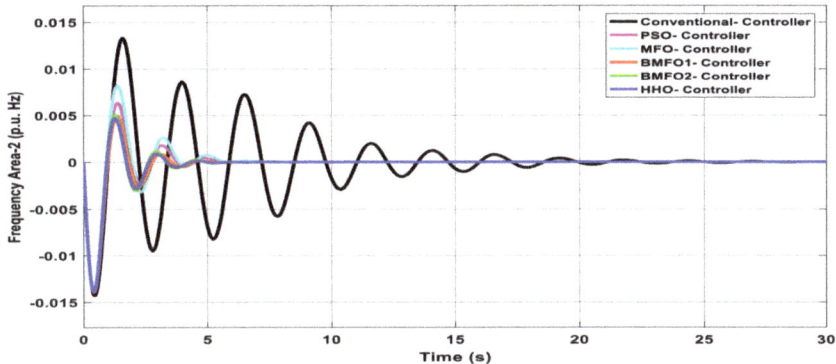

Figure 16. Dynamic response of Area2 frequency with various controller under contract violation case.

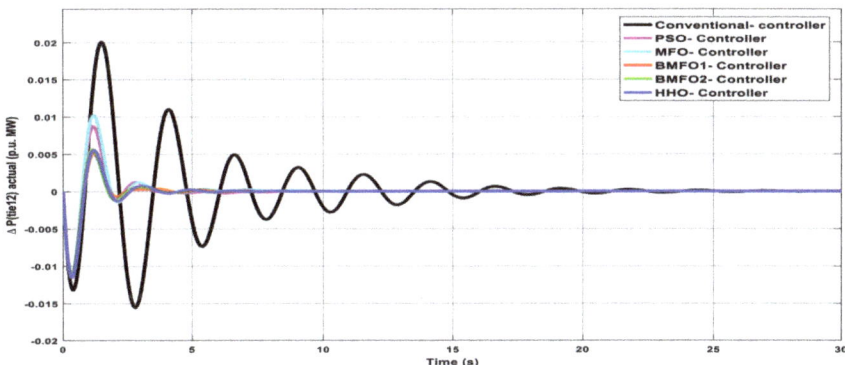

Figure 17. Deviation in actual tie-line power flow with various controllers under contract violation case.

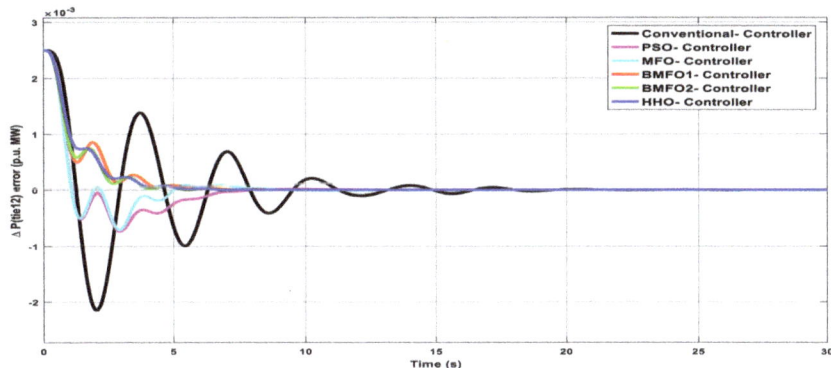

Figure 18. Deviation in tie-line error with various controllers under contract violation case.

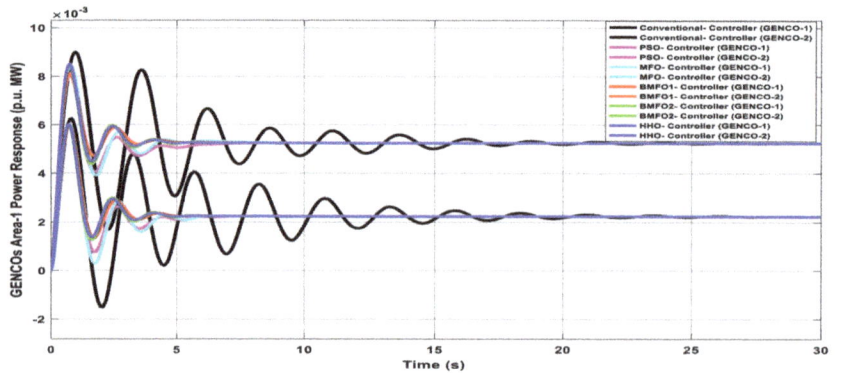

Figure 19. GENCOs generation response of Area1 with various controllers under contract violation case.

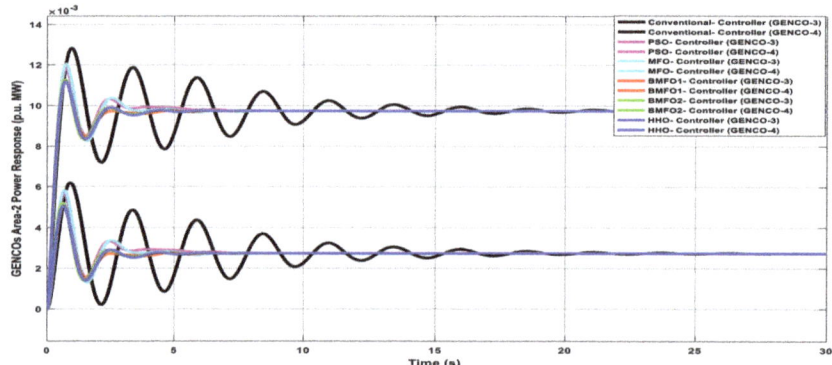

Figure 20. GENCOs generation response of Area2 with various controllers under contract violation case.

Dynamic response of Area1 frequency with respect to time in seconds with various controllers like conventional, PSO, MFO, BMFO1, BMFO2, and HHO under contract violation case are compared in Figure 15 and Table 15. The comparative outcomes obtained show improved results of the HHO controller with delay time −0.835 s and settling time of 20.86 s.

Table 15. Graph analysis of Area1 frequency with various controllers under contract violation case.

Controller	Delay Time	Rise Time	Peak Overshoot Time	Settling Time
Conventional	−2.414	−4.205	−4.668	23.39
PSO	−1.508	−2.715	−3.015	21.17
MFO	−1.503	−2.705	−3.005	21.01
BMFO1	−1.003	−1.805	−2.005	21.43
BMFO2	−0.978	−1.766	−1.955	21.43
HHO	−0.835	−1.499	−1.665	20.86

Dynamic response of Area2 frequency with respect to time in seconds with various controllers like conventional, PSO, MFO, BMFO1, BMFO2, and HHO under contract violation case are compared in Figure 16 and Table 16. The comparative outcomes obtained show improved results of the HHO controller with delay time 0.0023 s and settling time of 20.86 s.

Table 16. Graph analysis of Area2 frequency with various controller under contract violation case.

Controller	Delay Time	Rise Time	Peak Overshoot Time	Settling Time
Conventional	0.0066	0.0118	0.0131	23.39
PSO	0.0031	0.0047	0.0052	21.17
MFO	0.0045	0.0074	0.0082	21.01
BMFO1	0.0026	0.0046	0.0051	21.43
BMFO2	0.0027	0.0048	0.0053	21.43
HHO	0.0023	0.0041	0.0045	20.86

Dynamic response of deviation in actual tie-line power flow with respect to time in seconds with various controllers like conventional, PSO, MFO, BMFO1, BMFO2, and HHO under contract violation case are compared in Figure 17 and Table 17. The comparative outcomes obtained show improved results of the HHO controller with delay time 0.0029 s and settling time of 20.86 s.

Table 17. Graph analysis of deviation in actual tie-line power flow with various controller under contract violation case.

Controller	Delay Time	Rise Time	Peak Overshoot Time	Settling Time
Conventional	0.0106	0.0190	0.0211	23.39
PSO	0.0043	0.0077	0.0085	21.17
MFO	0.0063	0.0114	0.0125	21.01
BMFO1	0.0033	0.0059	0.0065	21.43
BMFO2	0.0031	0.0055	0.0061	21.43
HHO	0.0029	0.0053	0.0058	20.86

Dynamic response of deviation in tie-line error with respect to time in seconds with various controllers like conventional, PSO, MFO, BMFO1, BMFO2, and HHO under contract violation case are compared in Figure 18 and Table 18. The comparative outcomes obtained show improved results of the HHO controller with delay time 1.276 s and settling time of 19.8 s.

Table 18. Graph analysis of deviation in tie-line error with various controllers under contract violation case.

Controller	Delay Time	Rise Time	Peak Overshoot Time	Settling Time
Conventional	1.284	2.311	2.568	22
PSO	1.278	2.299	2.555	21.430
MFO	1.277	2.298	2.554	21.3
BMFO1	1.277	2.297	2.553	21.02
BMFO2	1.277	2.296	2.552	20.1
HHO	1.276	2.295	2.551	19.8

Dynamic response of GENCOs to DISCOs generation of Area1 with respect to time in seconds with various controllers like conventional, PSO, MFO, BMFO1, BMFO2, and HHO under contract violation case are compared in Figure 19 and Table 19a. The comparative outcomes obtained show improved results of the HHO controller with delay time 3.006 s and settling time of 20.86 s.

Table 19. (a). Graph analysis of GENCOs to DISCOs generation response of Area1 with various controllers under contract violation case. (b). Graph analysis of DISCOs to GENCOs generation response of Area1 with various controllers under contract violation case.

(a)				
Controller	Delay Time	Rise Time	Peak Overshoot Time	Settling Time
Conventional	3.130	5.630	6.255	23.39
PSO	3.116	5.608	6.231	21.17
MFO	3.056	5.514	6.112	21.01
BMFO1	3.056	5.514	6.112	21.43
BMFO2	3.044	5.478	6.088	21.43
HHO	3.006	5.411	6.012	20.86
(b)				
Controller	Delay Time	Rise Time	Peak Overshoot Time	Settling Time
Conventional	4.508	8.114	9.015	22
PSO	4.335	7.799	8.665	21.430
MFO	4.335	7.799	8.665	21.3
BMFO1	4.258	7.663	8.514	21.02
BMFO2	4.258	7.663	8.514	20.1
HHO	4.224	7.600	8.445	19.8

Dynamic response of DISCOs to GENCOs generation of Area1 with respect to time in seconds with various controllers like conventional, PSO, MFO, BMFO1, BMFO2, and HHO under contract violation case are compared in Figure 19 and Table 19b. The comparative outcomes obtained show improved results of the HHO controller with delay time 4.224 s and settling time of 19.8 s.

Dynamic response of GENCOs to DISCOs generation of Area1 with respect to time in seconds with various controllers like conventional, PSO, MFO, BMFO1, BMFO2, and HHO under contract violation case are compared in Figure 20 and Table 20a. The comparative outcomes obtained show improved results of the HHO controller with delay time 2.506 s and settling time of 20.86 s.

Table 20. (a). Graph analysis of GENCOs to DISCOs generation response of Area2 with various controllers under contract violation case. (b). Graph analysis of DISCOs to GENCOs generation response of Area2 with various controllers under contract violation case.

(a)				
Controller	Delay Time	Rise Time	Peak Overshoot Time	Settling Time
Conventional	3.114	5.591	6.212	23.39
PSO	2.943	5.297	5.885	21.17
MFO	2.998	5.397	5.996	21.01
BMFO1	2.857	5.143	5.714	21.43
BMFO2	2.663	4.792	5.324	21.43
HHO	2.506	4.511	5.012	20.86
(b)				
Controller	Delay Time	Rise Time	Peak Overshoot Time	Settling Time
Conventional	6.434	11.579	12.865	22
PSO	6.006	10.811	12.012	21.430
MFO	6.066	10.901	12.112	21.3
BMFO1	6.066	10.901	12.112	21.02
BMFO2	5.885	10.595	11.770	20.1
HHO	5.564	10.001	11.112	19.8

Dynamic response of DISCOs to GENCOs generation of Area1 with respect to time in seconds with various controllers like conventional, PSO, MFO, BMFO1, BMFO2, and HHO under contract violation case are compared in Figure 20 and Table 20b. The comparative outcomes obtained show improved results of the HHO controller with delay time 5.564 s and settling time of 19.8 s.

The relative resultsillustrate that an HHO-based PI controller fabricated finer conclusions when compared with structures including conventional PSO, MFO, BMFO1, and BMFO2-centered PI controllers. The peaks' overshoot/undershoot and settling period were compact and perturbations were quickly covered in the structure having a HHO-focused PI controller [28].

6. Conclusions

The power system's consistent operation necessitates a constant balancing of source and load as per recognized operating criteria. MFO is a very promising and interesting algorithm due to its advantages like fast searching speed and simplicity, but hasdrawbacks like getting stuck in bad local optima because it focuses on exploitation rather than exploration. Therefore, it is of great importance to research and put forward advanced optimization algorithms like BMFO1, BMFO2, and HHO with better performance to supplement the algorithm. This paper discussed major issues regarding the addition of RESs into frequency regulation power structure, which is most noticeable recently. This work briefly studied the utmost significant problems with the current accomplishments reported in the literature with different contract cases of unilateral, bilateral, and contract violation. The analysis results showed the improved consequences compared to conventional regulators with the help of modern soft computing techniques like the Harris hawks optimizer. The revised LFC model was also presented, which maintains the system frequency without any steady-state error, unlike conventional PI and moth flame optimizer. It instantaneously responds to different load disturbances and makes the system stable within a short time.

Author Contributions: Conceptualization, K.A. and A.K.; data curation, K.A., A.K. and V.K.K.; formal analysis, K.A.; investigation, V.K.K., B.S. and G.P.J.; methodology, D.P., B.S. and G.P.J.; software, A.K.; supervision, G.P.J.; validation, G.P.J.; visualization, K.A., D.P. and G.P.J.; writing—original draft, K.A.; writing—review and editing, G.P.J. All authors have read and agreed to the published version of the manuscript.

Funding: This research received no external funding.

Data Availability Statement: No new data were created or analyzed in this study. Data sharing is not applicable to this article.

Acknowledgments: We would like to thank Sudan Jha from the Department of Computer Science and Engineering, Chandigarh University, Punjab, India for his help in this work.

Conflicts of Interest: The authors declare no conflict of interest.

References

1. García, J.; Martí, J.V.; Yepes, V. The Buttressed Walls Problem: An Application of a Hybrid Clustering Particle Swarm Optimization Algorithm. *Mathematics* **2020**, *8*, 862. [CrossRef]
2. Ziegler, J.G.; Nichols, N.B. Optimum Settings for Automatic Controllers. *J. Dyn. Syst. Meas. Control.* **1993**, *115*, 220–222. [CrossRef]
3. Prashar, D.; Arora, K. Design of Two Area Load Frequency Control Power System under Unilateral Contract with the Help of Conventional Controller. *Int. J. Inf. Commun. Technol. Digit. Converg.* **2020**, *5*, 22–27.
4. Elgerd, O.I.; Fosha, C.E. Optimum Megawatt-Frequency Control of Multiarea Electric Energy Systems. *IEEE Trans. Power Appar. Syst.* **1970**, *89*, 556–563. [CrossRef]
5. Hossein, S.; Heider, A.; Shayanfar, A.J. Multi Stage Fuzzy PID Load Frequency Controller in a Restructured Power System. *J. Electr. Eng.* **2007**, *58*, 61–70.
6. *MATLAB 9.4, Version 9.4 (R2018a)*; The MathWorks Inc.: Natick, MA, USA, 2018.
7. Mirjalili, S. Moth-flame optimization algorithm: A novel nature-inspired heuristic paradigm. *Knowl. Based Syst.* **2015**, *89*, 228–249. [CrossRef]
8. Kothari, M.L.; Kaul, B.L.; Nanda, J. Automatic Generation Control of Hydro-Thermal System. *J. Inst. Eng. India* **1980**, *61*, 85–91.
9. Christie, R.; Bose, A. Load frequency control issues in power system operations after deregulation. *IEEE Trans. Power Syst.* **1996**, *11*, 1191–1200. [CrossRef]

10. Pan, C.-T.; Liaw, C.-M. An adaptive controller for power system load-frequency control. *IEEE Trans. Power Syst.* **1989**, *4*, 122–128. [CrossRef]
11. Bekhouche, N.; Feliachi, A. Decentralized estimation for the automatic generation control problem in power systems. In Proceedings of the First IEEE Conference on Control Applications, Dayton, OH, USA, 13–16 September 1992; Volume3, pp. 621–632.
12. Bid, A.P.; Sapclt, A.T., Jr.; bzvcm, C.S. An enhanced neural network load frequency control technique. In Proceedings of the International Conference on Control '94, Coventry, UK, 21–24 March 1994; Volume 389, pp. 409–415.
13. Elgerd, O.I.; Happ, H.H. Electric Energy Systems Theory: An Introduction. *IEEE Trans. Syst. Man Cybern.* **1972**, *2*, 296–297. [CrossRef]
14. Venter, G. Review of Optimization Techniques. In *Encyclopedia of Aerospace Engineering*; American Cancer Society: Atlanta, GA, USA, 2010.
15. Hopper, E.; Turton, B. An empirical investigation of meta-heuristic and heuristic algorithms for a 2D packing problem. *Eur. J. Oper. Res.* **2001**, *128*, 34–57. [CrossRef]
16. Trivedi, I.N.; Jangir, P.; Parmar, S.A.; Jangir, N. Optimal power flow with voltage stability improvement and loss reduction in power system using Moth-Flame Optimizer. *Neural Comput. Appl.* **2018**, *30*, 1889–1904. [CrossRef]
17. Konda, S.R.; Panwar, L.K.; Panigrahi, B.; Kumar, R. Solution to unit commitment in power system operation planning using binary coded modified moth flame optimization algorithm (BMMFOA): A flame selection based computational technique. *J. Comput. Sci.* **2018**, *25*, 298–317. [CrossRef]
18. Kamboj, V.K.; Nandi, A.; Bhadoria, A.; Sehgal, S. An intensify Harris Hawks optimizer for numerical and engineering optimization problems. *Appl. Soft Comput.* **2020**, *89*, 106018. [CrossRef]
19. Xu, D.; Liu, J.; Yan, X.; Yan, W. A novel adaptive neural network constrained control for a multi-area interconnected power system with hybrid energy storage. *IEEE Trans. Ind. Electron.* **2018**, *65*, 6625–6634. [CrossRef]
20. Xu, Y.; Li, C.; Wang, Z.; Zhang, N.; Peng, B. Load Frequency Control of a Novel Renewable Energy Integrated Micro-Grid Containing Pumped Hydropower Energy Storage. *IEEE Access* **2018**, *6*, 29067–29077. [CrossRef]
21. Yan, W.; Sheng, L.; Xu, D.; Yang, W.; Liu, Q. H∞ Robust Load Frequency Control for Multi-Area Interconnected Power System with Hybrid Energy Storage System. *Appl. Sci.* **2018**, *8*, 1748. [CrossRef]
22. Datta, A.; Bhattacharjee, K.; Debbarma, S.; Kar, B. Load frequency control of a renewable energy sources based hybrid system. In Proceedings of the 2015 IEEE Conference on Systems, Process and Control (ICSPC), Bandar Sunway, Malaysia, 18–20 December 2015; pp. 34–38.
23. Mehdi, N.; Reza, H.; Shoorangiz, S.S.A.F.; Sayed, M.S.B. Comparison of Artificial Intelligence Methods for Load Frequency Control Problem. *Aust. J. Basic Appl. Sci.* **2010**, *4*, 4910–4921.
24. Gulzar, M.M.; Rizvi, S.T.H.; Ling, Q.; Sibtain, D.; Din, R.S.U. Mitigating the Load Frequency Fluctuations of Interconnected Power Systems Using Model Predictive Controller. *Electronics* **2019**, *8*, 156. [CrossRef]
25. Kennedy, J.; Ebhart, R. Particle Swarm Optimization. In Proceedings of the IEEE International Conference on Neural Networks, Perth, Australia, 27 November–1 December 1995; Volume 1, pp. 1942–1948.
26. Swain, A.K. A Simple Fuzzy Controller for Single Area Hydropower System Considering Generation Rate Constraints. *J. Inst. Eng. India Part El Electr. Eng. Div.* **2006**, *87*, 12–17.
27. Arora, K.; Kumar, A.; Kamboj, V.K.; Prashar, D.; Jha, S.; Shrestha, B.; Joshi, G.P. Optimization Methodologies and Testing on Standard Benchmark Functions of Load Frequency Control for Interconnected Multi Area Power System in Smart Grids. *Mathematics* **2020**, *8*, 980. [CrossRef]
28. Alhelou, H.H.; Golshan, M.E.H.; Zamani, R.; Heydarian-Forushani, E.; Siano, P. Challenges and Opportunities of Load Frequency Control in Conventional, Modern and Future Smart Power Systems: A Comprehensive Review. *Energies* **2018**, *11*, 2497. [CrossRef]

Article

Grid Frequency and Amplitude Control Using DFIG Wind Turbines in a Smart Grid

José Antonio Cortajarena [1,*], Oscar Barambones [2], Patxi Alkorta [1] and Jon Cortajarena [3]

1. Engineering School of Gipuzkoa, University of the Basque Country, Otaola Hirib. 29, 20600 Eibar, Spain; patxi.alkorta@ehu.eus
2. Engineering School of Vitoria, University of the Basque Country, Nieves Cano 12, 01006 Vitoria, Spain; oscar.barambones@ehu.eus
3. Engineering School of Gipuzkoa, University of the Basque Country, Europa Plaza 1, 20018 Donostia, Spain; joncorta10@gmail.com
* Correspondence: josean.cortajarena@ehu.eus; Tel.: +34-943-033-041

Abstract: Wind-generated energy is a fast-growing source of renewable energy use across the world. A dual-feed induction machine (DFIM) employed in wind generators provides active and reactive, dynamic and static energy support. In this document, the droop control system will be applied to adjust the amplitude and frequency of the grid following the guidelines established for the utility's smart network supervisor. The wind generator will work with a maximum deloaded power curve, and depending on the reserved active power to compensate the frequency drift, the limit of the reactive power or the variation of the voltage amplitude will be explained. The aim of this paper is to show that the system presented theoretically works correctly on a real platform. The real-time experiments are presented on a test bench based on a 7.5 kW DFIG from Leroy Somer's commercial machine that is typically used in industrial applications. A synchronous machine that emulates the wind profiles moves the shaft of the DFIG. The amplitude of the microgrid voltage at load variations is improved by regulating the reactive power of the DFIG and this is experimentally proven. The contribution of the active power with the characteristic of the droop control to the load variation is made by means of simulations. Previously, the simulations have been tested with the real system to ensure that the simulations performed faithfully reflect the real system. This is done using a platform based on a real-time interface with the DS1103 from dSPACE.

Keywords: double feed induction generator; grid frequency and amplitude support; smart grid

1. Introduction

Wind energy is progressively gaining importance in the world's electricity production, with important engineering aspects to be addressed for its integration into conventional electricity grids. In fact, it contributes about 7% of total energy production worldwide and onshore and offshore wind energy together would generate more than a third of all electricity needs, becoming the main source of generation by 2050 [1].

Recently, some studies have been developed in order to analyze the installation of small wind turbines in urban areas. Installing wind turbines in all the possible extents can mitigate the rising energy demand. Built-up areas possess high potential for wind energy, including the rooftop of high-rise buildings, railway track, the region between or around multistoried buildings, and city roads. However, harnessing wind energy from these areas is quite challenging due to dynamic environments and turbulence for higher roughness on urban surfaces [2]. Some studies have been done in order to estimate the wind resource in an urban area [3,4]. These studies evaluate the urban wind resource by employing a physically-based empirical model to link wind observations at a conventional meteorological site to those acquired at urban sites. The approach is based on urban climate research that has examined the effects of varying surface roughness on the wind-field between and above buildings. These papers aim to provide guidance for optimizing the placement of small wind turbines in urban areas by developing an improved method of estimating the wind resource across a wide urban area.

There are different types of wind turbines like induction generator [5] or permanent magnet synchronous generators [6,7] based wind turbines, however the synchronous generators are usually used for a small-scale wind turbines.

In addition, the DFIM is the most commonly installed type of generator in wind turbines to date.

DFIG generators provide access to the rotor windings and regulating the rotor voltage, the generator active and reactive powers are fully controllable. Therefore, the design and implementation of a new control scheme for a DFIG based wind turbine system has attracted the attention of several authors in the last years [8–12].

The main reason for using the doubly fed induction generator is that the power converters have to manage only a fraction of the total system power, about 30%.

For this reason, there are fewer losses in the power electronics unit than in a full-power converter topology. The reduction in costs due to the use of a smaller inverter is another significant factor [13].

The regulation of the active and reactive powers in decoupled form with the DFIG is regulated using the field-oriented control (FOC) technique [14–19].

A smart grid has among its goals one dedicated to providing a more robust, efficient, and flexible electric power system [20,21]. In the last decade, the model predictive control (MPC) has been applied to microgrid systems to optimally schedule and control the microgrid, owing to the advantages of MPC such as fast response and robustness against parameter uncertainties. In the work presented in [22], a stochastic model predictive control framework to optimally schedule and control the microgrid with large scale renewable energy sources is proposed. This microgrid consists of fuel cell-based, wind turbines, PV generators, battery/thermal energy storage system gas fired boilers, and various types of electrical and thermal loads scheduled according to the demand response policy. In the work presented in [23], the authors propose a MPC for regulating frequency in stand-alone microgrids. This work analyzes the impact of system parameters on the control performance of MPC for frequency regulation, using a typical stand-alone microgrid, which consists of a diesel engine generator, an energy storage system, a wind turbine generator, and a load. A novel sensorless model predictive control (MPC) strategy of a wind-driven doubly fed induction generator (DFIG) connected to a dc microgrid is proposed in [24]. In this work, the MPC strategy has been used as a current controller to overcome the weaknesses of the inner control loop and to consider the discrete-time operation of the voltage source converter that feeds the rotor.

The extensive use of power generation using wind has forced countries to establish a set of rules for the operation and connection to the grid of wind generators. Amongst all the standards, those related to smart networks aim to guarantee a safe supply, and ensure the reliability and quality of the energy generated [25,26]. With the strong growth of grid-connected wind power plants, there is a need for grid-integrated wind farms with the ability to withstand grid voltage and frequency also during perturbations in the network [18].

To withstand the voltage and frequency, the wind generator must be able to change its operating point according to the needs. The most widely used and robust method is the well-known droop control. This control method is based on a concept known from the electrical networks and is based on the reduction of the frequency of the generator when its active power consumption increases [27].

This control strategy can be applied to different generators such as wind turbines in order to increase the reliability of the system.

In wind turbines, when control is carried out by means of the power drivers, as is the case of the DFIM, the speed is adjusted according to the wind speed in order to optimize energy production. This allows the regulation of the generator at the point of maximum power in a wide range of power. When using the droop technique, the generator must be running at a different speed than the maximum power speed. Consequently, when droop control is demanded, the control system will adjust the active and reactive powers to balance out deviations in frequency and amplitude of the grid voltage respectfully.

In the work presented in [11] the authors propose a Power Delta Control for a wind turbine in order to participate in the primary and secondary frequency regulation. This work presents a readily industrializable set of algorithms for torque and pitch control. However, the proposed control scheme are only validated by means of simulation using the NREL's FAST software.

The motivation of the present work is to validate the simulated algorithms in a real system based on wind generators to contribute to the compensation of the frequency and amplitude variations of the voltage of the microgrids.

This control scheme was initially proposed and simulated in the 2018 IEEE International Conference on Industrial Electronics for Sustainable Energy Systems [28]. However, in this previous work, the proposed control approach was only validated by means of some simulation results using a simple wind turbine model. This new work goes further, providing more simulations using a more detailed model, which accurately represents the dynamics of the real system. This will allow simulations to be made when real tests cannot be performed. Moreover, some real time experiments are presented over a test bench based on a commercial machine, typically used in industrial applications. Thus, the wind generator is a 7.5 kW Leroy Somer DFIM driven by a synchronous machine that emulates the desired wind profiles. Different experiments were developed using this test bench, designed and built ad hoc, and these real experiments validated the results previously obtained in the simulations. Thus, these experimental results can be used to demonstrate the applicability of this control scheme in industrial applications. It should be noted that the experimental validation of the new control schemes is a considerable research advance, since it facilitates its implementation in real industrial applications. This article deals with the process of controlling the DFIG in the contribution of active and reactive power for frequency and voltage regulation. The technique used is the well-known droop control.

The work is organized as follows; Sections 2 and 3 present the equations for the control of the DFIG. Sections 4–6 cover the droop control, the deloading process for frequency control and explain the voltage compensation and the calculation of the maximum reactive power based on the active working power. Section 7 introduces the laboratory results and the simulations. Lastly, the conclusions are presented.

2. DFIG Control Equations and Reference System

Figure 1 shows the stationary $\alpha\beta$ stator reference system, the rotor $\alpha'\beta'$ reference system and the dq reference system linked to the doubly fed induction machine stator flux vector.

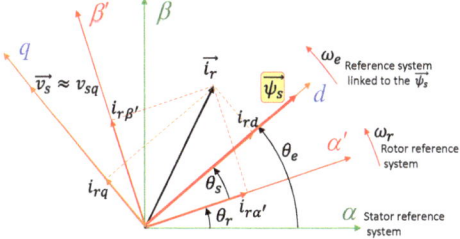

Figure 1. DFIG reference systems.

The doubly fed induction generator is controlled in the revolving reference frame aligned with the stator flux as shown in Figure 1. The next equations explain the operation of the DFIG in the dq frame [29,30].

$$\vec{v}_{s,dq} = R_s \vec{i}_{s,dq} + \frac{d\vec{\psi}_{s,dq}}{dt} + j\omega_e \vec{\psi}_{s,dq} \quad (1)$$

$$\vec{v}_{r,dq} = R_r \vec{i}_{r,dq} + \frac{d\vec{\psi}_{r,dq}}{dt} + j(\omega_e - \omega_r)\vec{\psi}_{r,dq} \quad (2)$$

$$\vec{\psi}_{s,dq} = L_s \vec{i}_{s,dq} + L_m \vec{i}_{r,dq} \text{ and } L_s = L_m + L_{ls} \quad (3)$$

$$\vec{\psi}_{r,dq} = L_r \vec{i}_{r,dq} + L_m \vec{i}_{s,dq} \text{ and } L_r = L_m + L_{lr} \quad (4)$$

$$T_e = \frac{3}{2} P_p (\psi_{sd} i_{sq} - \psi_{sq} i_{sd}) \quad (5)$$

$$T_e - T_L = J \frac{d\omega_m}{dt} + B\omega_m \quad (6)$$

$$P_s = \frac{3}{2}(v_{sd} i_{sd} + v_{sq} i_{sq}) \text{ and } Q_s = \frac{3}{2}(v_{sq} i_{sd} - v_{sd} i_{sq}) \quad (7)$$

where $\vec{v}_{s,dq}$, $\vec{i}_{s,dq}$, and $\vec{\psi}_{s,dq}$ are the stator voltage, current and flux vectors in the synchronous dq reference system. $\vec{v}_{r,dq}$, $\vec{i}_{r,dq}$, and $\vec{\psi}_{r,dq}$ are the rotor voltage, current, and flux vectors in the synchronous *dq* reference system. As can be seen in Figure 1, the stator flux q component is zero, and when operating with Equation (3) the next two equations are obtained,

$$i_{sd} = \frac{|\vec{\psi}_s|}{L_s} - \frac{L_m}{L_s} i_{rd} \quad (8)$$

$$i_{sq} = -\frac{L_m}{L_s} i_{rq} \quad (9)$$

Equations (8) and (9) shown that the stator current is controllable with the rotor current. With the omission of the stator resistance because it is so small, the stator flux can be assumed to be constant and its value is,

$$|\vec{\psi}_s| = \frac{|\vec{v}_s|}{\omega_e} \quad (10)$$

The stator voltage d component is almost zero because the reference system is oriented along the stator flux, so it can be obtained that,

$$P_s \approx \frac{3}{2} \omega_e \psi_s \frac{L_m}{L_s} i_{rq} \quad (11)$$

$$Q_s \approx \frac{3}{2} |\vec{v}_s| \left(\frac{|\vec{v}_s|}{\omega_e L_s} - \frac{L_m}{L_s} i_{rd} \right) \quad (12)$$

Equations (11) and (12) show that the stator active power is controlled with the q component of the rotor current and the stator reactive power with the rotor current d component.

Figure 2 shows the block diagram for the control of the DFIG from the rotor side using the rotor side converter (RSC). The current references i_{rq}^* and i_{rd}^* are calculated with (11) and (12) and with the required active and reactive power references. These current references are then compared with the real currents and the differences are the inputs signals of two PI regulators, obtaining in their outputs the references of the rotor voltage v_{rq}^* and v_{rd}^*. Lastly, the DC/AC inverter pulses SA, SB, and SC are produced using the seven segments space vector pulse width modulation (SVPWM).

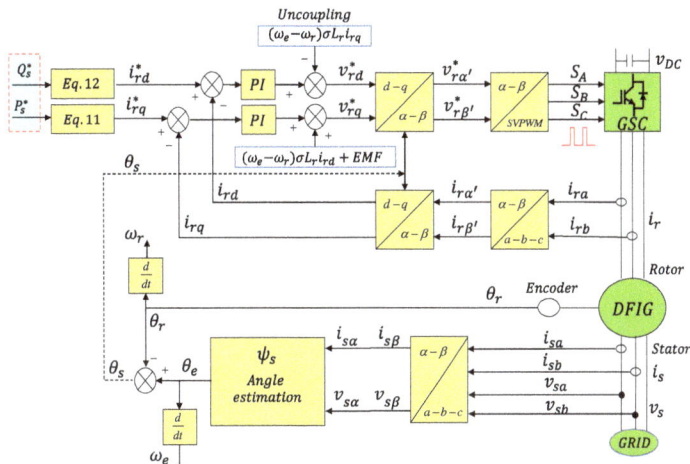

Figure 2. DFIG control structure.

3. Grid Side Converter Control Equations

For the control of the grid side converter (GSC), the revolving reference frame is aligned with the grid voltage vector as illustrated in Figure 3.

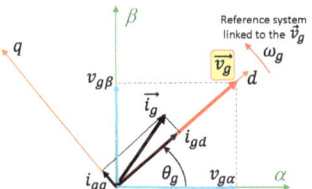

Figure 3. Grid voltage reference system.

The following equations describe the behavior of the grid side converter connected to the grid trough a line filter L_g and R_g in the mentioned rotating reference system,

$$\vec{v}_{conv,dq} = R_g \vec{i}_{g,dq} + L_g \frac{d\vec{i}_{g,dq}}{dt} + \vec{v}_{g,dq} \tag{13}$$

$$P_g = \frac{3}{2}\left(v_{gd}i_{gd} + v_{gq}i_{gq}\right) \text{ and } Q_g = \frac{3}{2}\left(v_{gq}i_{gd} - v_{gd}i_{gq}\right) \tag{14}$$

where $\vec{v}_{conv,dq}$, $\vec{i}_{g,dq}$, and $\vec{v}_{g,dq}$ are the GSC output voltage, the GSC current and grid voltage vectors. P_g and Q_g are the active and reactive powers regulated by the GSC.

Thus, the grid voltage q component is zero and the active power is regulated with the grid current d component and the reactive power is regulated with the grid current q component [31,32].

$$P_g = \frac{3}{2}v_{gd}i_{gd} \text{ and } Q_g = -\frac{3}{2}v_{gd}i_{gq} \tag{15}$$

Figure 4 illustrates a block diagram of the implemented grid side power converter control structure.

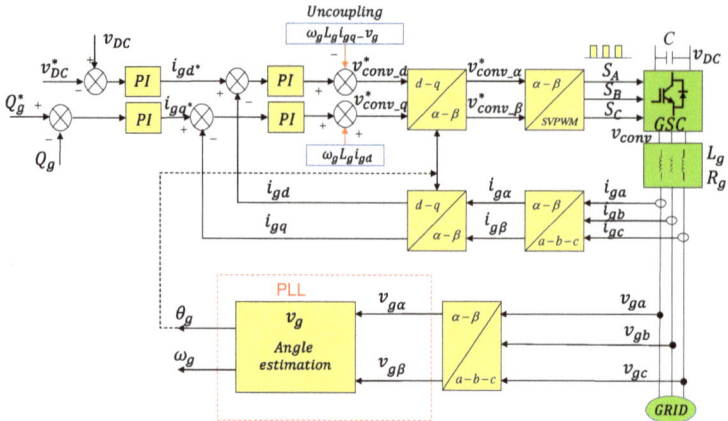

Figure 4. Grid side converter control structure.

The actual DC bus voltage is compared with the reference of the DC bus voltage and the difference will be the input of a PI regulator to obtain at its output the reference of the active current component i^*_{gd}. Then, i^*_{gd} is compared with the real i_{gd} current component to obtain at the output of the controller $v^*_{conv,d}$. The controller $v^*_{conv,q}$ is obtained from the reactive power controller section. Lastly, the DC/AC inverter pulses SA, SB, and SC are obtained using the seven segments space vector pulse width modulation (SVPWM).

4. Droop Control

One DFIM could be pictured as a voltage source that is connected to the principal or local network, through an impedance line Z as illustrated in Figure 5 [33,34].

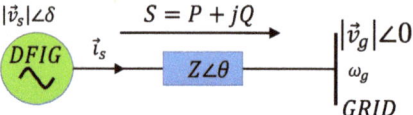

Figure 5. DFIG generator connected to the main grid trough line impedance.

In a system working in a sinusoidal steady state, the apparent power, S, which flows between the DFIG and the grid of Figure 5, could be defined as,

$$S = P + jQ = \vec{v}_s \vec{i}_s^{\,*} = \vec{v}_s \left(\frac{\vec{v}_s - \vec{v}_g}{Z\angle\theta} \right)^* = \frac{V_s^2}{Z} e^{j\theta} - \frac{V_s V_g}{Z} e^{j(\theta+\delta)} \quad (16)$$

whereas, V_s and V_g respectively are the voltage module of the DFIG and the main grid voltage vectors, δ is the angle between the stator voltage of the DFIG and the grid voltage, also known as the power angle, Z and θ are the magnitude and phase of the line impedance and $\vec{i}_s^{\,*}$ is the conjugated complex vector of the stator current. The active and reactive power of Equation (16) can be decomposed as,

$$P = \frac{V_s^2}{Z}\cos(\theta) - \frac{V_s V_g}{Z}\cos(\theta+\delta)$$
$$Q = \frac{V_s^2}{Z}\sin(\theta) - \frac{V_s V_g}{Z}\sin(\theta+\delta) \quad (17)$$

Supposing that the impedance of the line is mostly inductive, $|Z| = \omega_g L$ and $\theta = 90°$, and that δ is near to zero, $\sin(\delta) \approx \delta$ and $\cos(\delta) \approx 1$, therefore (17) can be simplified to,

$$P = \frac{V_s V_g}{\omega_g L}\delta \tag{18}$$

$$Q = V_s \frac{V_s - V_g}{\omega_g L} \tag{19}$$

Equations (18) and (19) show that regulating the active power the angle δ or the frequency of the network can be adjusted and the voltage difference $V_s - V_g$ is regulated by the reactive power. It should be stressed, that this is just true when the line impedance is mainly inductive, which is often the case. The aim of droop regulation is to adapt the frequency and amplitude of the grid voltage independently by regulating the active and reactive power.

The frequency and voltage droop characteristics are shown in Figure 6.

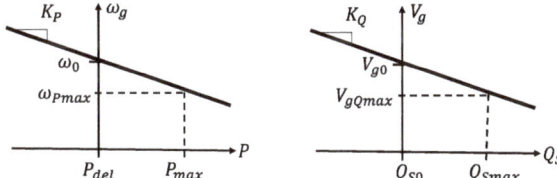

Figure 6. Droop characteristics. Left frequency, right voltage.

Where,

$$K_p = -\frac{\omega_0 - \omega_{P_{max}}}{P_{max} - P_{del}} \tag{20}$$

$$K_q = -\frac{V_{g0} - V_{gQmax}}{Q_{Smax} - Q_{S0}} \tag{21}$$

and, ω_0 is the rated frequency given to the power P_{del} and V_{g0} is the rated voltage amplitude given to the reactive power Q_{S0}. P_{max} is the maximum active power that the wind turbine can obtain from the wind speed and Q_{Smax} is the maximum reactive power of the stator. K_P and K_Q are the frequency and voltage droop coefficients, respectively. $\omega_{P_{max}}$ and V_{gQmax} are the admitted minimum grid frequency and amplitude for the possible P_{max} and Q_{Smax} respectively.

5. Getting Additional Active Power of the Wind Generator

In order to make a frequency regulation when the frequency of the system falls, some extra energy must be extracted from the wind turbine. Therefore, DFIG wind turbines must run with a deloaded power curve when the plant is running under regular frequency requirements [35,36]. The power transferred to the shaft of a wind generator is stated as

$$P_{turb} = \frac{1}{2} C_P(\lambda, \beta) \rho_{air} \pi R^2 v_w^3 \tag{22}$$

where ρ_{air}, is the mass density of the air, R is the radius of the propeller, $C_P(\lambda, \beta)$ is the power performance coefficient, v_w is the wind speed, β is the pitch angle and λ is the blade tip speed ratio and is defined as,

$$\lambda = \frac{R \omega_{pr}}{v_w} \tag{23}$$

Taking into account Equation (22), the wind turbine is deloaded by influencing the performance of the power coefficient. Cp depends on the tip speed ratio and pitch angle; thus, by adjusting one or both factors, the wind turbine could be deloaded.

The term applied to the C_p factor is described in Equation (24),

$$C_P(\lambda, \beta) = \frac{n_1(\lambda n_6 + (-n_4 - n_3(2.5+\beta) + n_2 A - n8B))}{e^{(n_5 A - n_8 B)}}$$
$$A = \frac{1}{\lambda + n_7(2.5+\beta)} \quad \text{and} \quad B = \frac{1}{1 + (2.5+\beta)^3} \tag{24}$$

where, $n_1 = 0.645$, $n_2 = 116$, $n_3 = 0.4$, $n_4 = 5$, $n_5 = 21$, $n_6 = 9.12 \times 10^{-3}$, $n_7 = 0.08$ and $n_8 = 0.035$ [37].

The power of the turbine produced with the C_p coefficient, for a zero pitch angle and for various wind speeds in relation to the speed of the DFIG rotor is illustrated in Figure 7.

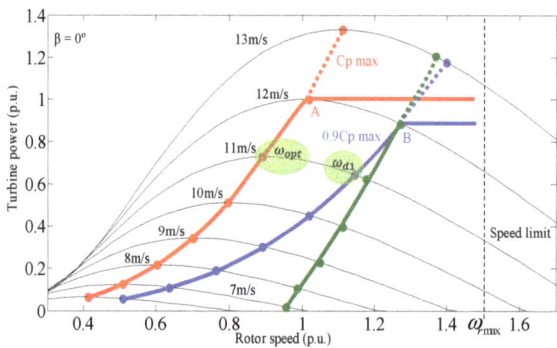

Figure 7. Turbine deloaded and maximum power curves.

With Equations (24) and (22), various turbine power characteristics are determined as illustrated in Figure 7. The red one belongs to the maximum achievable power, in which the power coefficient is the highest available. The blue and green curves are the deloaded power curves. The blue curve represents a 10% deloaded power curve at each given wind speed, being able to extract when necessary a 10% of the maximum power for that wind speed. However, the green curve allows 10% of the maximum power to be extracted at any wind speed. The deload procedure decreases the efficiency of a wind turbine by raising the tip speed ratio. Thus, the deload curves shift to the right of the maximum power curve. Because of the increase in speed, kinetic energy is stored in the inertia of the wind turbine and is expressed as

$$E_{kinetic} = \frac{1}{2} J_T \left(\omega_{d1}^2 - \omega_{opt}^2 \right) \tag{25}$$

where J_T is the wind turbine inertia, ω_{d1} is the turbine deloaded speed and ω_{opt} is the maximum power speed or optimum speed as illustrated in Figure 7. This energy could be employed for short-term frequency control [38]. The additional active power that can be achieved for the deloading operation can be derived from Equation (22) and is dependent on the wind speed and the deloading percentage used as

$$\Delta P_{turb} = \frac{1}{2} \Delta C_P(\lambda, \beta) \rho_{air} \pi R^2 v_w^3 \tag{26}$$

ΔP_{turb} is the power reserve, ΔC_P is the increase in the power coefficient when the turbine moves from the deloaded working point to a greater power line. This power increase is utilized for long-term frequency adjustment [30].

The reference calculation of the downloaded DFIG speed for every wind speed starts with the specification of the optimal tip speed relation. Thereafter, with the extra active power demanded by the smart grid supervisor, the new tip speed ratio is computed based on Equations (23) and (24).

Observing Figure 7, for 12 m/s of wind speed, the wind turbine can run 10% deloaded as in point B. When the grid frequency decreases, due to a sudden increase in the load, the

DFIG must boost the set point of the desired active power. Consequently, the operating power coefficient of the wind turbine moves from the point B of the deloaded curve to the point A of the higher power curve.

In this way, the increase in active power supports the control of the grid frequency. If the highest deloaded power is achieved, (point B) due to increased wind speed, the pitch controller will be required to maintain the required power reserve, which is the difference between the red and blue horizontal lines.

The pitch controller can be used also to avoid overstress or to limit the speed of the wind turbine while regulating the frequency [20].

The major issue of the wind generator is the uncontrollability of the wind as its speed variations affect the amount of energy stored. The manner of saving the desired amount of power, regardless of the wind speed, is to deload the wind turbine for a specified power value as shown in Figure 7. For instance, the green curve shows that the reserved power is 10% of the maximum power for all speeds.

However, working in this way wastes a large amount of wind energy at low speeds. In Figure 8, a diagram is shown for the generation of the active power reference to be introduced in the P^* input of the DFIG control scheme presented in Figure 2. Figure 8 also includes droop control, inertial energy, and the pitch controller. The pitch controller will act when the active power surpasses the maximum deloaded power, the peak mechanical power or when the wind turbine speed is greater than the higher limit [39]. The control command $\Delta P_{supervisor}$ is applied by the network supervisor for other control strategies such as automatic control of generation and energy flow or to limit the energy level to a specified level [20,40].

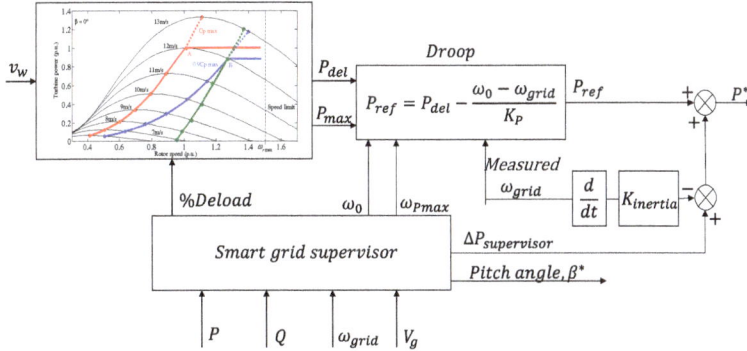

Figure 8. DFIG reference generation of active power. Droop and smart grid supervisor defined control.

6. Reactive Power and Voltage Amplitude Control

The amplitude of the stator voltage is regulated by adjusting the reactive power as described in the Section 4.

The reactive energy is controlled by the GSC and by the RSC [17]. Both converters are dimensioned to handle about 30% of the nominal power of the generator. Converters are principally utilized to supply the active energy from the rotor to the stator or vice versa. The GSC can be employed to supply reactive energy together with the stator to the grid.

It should be noted that the maximum grid-side converter current cannot be exceeded and therefore the amount of reactive power that can be injected by the grid-side converter will depend on the amount of active power flowing through the rotor. Thus, when the speed of the rotor is synchronous, the active power through the rotor is approximately zero and therefore the reactive power could be maximum. However, as the power through the rotor increases due to an increase in slip, the reactive power must be reduced.

As shown in Equation (12), the reactive power through the stator is regulated by the rotor current d-component. When the load increases suddenly, not only does the frequency decrease, but also the amplitude of the grid voltage tends to be reduced. Because of that, the active and reactive power increases trying to correct the frequency and the amplitude of the grid voltage.

The rotor current must be limited to its maximum value taking into account the active and reactive power values. Equation (27), derived from Equations (11) and (12), provides the maximum active and reactive powers permitted considering the rotor maximum current,

$$(P_s)^2 + \left(Q_s - \frac{3}{2}\frac{|\vec{v}_s|^2}{\omega_e L_s}\right)^2 = \left(\frac{3}{2}|\vec{v}_s|\frac{L_m}{L_s}\right)^2 \left(i_{rd}^2 + i_{rq}^2\right) \tag{27}$$

A graphical representation of Equation (27) in Figure 9, illustrates the full range of $P_s - Q_s$ generation in steady state. If Q_s is greater than zero, the DFIG absorbs reactive energy due to its inductive feature [41], however, Q_s values lower than zero correspond to the reactive power that the DFIG is able to supply to the grid.

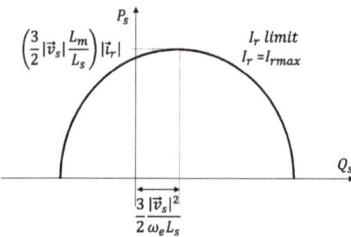

Figure 9. Stator $P_s - Q_s$ boundaries for rated current of the rotor.

Figure 10 illustrates the red curve of maximum mechanical power that can be obtained for the specified wind speeds. The blue and green curves show the reactive power that the DFIG can provide to the grid when the generator is operating at rated rotor current.

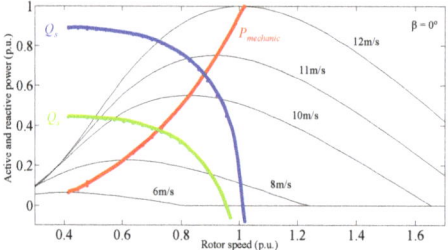

Figure 10. Maximum mechanical power and reactive powers obtainable for maximum rotor current. DFIG 7.5 kW, green curve. DFIG 1.2 MW, blue curve.

The blue Q_s line corresponds to a 1.2 MW DFIG and the green line to a 7.5 kW DFIG. When the DFIG is taking the maximum power of the wind, all the rotor current is formed by the q component and the value of the d component will be zero. This implies that the reactive power is absorbed from the grid and its value will be defined by Equation (12). As shown in the figure, the lower the mechanical power, the more reactive power the DFIG can provide to the grid to compensate for the voltage drop in the grid.

The selection of the maximum active power in the grid frequency compensation does not provide any choice to assist in the correction of the grid amplitude from the stator, but the GSC could be used if their current limit is maintained. This is why certain criteria must be detailed by the smart grid controller in the distribution of the maximum active and reactive powers while respecting the current limits of the stator, rotor and the GSC.

Figure 11 provides a diagram for generating the reactive power references of the stator and the GSC converter. The value of K_Q depends on the capacity of the DFIG to supply reactive power and this value changes as the active power varies as shown in Figure 10. The $\Delta Q_{supervisor}$ control signal is employed by the grid controller for additional regulation strategies, like for example power factor compensation.

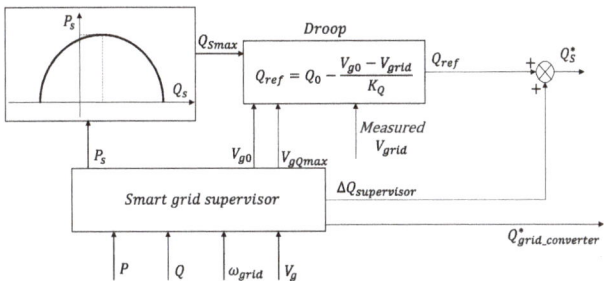

Figure 11. Diagram for generating the reactive power references of the stator and the GSC converter.

7. Experimental Results

Figure 12 shows the picture of the real test bench for testing the suggested controlling solutions. The system consists of a 7.5 kW DFIM and a 10 kW synchronous machine (PMSM) that performs the tasks of a household wind turbine [42–45]. The probes for monitoring all the currents and voltages of the system are matched and wired to a DS1103 dSPACE [44] control system.

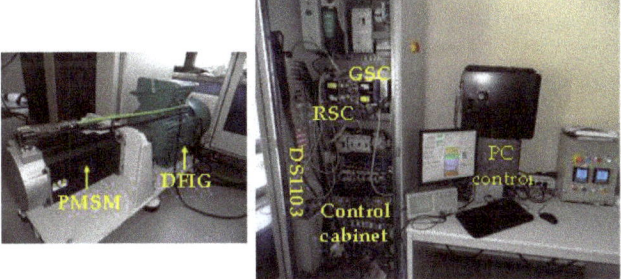

Figure 12. DFIG platform used to test the proposed control strategies.

The gate signals of the RSC and GSC IGBTs are generated by the DS1103 and the rotor position and speed are determined by a 4096 encoder wired to the DS1103. The main characteristics of the double feed induction machine are given in Table 1. The RSC and GSC are two NFS-200 converters with Mitsubishi IGBTs fabricated by Dutt [46]. Three 2 mH @ 15A inductors form the line filter between the GSC and the grid.

Table 1. Leroy Somer DFIM main parameters

Parameter	Value
Stator voltage	380 V
Rotor voltage	190 V
Rated stator current	18 A
Rated rotor current	24 A
Rated speed	1447 rpm @ 50 Hz
Rated torque	50 Nm
Stator resistance	0.325 Ω
Rotor resistance	0.275 Ω
Magnetizing inductance	0.0664 H
Stator leakage inductance	0.00264 H
Rotor leakage inductance	0.00372 H
Inertia moment	0.07 Kg*m^2

Before connecting the DFIM to the grid, the DC bus formed by the RSC and GSC capacitors in a back-to-back configuration must be charged. Once the DC Bus is charged and if the output voltage vector of the GSC is aligned with the grid voltage vector, the GSC connects to the grid and starts the regulation of the DC Bus voltage to a set value of 580 V.

When the wind (emulated by the PMSM) reaches the minimum speed of 5 m/s, the double fed induction machine starts the process of hooking up to the grid. This is done in two stages. In the first stage, the encoder offset with respect to the stator flux must be obtained. In the second stage, the voltage vector generated by the stator must be aligned with the grid voltage vector. When the last process is complete, the DFIM's stator is connected to the grid and the actual regulation process begins.

Figure 13 shows the regulation of the DFIG when the wind speed changes from 7 to 12 and then to 15 m/s. This means a rotor speed from 900 (sub-synchronous speed) to 1500 (synchronous speed) and then to 1900 rpm (super-synchronous speed) respectively.

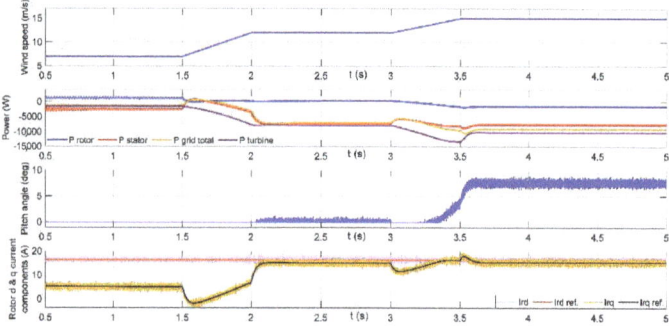

Figure 13. DFIG signals for a wind speed from 7 to 15 m/s. Upper graph, wind speed reference. Second graph, stator, rotor, turbine and total electrical power. Third graph, pitch angle. Fourth graph, references and actual values of rotor current d and q components.

In the DFIG the energy from the stator always flows into the grid. The flow of energy via the rotor changes sense depending on the speed of the DFIG. Thus, when the speed of the DFIG is lower than the synchronous speed, the energy of the rotor is absorbed from the grid and goes from the rotor to the stator, in the figure the interval from 0.5 to 1.5 s at 7 m/s of wind speed. Because of this, the total power is inferior to the power of the stator. When the DFIG speed is higher than the synchronous speed, the rotor energy changes sense and goes from the rotor to the grid.

As soon as the total power exceeds the nominal power (7.5 kW) and the wind speed increases, the pitch regulator forces a modification of the pitch angle to ensure that the generator does not exceed the nominal power. As it can be observed in the figure, since no changes are made to the reactive power reference, the d component of the rotor current is constant. However, due to changes in wind speed and therefore in mechanical power, the q component of the rotor changes as does the power of the stator.

Figure 14 shows the implementation made on the real platform to experimentally check the contribution of the DFIG on the voltage amplitude when the load is connected to the grid.

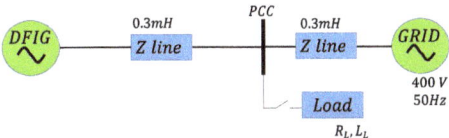

Figure 14. Single line diagram of the system implemented and used for testing.

To observe the voltage variation at the PCC (point of common coupling), an inductive load is set where $R_L = 0$ and $L_L = 20\ mH$. The wind speed is fixed to 6 m/s and Figure 15 shows the grid phase voltage amplitude reduction when the load is connected at t = 0.5 s. Due to the inductive characteristic of the line and the inductive load, there is a 5 V reduction in the PCC. At 0.75 s the voltage compensation is activated, supplying the DFIG through the stator with a capacitive reactive power of 7 KVAR (maximum rotor current). This results in a voltage increase on the PCC of 1.5 V. Since the GSC (grid side converter) is able to provide reactive power, 14 KVAR of capacitive reactive power is injected at 1.25 s ordered by the grid supervisor, which causes an additional 3 V increase bringing the PCC voltage closer to the grid voltage.

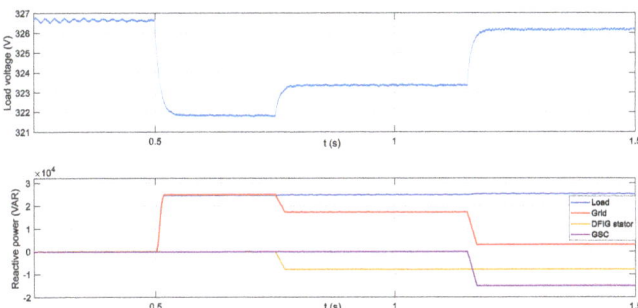

Figure 15. Load voltage for inductive load and voltage compensation with the controlled DFIG reactive power for Figure 14 structure.

Due to the impossibility of varying the frequency on the real platform, a simulation is performed to observe the frequency variation on the PCC. The grid is configured so that its frequency varies as a function of the active power consumed, as occurs in a generator. Thus, the grid decreases by 1 Hz when there is an active power consumption of 10 kW. Therefore, in our test the frequency value drops from 50 Hz when there is no consumption to 49 Hz when the consumption in the grid is 10 kW.

If the power is negative (the grid absorbs energy), the frequency increases in the same ratio. The active power is consumed when the load is connected in the line of Figure 14 with a resistive load where $R_L = 29\ \Omega$ and $L_L = 0$, that is 5.5 kW are consumed for the load.

Figure 16 illustrates the behavior of the system by showing the frequency of the PCC, the wind speed and the active powers of the load, the grid and the DFIG. The wind speed has been set at 7 m/s up to 1 s. With this wind the power that the DFIG delivers is 1.5 kW. Up to 0.5 s the grid absorbs this power and the grid frequency at the PCC point is 50.15 Hz. At 0.5 s, the load that absorbs 5.5 kW is connected producing a decrease in frequency of

0.55 Hz, until it reaches 49.6 Hz. From 1 s, the wind speed increases as shown in the figure until it reaches 10 m/s, producing an increase in the contribution of active power to the grid, up to 3.9 kW and reducing the contribution to be made by the grid to 1.6 kW.

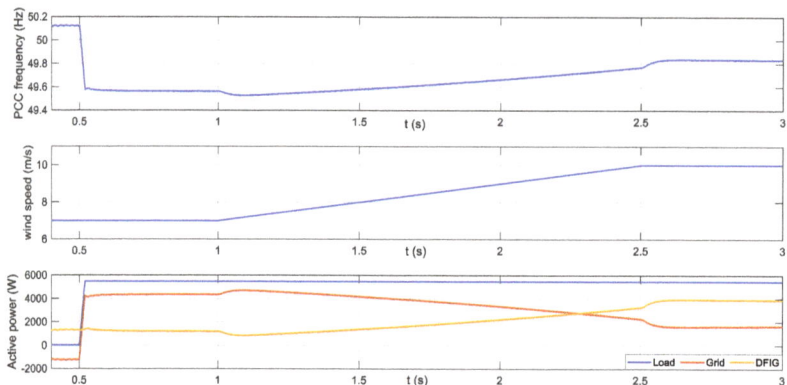

Figure 16. PCC frequency, wind speed and active power. Frequency drop compensation for the resistive load in the scheme of Figure 14.

This reduction of the active power contribution by the grid means a recovery of the frequency in the grid at 49.83 Hz.

Figure 17 presents the behavior of the system by showing the frequency of the PCC, the power coefficient CP, the DFIG mechanical speed and the active powers of the load, the grid and the DFIG. The wind speed has been set at 12 m/s for all the time. From the beginning the DIFG is running to 2300 rpm, this speed is higher than the optimum speed, thus the DFIG is working 35% deloaded generating 3.5 kW when the obtainable maximum power is 5.4 kW. The load is connected at 0.75 s, which means a drop in frequency from 50.16 to 49.6 Hz. In the instant of 1 s, the energy is extracted from the wind, going from the deloaded working speed to the maximum power speed. This speed deceleration (from 2300 to 1780 rpm) provides extra kinetic energy which produces an increase in frequency until it stabilizes at 49.8 Hz when the maximum possible power is extracted from the wind. Obviously, if it is not possible to extract more energy from the wind it is not possible to reduce the frequency deviation further. The important currents that reflect the process shown in Figure 17, can be seen in Figure 18. In this figure, the grid current, the current consumed by the load, the stator current of the DFIG, the current of the grid side converter and finally the current through the rotor of the DFIG can be seen.

Until 0.75 s the load is disconnected and the DFIG works in deloaded mode. The generator works in super-synchronous speed and the power is delivered to the grid by the stator and the rotor or GSC converter. The generated power of 3.5 kW produces a stator peak current of 4.4 A and a GSC peak current of 1.7 A with the imposed reactive power of 0 VAR. When the load is connected at 0.75 s the power consumption is 5.54 kW and the load peak current is 11.2 A. Since the power delivered by the DFIG is less than the power absorbed by the load, the grid must supply the difference and therefore the grid current peak becomes 5.1 A. In the instant 1 s it passes from the state of deloaded to the point of maximum power. As the deceleration from 2300 to 1780 rpm occurs, the power delivered by the DFIG increases as the CP of the turbine increases and also recovers kinetic energy in deceleration. For this reason, the stator and grid side converter currents increase while the grid current decreases. Finally, when the DFIG reaches full power speed, the current through the stator is 8.4 A and through the GSC is 1.3 A. The rest of the current, until reach the 11.2 A that the load absorbs, is provided by the grid. The amplitude variation of the rotor current is small since the d component of the current predominates to impose a reactive power of 0 VAR. The Ird value is 16.5 A and the amplitude variations are due to the variation of the Irq component to manage the active power. The frequency reduction of the current is observed as it approaches the synchronism speed.

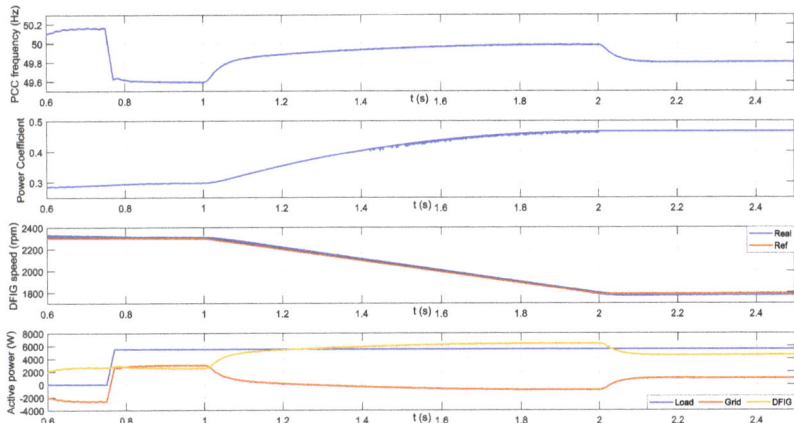

Figure 17. PCC frequency, CP coefficient, DFIG speed, and active power. Frequency drop compensation for the resistive load in the scheme in Figure 14 with the DFIG working deloaded.

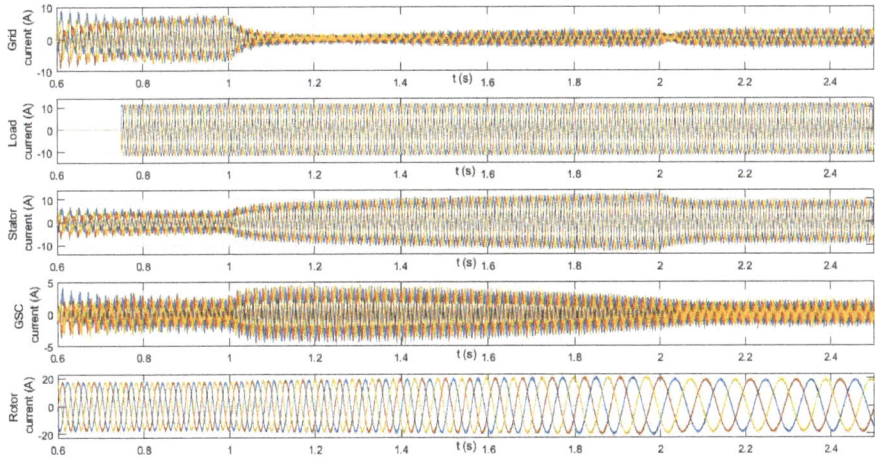

Figure 18. Grid, load, stator, grid side converter, and rotor currents for the process shown in Figure 17.

8. Conclusions

This article proposes the use of the double feed induction generator in wind power systems to control the frequency and the amplitude of the point of common coupling voltage. It has been demonstrated that the DFIG must be deloaded to extract the active power from the wind that is required to compensate for the frequency of the grid, and the smart grid controller must establish the quantity deloaded. In addition, the smart grid controller has to choose the required maximum reactive power, taking into account the electrical limits of the double fed induction generator, to balance out the amplitude of the grid voltage. The capacity given to a smart grid to handle frequency and voltage amplitude enhances the reliability of the installation.

The efficiency of the wind turbine is reduced when it is desired to reserve the wind energy in a wind turbine to compensate for the frequency. Since the maximum current value of the DFIG limits the active and reactive power set, by using the converter on the grid side, the reactive power injection can be increased to compensate the grid voltage. Frequency compensation has been tested by performing simulations creating a weak grid. However, the voltage compensation has been implemented in a real platform with a commercial DFIG that has successfully validated the algorithms presented, allowing the commercial implementation of such a system.

Author Contributions: Methodology, J.A.C. and O.B.; Software, O.B., P.A. and J.C.; Hardware and tests, J.A.C., P.A. and J.C. Writing—original draft, J.A.C.; Writing—review editing, O.B., P.A. and J.C. All authors have read and agreed to the published version of the manuscript.

Funding: This research was funded by the Basque Government through the project SMAR3NAK (ELKARTEK KK-2019/00051), the Diputación Foral de Álava (DFA) through the project CONAVAUTIN 2 and the University of the Basque Country (UPV/EHU) through (PPGA20/06).

Institutional Review Board Statement: Not applicable.

Informed Consent Statement: Not applicable.

Data Availability Statement: Not applicable.

Conflicts of Interest: The authors declare no conflict of interest.

Abbreviations

RSC	Rotor side converter
GSC	Grid side converter
α, β	Direct and quadrature axes expressed in the stationary reference frame
α', β'	Rotor direct and quadrature axes expressed in the rotor reference frame
d, q	Direct and quadrature axes expressed in the synchronous rotating reference frame
$\vec{v}_s$	Stator voltage vector
$\vec{i}_s$	Stator current vector
$\vec{\psi}_s$	Stator flux vector
L_m	DFIG mutual inductance
L_s	DFIG stator inductance
L_r	DFIG rotor inductance
L_{ls}	DFIG stator leakage inductance
L_{lr}	DFIG rotor leakage inductance
R_s	DFIG stator resistance
R_r	DFIG rotor resistance
ω_e	Synchronous speed
ω_r	Rotor electrical speed
ω_m	Rotor mechanical speed
T_e	DFIG electromagnetic torque
T_L	DFIG load torque
J	DFIG inertia
B	DFIG friction coefficient
R_L	Load resistance
L_L	Load inductance
P_s	DFIG stator active power
Q_s	DFIG stator reactive power
P_p	Pair of poles
$\vec{v}_g$	Grid voltage vector
$\vec{i}_g$	Grid current vector
R_g	Grid filter resistance

L_g	Grid filter inductance
ω_g	Grid frequency
$\vec{v}_{conv}$	Grid side converter output voltage vector
P_g	Grid side converter active power
Q_g	Grid side converter reactive power
P	DFIG total active power
Q	DFIG total reactive power
S	DFIG total apparent power
Z	Line impedance
K_P	Voltage droop coefficient
K_Q	Frequency droop coefficient
V_s	$\lvert\vec{v}_s\rvert$
V_g	$\lvert\vec{v}_g\rvert$
δ	Angle between the DFIG stator voltage and the grid voltage
ω_{pr}	Propeller speed
ω_{d1}	Deloading propeller speed
ω_{opt}	Propeller optimum speed
L	Line inductance
IGBT	Insulated gate bipolar transistor

References and Note

1. *Future of Wind: Deployment, Investment, Technology, Grid Integration and Socio-Economic Aspects*; A Global Energy Transformation Paper; International Renewable Energy Agency (IRENA): Abu Dhabi, UAE, 2019.
2. Tasneem, Z.; Al Noman, A.; Das, S.K.; Saha, D.K.; Islam, R.; Ali, F.; Badal, F.R.; Ahamed, H.; Moyeen, S.I.; Alam, F. An analytical review on the evaluation of wind resource and wind turbine for urban application: Prospect and challenges. *Dev. Built Environ.* **2020**, *4*, 100033. [CrossRef]
3. Sunderland, K.M.; Mills, G.; Conlon, M.F. Estimating the wind resource in an urban area: A case study of micro-wind generation potential in Dublin, Ireland. *J. Wind Eng. Ind. Aerodyn.* **2013**, *118*, 44–53. [CrossRef]
4. Drew, D.R.; Barlow, J.F.; Cockerill, T.T. Estimating the potential yield of small wind turbines in urban areas: A case study for Greater London, UK. *J. Wind Eng. Ind. Aerodyn.* **2013**, *115*, 104–111. [CrossRef]
5. Tanvir, A.A.; Merabet, A. Artificial Neural Network and Kalman Filter for Estimation and Control in Standalone Induction Generator Wind Energy DC Microgrid. *Energies* **2020**, *13*, 1743. [CrossRef]
6. Guerrero, J.M.; Lumbreras, C.; Reigosa, D.; Garcia, P.; Briz, F. Control and emulation of small wind turbines using torque estimators. *IEEE Trans. Industy Appl.* **2011**, *53*, 4863–4876. [CrossRef]
7. Calabrese, D.; Tricarico, G.; Brescia, E.; Cascella, G.L.; Monopoli, V.G.; Cupertino, F. Variable Structure Control of a Small Ducted Wind Turbine in the Whole Wind Speed Range Using a Luenberger Observer. *Energies* **2020**, *13*, 4647. [CrossRef]
8. Liu, T.; Gong, A.; Song, C.; Wang, Y. Sliding Mode Control of Active Trailing-Edge Flap Based on Adaptive Reaching Law and Minimum Parameter Learning of Neural Networks. *Energies* **2020**, *13*, 1029. [CrossRef]
9. Bashetty, S.; Guillamon, J.I.; Mutnuri, S.S.; Ozcelik, S. Design of a Robust Adaptive Controller for the Pitch and Torque Control of Wind Turbines. *Energies* **2020**, *13*, 1195. [CrossRef]
10. Kong, X.; Ma, L.; Liu, X.; Abdelbaky, M.A.; Wu, Q. Wind Turbine Control Using Nonlinear Economic Model Predictive Control over All Operating Regions. *Energies* **2020**, *13*, 184. [CrossRef]
11. Elorza, I.; Calleja, C.; Pujana-Arrese, A. On Wind Turbine Power Delta Control. *Energies* **2019**, *12*, 2344. [CrossRef]
12. Barambones, O.; Cortajarena, J.A.; Calvo, I.; Gonzalez de Durana, J.M.; Alkorta, P.; Karami, A. Variable speed wind turbine control scheme using a robust wind torque estimation. *Renew. Energy* **2018**, *133*, 354–366. [CrossRef]
13. Hansen, A.D.; Iov, F.; Blaabjerg, F.; Hansen, L.H. Review of contemporary wind turbine concepts and their market penetration. *J. Wind Eng.* **2004**, *28*, 247–263. [CrossRef]
14. Ananth, D.V.N.; Nagesh Kumar, G.V. Tip Speed Ratio Based MPPT Algorithm and Improved Field Oriented Control for Extracting Optimal Real Power and Independent Reactive Power Control for Grid Connected Doubly Fed Induction Generator. *Int. J. Electr. Comput. Eng.* **2016**, *6*, 1319–1331.
15. Amrane, F.; Chaiba, A.; Francois, B.; Babes, B. Experimental design of stand-alone field oriented control for WECS in variable speed DFIG-based on hysteresis current controller. In Proceedings of the 2017 15th International Conference on Electrical Machines, Drives and Power Systems, Sofia, Bulgaria, 1–3 June 2017; pp. 1–5.
16. Djilali, L.; Sanchez, E.N.; Belkheir, M. Real-time implementation of sliding-mode field-oriented control for a DFIG-based wind turbine. *Int. Trans. Electr. Energy Syst.* **2018**, *28*, e2539. [CrossRef]
17. Kaloi, G.S.; Wang, J.; Baloch, M.H. Active and reactive power control of the doubly fed induction generator based on wind energy conversion system. *Energy Rep.* **2016**, *2*, 194–200. [CrossRef]

18. Soued, S.; Ramadan, H.S.; Becherif, M. Dynamic Behavior Analysis for Optimally Tuned On-Grid DFIG Systems. Special Issue on Emerging and Renewable Energy: Generation and Automation. ScienceDirect. *Energy Procedia* **2019**, *162*, 339–348. [CrossRef]
19. Shenglong, Y.; Kianoush, E.; Tyrone, F.; Herbert, H.C.J.; Kit, P.W. State Estimation of Doubly Fed Induction Generator Wind Turbine in Complex Power Systems. *IEEE Trans. Power Syst.* **2016**, *31*, 4935–4944.
20. Palensky, P.; Dietrich, D. Demand side management: Demand response, intelligent energy systems and smart loads. *IEEE Trans. Ind. Inform.* **2011**, *7*, 381–389. [CrossRef]
21. Güngör, V.C.; Sahin, D.; Kocak, T.; Ergüt, S.; Buccella, C.; Cecati, C.; Hancke, G.P. Smart grid technologies: Communication technologies and standards. *IEEE Trans. Ind. Inform.* **2011**, *7*, 529–539. [CrossRef]
22. Zhang, Y.; Meng, F.; Wang, R.; Kazemtabrizi, B.; Shi, J. Uncertainty-resistant stochastic MPC approach for optimal operation of CHP microgrid. *Energy* **2019**, *179*, 1265–1278. [CrossRef]
23. Nguyen, T.T.; Yoo, H.J.; Kim, H.M. Analyzing the Impacts of System Parameters on MPC-Based Frequency Control for a Stand-Alone Microgrid. *Energies* **2017**, *10*, 417. [CrossRef]
24. Bayhan, H.A.; Ellabban, O. Sensorless model predictive control scheme of wind-driven doubly fed induction generator in dc microgrid. *IET Renew. Power Gener.* **2016**, *10*, 514–521. [CrossRef]
25. Valencia-Rivera, G.H.; Merchan-Villalba, L.R.; Tapia-Tinoco, G.; Lozano-Garcia, J.M.; Avina-Cervantes MA, I.M.; Gabriel, J. Hybrid LQR-PI Control for Microgrids under Unbalanced Linear and Nonlinear Loads. *Mathematics* **2020**, *8*, 1096. [CrossRef]
26. Thounthong, P.; Mungporn, P.; Pierfederici, S.; Guilbert, D.; Bizon, N. Adaptive Control of Fuel Cell Converter Based on a New Hamiltonian Energy Function for Stabilizing the DC Bus in DC Microgrid Applications. *Mathematics* **2020**, *8*, 2035. [CrossRef]
27. Qiao, W.; Harley, R.G. *Grid Connection Requirements and Solutions for DFIG Wind Turbines*; IEEE: Piscataway, NJ, USA, 2008.
28. Cortajarena, J.A.; De Marcos, J.; Alkorta, P.; Barambones, O.; Cortajarena, J. DFIG wind turbine grid connected for frequency and amplitude control in a smart grid. In Proceedings of the 2018 IEEE International Conference on Industrial Electronics for Sustainable Energy Systems (IESES), Hamilton, New Zealand, 31 January–2 February 2018.
29. Li, S.; Challoo, R.; Nemmers, M.J. Comparative Study of DFIG Power Control Using Stator Voltage and Stator-Flux Oriented Frames. In Proceedings of the 2009 IEEE Power & Energy Society General Meeting, Calgary, AB, Canada, 26–30 July 2009; pp. 1–8.
30. Ghosh, S.; Isbeih, Y.J.; Bhattarai, R.; El Moursi, M.S.; El-Saadany, E.F. A Dynamic Coordination Control Architecture for Reactive Power Capability Enhancement of the DFIG-Based Wind Power Generation. *IEEE Trans. Power Syst.* **2020**, *35*, 3051–3064. [CrossRef]
31. Cortajarena, J.A.; Barambones, O.; Alkorta, P.; De Marcos, J. Sliding mode control of grid-tied single-phase inverter in a photovoltaic MPPT application. *Solar Energy* **2017**, *155*, 793–804. [CrossRef]
32. Kazmierkowski, M.P.; Jasinski, M.; Wrona, G. DSP-Based control of grid-connected power converters operating under grid distortions. *IEEE Trans. Ind. Inform.* **2011**, *7*, 204–211. [CrossRef]
33. Guerrero, J.M.; Matas, J.; Vicuña, L.G.; Castilla, M.; Miret, J. Decentralized control for parallel operation of distributed generation inverters using resistive output impedance. *IEEE Trans. Ind. Electron.* **2007**, *54*, 994–1004. [CrossRef]
34. Guerrero, J.M.; Vicuña, L.G.; Matas, J.; Castilla, M.; Miret, J. A wireless controller to enhance dynamic performance of parallel inverters in distributed generation systems. *IEEE Trans. Power Electron.* **2004**, *19*, 1205–1213. [CrossRef]
35. Mun-Kyeom, K. Optimal Control and Operation Strategy for Wind Turbines Contributing to Grid Primary Frequency Regulation. *Appl. Sci.* **2017**, *7*, 1–23.
36. Aho, J.; Buckspan, A.; Laks, J.; Fleming, P.; Jeong, Y.; Dunne, F.; Churchfield, M.; Pao, L.; Johnson, K. Tutorial of Wind Turbine Control for Supporting Grid Frequency through Active Power Control. In Proceedings of the 2012 American Control Conference (ACC), Montréal, QC, Canada, 27–29 June 2012; pp. 1–12.
37. Matlab/Simulink. Wind turbine model. 2009.
38. Ramtharan, G.; Ekanayake, J.B.; Jenkins, N. Frequency support from doubly fed induction generator wind turbines. *IET Renew. Power Gener.* **2007**, *1*, 3–9. [CrossRef]
39. Saenz-Aguirre, A.; Zulueta, E.; Fernandez-Gamiz, U.; Teso-Fz-Betoño, D.; Olarte, J. Kharitonov Theorem Based Robust Stability Analysis of a Wind Turbine Pitch Control System. *Mathematics* **2020**, *8*, 964. [CrossRef]
40. North American Electric Reliability Corporation. *Accommodating High Levels of Variable Generation*; North American Electric Reliability Corporation: Princeton, NJ, USA, 2009.
41. Kayikc, M.; Milanovic, J.V. Reactive power control strategies for DFIG-based plants. *IEEE Trans. Energy Convers.* **2007**, *22*, 389–396.
42. Gallardo, S.; Carrasco, J.M.; Galván, E.; Franquelo, L.G. DSP-based doubly fed induction generator test bench using a back-to-back PWM converter. In Proceedings of the 2004 30th Annual Conference of the IEEE Industrial Electronics Society, Busan, Korea, 2–6 November 2004.
43. dSPACE. Real–Time Interface. In *Implementation Guide. Experiment Guide*; For Realese 5.0.; GmbH: Paderborn, Germany, 2005.
44. Cortajarena, J.A.; De Marcos, J.; Alvarez, P.; Vicandi, F.J.; Alkorta, P. Start up and control of a DFIG wind turbine test rig. In Proceedings of the 2011 37th Annual Conference of the IEEE Industrial Electronics Society, Melbourne, VIC, Australia, 7–10 November 2011.
45. Junseon, P.; Seungjin, L.; Joong Yull, P. Effects of the Angled Blades of Extremely Small Wind Turbines on Energy Harvesting Performance. *Mathematics* **2020**, *8*, 1295.
46. Dutt. Power Electronics & Control. Available online: http://www.duttelectronics.com (accessed on 2 October 2020).

Article

Multi-Stack Lifetime Improvement through Adapted Power Electronic Architecture in a Fuel Cell Hybrid System

Milad Bahrami [1],*, Jean-Philippe Martin [1], Gaël Maranzana [1], Serge Pierfederici [1], Mathieu Weber [1], Farid Meibody-Tabar [1] and Majid Zandi [2]

[1] University of Lorraine, CNRS, LEMTA, 54000 Nancy, France; jean-philippe.martin@univ-lorraine.fr (J.-P.M.); gael.maranzana@univ-lorraine.fr (G.M.); serge.pierfederici@univ-lorraine.fr (S.P.); mathieu.weber@univ-lorraine.fr (M.W.); farid.meibody-tabar@univ-lorraine.fr (F.M.-T.)

[2] Renewable Energies Engineering Department, Shahid Beheshti University, Tehran 1983969411, Iran; m_zandi@sbu.ac.ir

* Correspondence: milad.bahrami@univ-lorraine.fr

Received: 31 March 2020; Accepted: 2 May 2020; Published: 7 May 2020

Abstract: To deal with the intermittency of renewable energy resources, hydrogen as an energy carrier is a good solution. The Polymer Electrolyte Membrane Fuel Cell (PEMFC) as a device that can directly convert hydrogen energy to electricity is an important part of this solution. However, durability and cost are two hurdles that must be overcome to enable the mass deployment of the technology. In this paper, a management system is proposed for the fuel cells that can cope with the durability issue by a suitable distribution of electrical power between cell groups. The proposed power electronics architecture is studied in this paper. A dynamical average model is developed for the proposed system. The validation of the model is verified by simulation and experimental results. Then, this model is used to prove the stability and robustness of the control method. Finally, the energy management system is assessed experimentally in three different conditions. The experimental results validate the effectiveness of the proposed topology for developing a management system with which the instability of cells can be confronted. The experimental results verify that the system can supply the load profile even during the degradation mode of one stack and while trying to cure it.

Keywords: multi-stack; Polymer Electrolyte Membrane Fuel Cell (PEMFC); energy management; power electronics; stability analysis

---

1. Introduction

Based on the energy and environmental crisis in the world, the approach is toward using renewable and clean energy [1]. The most important issue of renewable energies is the intermittency of the resources [2–4]. Hydrogen as an energy carrier can cope with this problem [5]. This gas can be directly converted to electricity by a Polymer Electrolyte Membrane Fuel Cell (PEMFC) [4,6]. The nominal voltage of one cell is near 0.7 V. Therefore, a number of cells are connected in series inside of one stack to increase the output voltage. The basic disadvantage of such connections is the lifetime dependency of the stack on each cell lifetime. The other disadvantage of this connection is the probability of the fault propagation from one cell to the adjacent ones because of thermal coupling. The cell management can cope with these problems.

Currently, the performance of PEMFC, in terms of power density (3.1 kW/L) and electrical energy efficiency, is sufficient to allow large-scale deployment of the technology [7,8]. On the other hand, lifetime and cost are two points that need to be improved [9]. To improve the durability, it is possible to develop new materials more resistant but also to better manage the operating conditions to avoid the electrochemical instabilities that lead to irreversible damages [10]. For instance, a novel converter was proposed in [11] to reduce the current ripple for fuel cell applications. A specific management system like the battery management systems can deal with the instabilities of the cells inside a stack. For instance, if one cell is more degraded than other cells, its voltage decreases and it dissipates more heat. The direct consequence of the higher temperature of the degraded cell is the drying of its membrane and the subsequent increase of its ionic resistance that at the end amplifies the voltage loss. This snowball effect can destroy the cell. If it was possible to separately change the current of the damaged cells to produce more water or to reduce the heat production, it would deal with the mentioned snowball effect.

Developing such a management system for a single stack needs a special stack, which allows modifying the current of any number of cells inside the stack, and a power electronic structure that allows controlling the power flow of different cells. A new patent obviates the first requirement and allows access to the current of any cells [12]. Multi-stack can be considered as an attempt to increase the lifetime and durability of the fuel cell system at the cost of compactness loss. The multi-stack can be used for high power [13–15], vehicular and transportation [16–24], and stationary [16,25–27] applications due to its higher reliability and efficiency and its optimized fuel consumption. This technology has been already used by a Mercedes bus, power supplies of space exploration vehicles, and air-independent propulsion for submarines [28]. In [21], the multi-stack was investigated in the fault mode and the faulty stack could partially or totally be disconnected by a diode by-pass circuit while the other stacks supply the load. A model was also developed for the proposed architecture. Two PEMFC stacks with the rated power of 20 kW (total rated power of 40 kW) were used in [22] to totally supply a manned aircraft. Ten PEMFC stacks with the rated power of 480 kW were used in [27] to develop a stationary distributed generation system. In [23], a hybrid system of two PEMFC stacks with the rated power of 150 kW (total 300 kW) and a battery bank were used to supply a locomotive. The battery bank was used to supply the load profile over transient conditions.

As seen in Figure 1., four basic topologies are used for the multi-stack or segmented FCs: series, parallel, cascade, series-parallel [21,28]. The series topology (Figure 1a) requires a low voltage ratio converter. In such a connection, the failure of a single cell means losing the whole system. Furthermore, there is no freedom degree in controlling the cells separately. The second topology, which the cells or stacks separately connect to the DC link by individual converters, provides the freedom degree in controlling the cells [18]. However, the high conversion ratio converters, which are required to increase the output voltage, provide higher stress on the semiconductor devices. This architecture is the most expensive topology due to a great requirement of the passive energy storage components. The cascade topology resolves the problem of the parallel topology. In this topology, the DC-link voltage is divided between the cells. This leads to lower stress on semiconductor devices. The series-parallel topology (Figure 1d) is identical to the parallel topology except that more cells are connected to the input of each converter. This topology inherits the advantages and disadvantages of series and parallel topologies. In such a topology, the converters with lower voltage ratio can be used.

Considering the ability of separately controlling the cells or stacks, the cascade topology (Figure 1c) can be used to manage the cells. However, connecting the cells to the high voltage DC link while maintaining the controllability of each converter is challenging. It should be noted that the proposed structure can be used for a single stack that allows accessing the current of different cell groups or a multi-stack. As a result, the word stack is used instead of cell groups for the sake of simplicity in the rest of this paper.

Due to the low voltage of a cell or a small group of cells, a converter with a high voltage ratio has to be used [29,30]. In this paper, the classical DC–DC boost converters are used while their output capacitors are connected in series, inspired by the cascade topology. In such a connection, the same load current passes through all output capacitors. Therefore, if the input power of one cell becomes lower than the required amount, the voltage of the corresponding capacitor will decrease. In this case, the controllability will be lost if the output voltage becomes lower than the input voltage.

As mentioned before, the objective is to develop a management system that separately controls the current of cells or stacks. Considering the cascade topology with N groups (Figure 2), if the load power (P_{load}) is constant and the DC-link voltage is regulated at V_{dc}, then the load current has a constant value of i_{load} in steady-state and, as a result, the supplied power by cells can be calculated as $P_k = i_{load} V_{C_k} \ \forall \ k \in \{1,2\ldots,N\}$. Assuming that the total supplied power by stacks is equal to the load power, and the first group of cells should inject a part (x) of the nominal power (P_{load}/N), whereas the other groups inject an identic value of power, the following equation can be obtained in steady-state as follows:

$$\begin{cases} P_1 = x\frac{P_{load}}{N} = i_{load} V_{C1} \rightarrow V_{C1} = \frac{x}{N} V_{dc} \\ P_k = \frac{N-x}{N(N-1)} P_{load} = i_{load} V_{Ck} \rightarrow V_{Ck} = \frac{N-x}{N(N-1)} V_{dc} \end{cases} \quad (1)$$

where k in this equation can be $\in \{2,3\ldots,N\}$. The voltage of the first capacitor must be greater than the input voltage of the corresponding converter. Thus, considering the voltage ratio of the boost converter (R_v) and (1), $x > \frac{N V_{FC}}{V_{dc}} > \frac{1}{R_v}$. Therefore, the supplied power of the first group cannot be less than $\frac{1}{R_v N} P_{load}$. Otherwise, the controllability is lost. In such conditions, a voltage equalizer or a balancing system has to be added to ensure the controllability of converters.

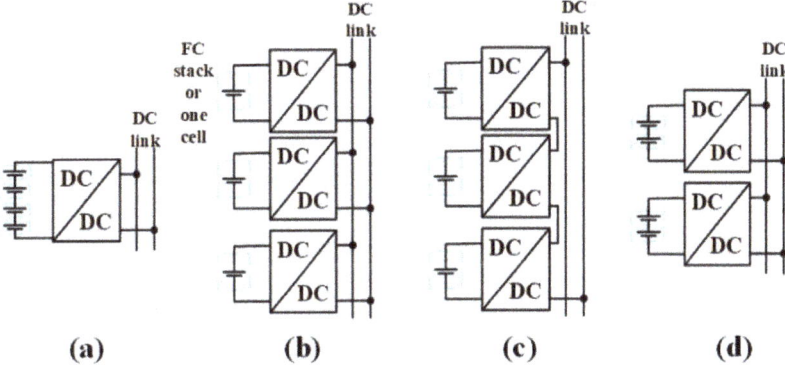

Figure 1. Multi stack topologies: (**a**) Series; (**b**) Parallel; (**c**) Cascade; (**d**) Series-parallel.

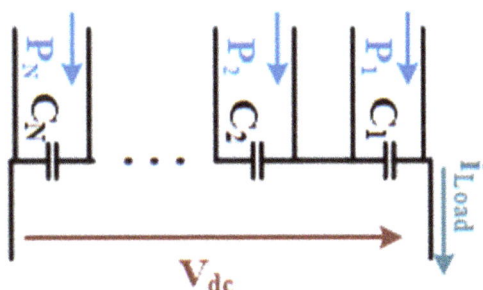

Figure 2. Cascade topology with different injected powers by stacks.

A new voltage equalizer, which was proposed in [31], is modified and used in this paper. In [32], the equalizer was controlled according to the voltage difference and a hysteresis method. The proportional gain controller, as used in [31], is improved in this paper to enhance the dynamical behavior of the equalization. The improved method can cope with the overvoltage of the output capacitors due to operating conditions of the fuel cell management system in some particular cases.

Due to dynamic constraints, the hybridization of a fuel cell with an energy storage system is required to supply a given load, especially for transport or stationary applications [23,33–36]. The battery has a high energy density and a higher power density than the FC. Therefore, the FC/Battery hybridization has been used in some papers [34,37–39]. Contrary to the batteries, supercapacitors (SCs) have a higher power density and a lower energy density than the batteries. As a result, the FC/SC hybrid system has been widely used in the literature due to the ability of the SC in compensating for the slow dynamic of the FC [34,40]. In [35,41,42], an FC/SC/Battery structure was proposed for vehicular application and an energy management strategy was proposed to supply the load. The main objective in [38,41,42] was to develop an energy management strategy to improve fuel consumption. In this paper, the main objective is to develop a system in which energy management can be implemented while curing the defective stacks.

The conventional hybridizing fuel cell with SC is used in this paper to respond to the high dynamic load profile. To regulate the output DC bus voltage, the SC is connected to the DC link through a bidirectional DC–DC boost converter.

The hybridized system with the equalizer system needs a control method that can be implemented easily and ensure the stability of the system. In this paper, a control method is proposed for the whole system.

An accurate model of the system has to be realized to perform the sizing of the passive elements and control parameters or/and dynamical stability analysis. There are three basic approaches to investigate the dynamic behavior of the system: Discrete-time model, Switching model, and Average model. In [43], these three approaches were used to investigate the dynamic behavior of a boost converter with a sliding mode current controller. In [44], the stability of a boost converter based on the discrete-time model and the average model was investigated when a hysteresis current controller was used. The dynamic behavior of a boost converter with an LC filter was investigated based on the discrete-time model approach in [45]. In [45], the output voltage and current of the boost converter were controlled based on the dynamic separation. The discrete-time model cannot be used especially for some complicated systems. For such systems, the average model can be used to study the dynamic behavior of the system. In [46], an interleaved-boost full-bridge converter was modeled by obtaining the state-space average models of the system over different operation modes. This kind of modeling approach can involve many equations for complex systems. Therefore, this model can be very complex. In [47], an average model was proposed to perform the stability analysis for a hybrid photovoltaic and wave power generation system. The average model of each system and control method were used to obtain an overall dynamic average model for the hybrid system. Then, the stability analysis, based on the eigenvalues of the proposed average state-space model, has been used to verify the stable operation of the system under various operating conditions. An average model based on the state-space equations of the system was introduced for a balancer in [48]. This balancer can equalize the voltage between two SC banks. Indeed, the proposed balancer was a synchronous buck-boost topology that can transfer the stored energy in the inductor of the converter to the banks and vice versa. In such a connection, there is no intermediary AC power stage. For systems with AC power stage, the average values of pure AC variables are equal to zero. This last property involves an order reduction to establish the average model. To deal with this phenomenon, the DC terms and the first-order terms of Fourier series of state variables were used in [49]. Since the transformer current of a dual active bridge converter is purely AC, its DC component is equal to zero. Therefore, the switching frequency terms in the Fourier series of state variables were used to capture the effect of the transformer current on the system dynamics in [49]. In [50], a dynamical average model was proposed for an isolated boost converter. In this converter, the average value of the transformer current over each period of switching is equal to zero. Therefore, the energy cannot be transferred through the transformer by using this average value in the average state-space model. The equation of the average value of the leakage current was obtained over a half period. Since the leakage current of the isolated boost converter is symmetric, this equation is valid for any half periods. As a result, the obtained average value is not equal to zero and the energy can be transferred through the transformer even in the average model. The main disadvantage of this approach is the assumption in which the transformer current waveform must be symmetrical.

In this paper, the proposed equalizer includes an AC power stage but the current waveforms can be asymmetrical. Thus, a reduced-order average model is proposed in which the symmetrical waveform of the transformer current is not mandatory. Moreover, the proposed average model takes into account the cross-coupling effect due to the serial connection of output capacitors. Dynamic stability analysis of the system is also performed based on the proposed model. This proposed approach is validated by simulation and experimental results.

The rest of this paper is organized as follows: The proposed system configuration, the improved control method, and the definition of the dynamical average model for any operating conditions are detailed in Section 2. The SC and DC-link voltage control methods are described in Section 3. The simulation and experimental results are presented in Section 4. The stability analysis and energy management are performed in Section 5. Finally, conclusions are presented in Section 6.

2. Proposed System Configuration

The proposed power electronic architecture used to realize the fuel cell management system is shown in Figure 3. As seen in this figure, the N fuel cell stacks are used and connected to the DC link through the DC–DC boost converters. Inspired by the cascade topology, the output capacitors of the boost converters are connected in series. These boost converters allow the separate management of those stacks. To ensure the controllability of different boost converters, an equalizer architecture based on the multi-winding transformer is used in this paper. Indeed, the input of the H-bridge inverter is connected to the DC-link where a capacitor C_H is used to stabilize the voltage. The H-bridge inverter converts this DC voltage to an AC voltage at the input of the HF transformer. The diode structure at the output of the HF transformer allows the transmission of energy from AC secondary windings to the lower voltage capacitors. The equalizer architecture was deeply investigated and verified by experimental results in [31] without hybridization. To regulate the DC bus voltage, the SC with the conventional bidirectional boost converter is used. In order to realize the whole system, each part is modeled separately.

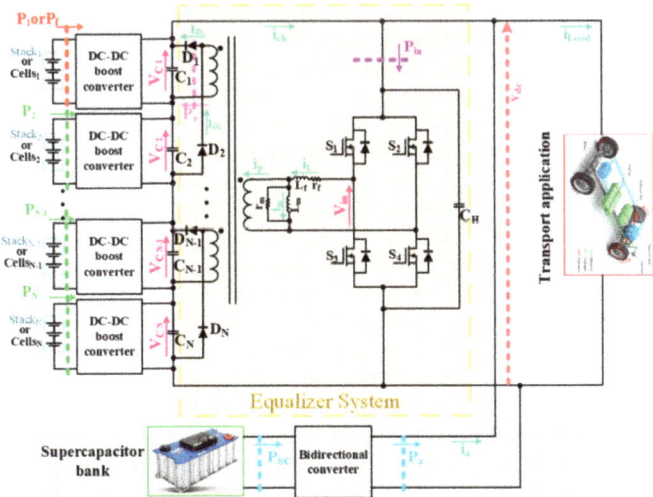

Figure 3. Proposed power electronic architecture to realize fuel cell management system.

2.1. Model of the Equalizer

As shown in Figure 3 H-bridge inverter is connected to the primary side of a High Frequency (HF) transformer. To stabilize the input voltage of the H-bridge inverter, a capacitor (C_H) is used. A special diode structure is used on the secondary side of the transformer. This topology allows the energy to be transferred from the DC-link to the lower voltage capacitors.

The following assumptions are considered for the purpose of simplicity:

(1) Turn ratio for all secondary windings are the same and equal to $m = \frac{N_2}{k N_1}$ and k is the coupling coefficient.
(2) The coupling coefficient between the primary and secondary windings are identical and equal to k.
(3) The coupling coefficient between the secondary windings is unitary.
(4) The DC bus voltage is controlled to a reference constant value.
(5) The switches are considered as the ideal devices. The diode voltage drops are taken into account but their dynamical resistance is assumed zero.

The theoretical waveforms of the proposed equalizer in steady-state are shown in Figure 4a. when C_1 (and C_2) are the lower voltage capacitors between odd and even-numbered capacitors respectively. As seen in this figure, based on the proposed switching commands for the H-bridge inverter, a symmetrical square waveform (V_{in}) is imposed on the primary side of the HF transformer with a variable duty cycle (d). Therefore, the odd\even numbered diodes can be naturally turned on during the positive\negative part of this square wave because of the diode structure and the polarity of secondary windings. The diode corresponding to the lower voltage capacitor between odd\even numbered diodes is automatically turned on during the positive\negative part of the input voltage of the HF transformer. Due to the same polarity of the secondary windings, a voltage equal to the lower voltage between the odd\even numbered capacitor voltages is induced to all windings when an odd\even numbered diode starts to conduct. Therefore, other odd\even numbered diodes are negatively biased during the conduction of one diode.

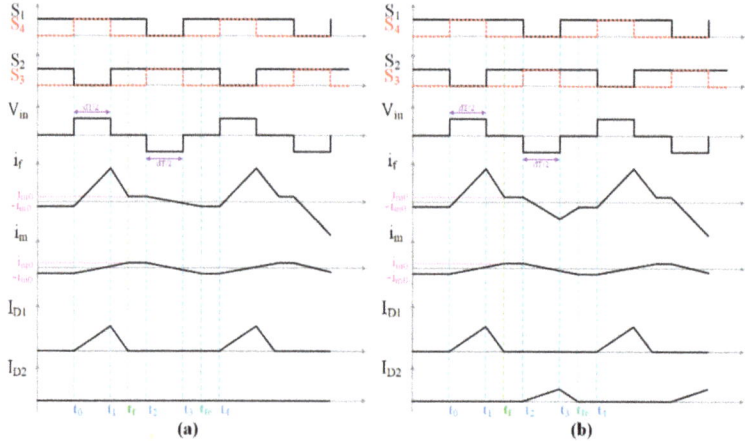

Figure 4. Theoretical waveforms in the steady-state of the proposed equalizer whereas: (a) V_{c1} is the lowest voltage between all capacitors; (b) $V_{c1}\backslash V_{c2}$ is the lowest voltage between the odd\even numbered capacitors.

Theoretical waveforms of the proposed equalizer are shown in Figure 4a when the first capacitor has the lowest voltage and in Figure 4b when the first and second capacitors have the lower voltage between odd\even numbered capacitors and other capacitors have the same voltage. As seen in those figures, the transformer current waveform can be symmetrical or not. To be more general, the second case is presented as the operation modes as follows:

(1) Mode 1 [t_0–t_1: Figure 5a]: Based on the switching command of the H-bridge inverter, a positive voltage is imposed on the primary side of the transformer. Regarding the assumption that the first capacitor has the lower voltage between the odd-numbered capacitor, the first diode is turned on and the derivative equation of the system is as follows:

$$\begin{cases} L_m \frac{di_m}{dt} = \frac{V_{C_1}+V_d}{m} \\ L_f \frac{di_f}{dt} = V_{in} - \frac{V_{C_1}+V_d}{m} - r_f i_f \\ C_1 \frac{dV_{C_1}}{dt} = \frac{\left(i_f - i_m - \frac{V_{C_1}+V_d}{mr_m}\right)}{m} - i_{load} + \frac{P_1}{V_{C_1}} \\ C_j \frac{dV_{C_j}}{dt} = \frac{P_j}{V_{C_j}} - i_{load} \quad j = 2, 3, 4 \end{cases} \quad (2)$$

where L_m and L_f are, respectively, the magnetic and leakage inductances, r_m and r_f are the magnetic and leakage resistances, i_m is the magnetizing current, i_f is the leakage current, V_d is the diode drop voltage, V_{in} is the input voltage at the primary side of the transformer, V_{cj} is the voltage of C_j, P_j is the injected power by the boost converter corresponding to the jth fuel cell, and i_{load} is the load current.

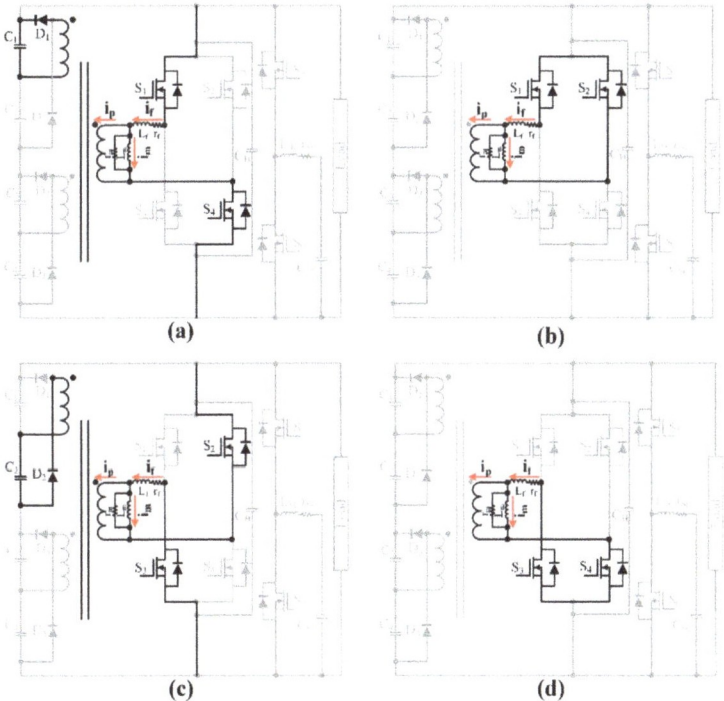

Figure 5. Different operation modes of the proposed equalizer. (a) Mode 1: $t_0 < t < t_1$. (b) Mode 2: $t_1 < t < t_2$. (c) Mode 3: $t_2 < t < t_3$. (d) Mode 4: $t_3 < t < t_4$.

(2) Mode 2 [t_1–t_2: Figure 5b]: The input voltage of the HF transformer is equal to zero in this mode. D_1 can continue to conduct during the interval of [t_1,t_f] and, as a result, the derivative equations of the system are as follows:

$$\begin{cases} L_m \frac{di_m}{dt} = \frac{V_{C_1}+V_d}{m} \\ L_f \frac{di_f}{dt} = -\frac{V_{C_1}+V_d}{m} - r_f i_f \\ C_1 \frac{dV_{C_1}}{dt} = \frac{\left(i_f - i_m - \frac{V_{C_1}+V_d}{m r_m}\right)}{m} - i_{load} + \frac{P_1}{V_{C_1}} \\ C_j \frac{dV_{C_j}}{dt} = \frac{P_j}{V_{C_j}} - i_{load} \quad j = 2,3,4 \end{cases} \quad (3)$$

The switching frequency of the H-bridge inverter is low enough that the diode can be turned off before the end of this mode. As a result, the derivative equations of the system during the interval of [t_f,t_2] are as following when the diode is off:

$$\begin{cases} L_m \frac{di_m}{dt} = r_m(i_f - i_m) \\ L_f \frac{di_f}{dt} = -r_f i_f - r_m(i_f - i_m) \\ C_j \frac{dV_{C_j}}{dt} = \frac{P_j}{V_{C_j}} - i_{load} \quad j = 1,2,3,4 \end{cases} \quad (4)$$

(3) Mode 3 [t_2–t_3: Figure 5c]: A negative voltage is imposed on the primary side of the HF transformer in this mode. As a result, the second diode corresponding to the lower voltage capacitor among even-numbered capacitors starts to conduct. The derivative equations of the system are as follows:

$$\begin{cases} L_m \frac{di_m}{dt} = -\frac{V_{C_2}+V_d}{m} \\ L_f \frac{di_f}{dt} = V_{in} + \frac{V_{C_2}+V_d}{m} - r_f i_f \\ C_2 \frac{dV_{C_2}}{dt} = -\frac{\left(i_f - i_m + \frac{V_{C_2}+V_d}{m r_m}\right)}{m} - i_{load} + \frac{P_2}{V_{C_2}} \\ C_j \frac{dV_{C_j}}{dt} = \frac{P_j}{V_{C_j}} - i_{load} \quad j = 1,3,4 \end{cases} \quad (5)$$

It can be noted that this mod does not exist for Figure 4a.

(4) Mode 4 [t_3–t_4: Figure 5d]: This mode is similar to mode 2 but due to negative voltage imposed on the primary side of the HF transformer, the even-numbered diodes can be turned on. Since the second capacitor is assumed to be the lower voltage capacitor between the even-numbered capacitors, the second diode begins to conduct the current. The derivative equation of the system before the diode stops to conduct is as following during the interval of [t_3,t_{fe}]:

$$\begin{cases} L_m \frac{di_m}{dt} = -\frac{V_{C_2}+V_d}{m} \\ L_f \frac{di_f}{dt} = \frac{V_{C_2}+V_d}{m} - r_f i_f \\ C_2 \frac{dV_{C_2}}{dt} = -\frac{\left(i_f - i_m + \frac{V_{C_2}+V_d}{m r_m}\right)}{m} - i_{load} + \frac{P_2}{V_{C_2}} \\ C_j \frac{dV_{C_j}}{dt} = \frac{P_j}{V_{C_j}} - i_{load} \quad j = 1,3,4 \end{cases} \quad (6)$$

Over the interval of [t_{fe},t_4] the derivative equations of the system are changed as (4) when the diode D_2 is off in this mode. Moreover, this mod does not take place for Figure 4a.

Based on these modes, the power can be transferred from the series connection of capacitors to the lower voltage capacitors. Further information and details about this equalizer can be found in [31].

2.2. Improved Control Method for the Equalizer

As seen in Figure 6, the maximum voltage among the output capacitors connected to the boost converters of different stacks is compared with the lowest voltage and this difference is multiplied by the proportional gain to obtain the duty cycle. Compared to [31], this controller is also able to decrease the maximum voltage on each output capacitor and can consequently reduce the stress on the capacitors and semiconductor devices. A low-pass filter is used in this controller to optimize the dynamical behavior of the equalizer. Due to the possibility of a large external perturbation during the transient states, this controller can impose a high value of duty cycle that leads to conduct a high value of current in semiconductors. Using a dynamic saturation can cope with this issue. To model such a control system, the derivative equation of the filter is used as follows:

$$\frac{dy}{dt} = \omega_f \left(V_{C_{max}} - V_{C_{min}} - y \right) \tag{7}$$

where y is the output of the filter, ω_f is the cut-off frequency of the filter, $V_{C_{min}}$ and $V_{C_{max}}$ are the minimum and maximum voltages among the capacitor voltages, respectively. The difference between the maximum and the minimum voltages is used as the input of the filter. As a result, the duty cycle of the H-bridge inverter is calculated as follows:

$$d = K_p y \tag{8}$$

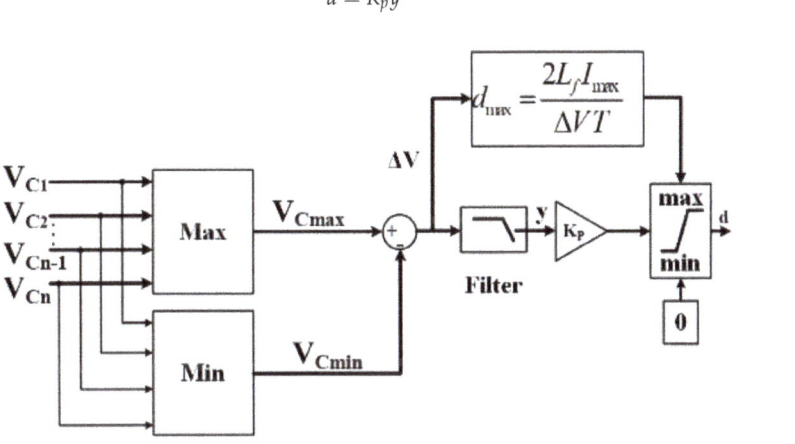

Figure 6. Schematic diagram of the control method used to determine the duty cycle (d) of the H-bridge inverter.

In this paper, a protection circuit is also proposed to dynamically control the maximum value of the duty cycle and indirectly control the maximum value of the current. As seen in Figure 6, this protection is implemented by considering the derivative equation of leakage inductance of the transformer. Based on Figure 4, the leakage current of the transformer can be positively\negatively reached to its maximum during the positive\negative part of the input voltage of the transformer. The time interval of $[t_0,t_1]$ or $[t_2,t_3]$ is equal to dT/2 where T is 1/F and F is the switching frequency. Therefore, the maximum duty cycle can be deduced from the model as following regardless of the losses:

$$L_f \frac{I_{max}}{dT/2} = \Delta V \tag{9}$$

2.3. Average Model

Considering the supplied power by the first stack is lower than the others, the first capacitor voltage has the lowest voltage. The other stacks inject the nominal power and as a result, the output capacitors of their boost converters have the same voltage. A dynamical average model can be considered for this system by calculating the transmitted power through the transformer. To calculate this power, the diode current equation can be used. Based on the approach in [31], an equation for the transmitting power through the transformer can be obtained as follows:

$$P_{e1} = \frac{dV_{C_1}}{8F^2 m^2}\left(\frac{dF(mV_{dc}-B_1)}{L_f} - \frac{dFB_1}{L_m} - \frac{16F^3 L_m (mr_m V_{dc} - 2L_f B_1)}{r_m(-4FL_f L_m + 4FL_C r_m - dr_f r_m)}\right.$$
$$\left. + \frac{d^2 F r_f (-mr_m V_{dc} + 2L_f B_1)}{L_f (4FL_f L_m - 4FL_C r_m + dr_f r_m)} - \frac{d(-4FL_C + 2dr_f) A_1}{4L_f L_m B_1 (4FL_f L_m - 4FL_C r_m + dr_f r_m)^2}\right) \quad (10)$$

This power is received by the first capacitor. A similar equation can be obtained for the transferred power to one of the even-numbered capacitors. Considering the second capacitor as a lower voltage capacitor among the even-numbered capacitors, the transferred power can be calculated as follows:

$$P_{e2} = \frac{dV_{C_2}}{8F^2 m^2}\left(\frac{dF(mV_{dc}-B_2)}{L_f} - \frac{dFB_2}{L_m} - \frac{16F^3 L_m (mr_m V_{dc} - 2L_f B_2)}{r_m(-4FL_f L_m + 4FL_C r_m - dr_f r_m)}\right.$$
$$\left. + \frac{d^2 F r_f (-mr_m V_{dc} + 2L_f B_2)}{L_f (4FL_f L_m - 4FL_C r_m + dr_f r_m)} - \frac{d(-4FL_C + 2dr_f) A_2}{4L_f L_m (4FL_f L_m - 4FL_C r_m + dr_f r_m)^2 B_2}\right) \quad (11)$$

This power is received by the second capacitor. It can be noted that this power is negligible in the case of Figure 4a.

To obtain a dynamical average model, the power that is consumed by the equalizer is also required. To calculate this power, the losses inside the system should be added to $P_{e1} + P_{e2}$. The injected power to the equalizer can be calculated as follows:

$$P_{in} = P_{e1} + P_{e2} + \frac{\overline{V}_{p1}^2 + \overline{V}_{p2}^2}{r_m} + r_f\left(\overline{I}_{f1}^2 + \overline{I}_{f2}^2\right) \quad (12)$$

where $\overline{I}_{f1}$ is the RMS value of the leakage current of the transformer due to the transmitting energy to the first capacitor, and $\overline{V}_{p1}$ is the RMS voltage on the primary side of the transformer due to the transmitting energy to the first capacitor that is the RMS value of $\frac{V_d + V_{C_1}}{m}$ over the interval of t_f-t_0 in the switching period as follows:

$$\begin{cases} \overline{I}_{f1}^2 = \frac{1}{T}\int_{t_0}^{t_f}(i_f(t))^2 dt = \frac{1}{T}\left(\int_{t_0}^{t_1}(i_f(t))^2 dt + \int_{t_1}^{t_f}(i_f(t))^2 dt\right) \\ \overline{V}_{p1}^2 = \frac{1}{T}\int_{t_0}^{t_f}\left(\frac{V_d+V_{C_1}}{m}\right)^2 dt = \frac{(t_f - t_0)}{T}\left(\frac{V_d+V_{C_1}}{m}\right)^2 \end{cases} \quad (13)$$

where $\overline{I}_{f2}$ and $\overline{V}_{p2}$ are the RMS values of the leakage current of the transformer and the RMS voltage on the primary side of the transformer due to the transmitting energy to the second capacitor that can be calculated in a similar way in the interval of $[t_2, t_{fe}]$. It can be noted that this current is negligible in the case of Figure 4a. The load current that is seen by the series connection of the capacitors can be calculated regarding the injecting power to the equalizer as follows:

$$i_{ch} = i_{load} + \frac{P_{in}}{V_{dc}} - i_a \quad (14)$$

where i_{load} is the load current, and i_a is the current that is injected by the SC.

Calculating the received power by the capacitors, a dynamical average model based on the derivative equations of the system can be obtained as follows:

$$\begin{cases} C_1 \frac{dV_{C_1}}{dt} = \frac{P_1 + P_{e1}}{V_{C_1}} - i_{ch} \\ C_2 \frac{dV_{C_2}}{dt} = \frac{P_2 + P_{e2}}{V_{C_2}} - i_{ch} \\ C_j \frac{dV_{C_j}}{dt} = \frac{P_j}{V_{C_j}} - i_{ch} \ \forall \ j \in \{3, 4, \ldots, n\} \end{cases} \quad (15)$$

To control the power delivered by each stack, the sliding mode controller is used. This control is explained in the following section. The parameter of this controller is chosen in such a way that this controller is only as fast as required. Therefore, the reference power is always followed by this controller.

3. Hybridization

The diagram of the overall control structure is shown in Figure 7. As seen in this figure, each part of the system is controlled by its own controller. In this section, the method of regulating the DC-link and the SC voltages will be detailed. This method is based on an energy regulator (extern loop) and an indirect sliding mode control. This simple method can ensure that the dynamic performances are independent of the operating points.

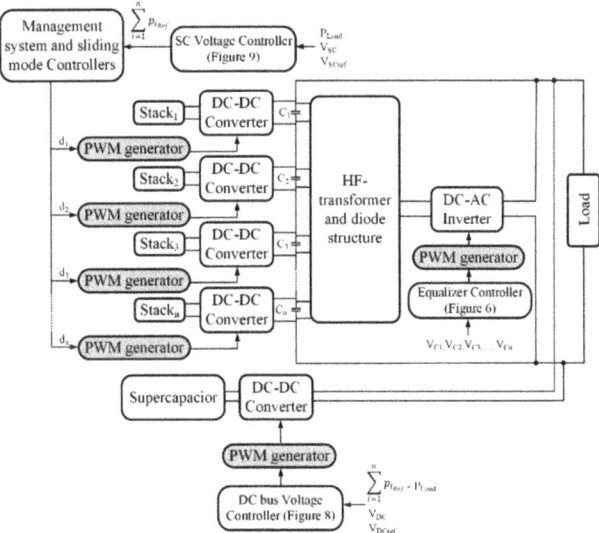

Figure 7. Diagram of the overall control structure.

3.1. DC Bus Voltage Controller

To regulate the DC bus voltage, an SC is connected to the DC bus by a bidirectional boost converter. The state-space model of the super-capacitor with a boost converter is as follows:

$$\begin{cases} L_{SC} \frac{di_{SC}}{dt} = v_{sc} - (1 - d_{SC})V_{dc} - r_{SC}i_{SC} \\ C_{SC} \frac{dv_{SC}}{dt} = -i_{SC} \end{cases} \quad (16)$$

where L_{sc} is the inductance connected to the boost converter of the SC and r_{sc} is its resistance, v_{sc} is the SC voltage, and i_{sc} is its current. Therefore, the injected current to the DC bus by the SC can be calculated as follows:

$$i_a = (1 - d_{SC})i_{SC} \tag{17}$$

To control the DC bus voltage, a controller based on the flatness theory is used in this paper. This controller is deeply studied in [51,52]. The stored energy in the DC link is used as the output variable of this controller. This energy can be calculated as follows:

$$y_{dc} = \frac{1}{2}C_{eq}V_{dc}^2 \tag{18}$$

where C_{eq} is the equivalent capacitor of series-connected capacitors. The derivation of this energy is as follows:

$$\dot{y}_{dc} = \sum_{j=1}^{n} P_j + P_a - P_{load} \tag{19}$$

where P_a is the power injected to DC bus by the SC. Hence:

$$P_a = \dot{y}_{dc} + P_{load} - \sum_{j=1}^{n} P_j \tag{20}$$

Considering the losses of the boost converter in the resistance of its inductance, this power can be calculated as follows:

$$P_a = P_{SC} - r_{SC}\left(\frac{P_{SC}}{v_{sc}}\right)^2 - L_{SC}\left(\frac{P_{SC}}{v_{sc}}\right)\frac{d_{SC}\left(\frac{P_{SC}}{v_{sc}}\right)}{dt} \tag{21}$$

where P_{sc} is the injected power by the SC. As seen in Figure 8, this control is realized by two loops. The inner loop is the power loop and the energy loop is an outer loop. Based on the energy stored in the DC bus, the energy loop can control the voltage of the DC bus. Assuming that the outer loop (energy loop) is slower enough than the inner loop (power loop), the variation of the magnetic energy can be neglected. Furthermore, the injected power by SC can be rewritten as follows:

$$P_{SC} = 2P_{max}\left(1 - \sqrt{1 - \frac{P_a}{P_{max}}}\right) \tag{22}$$

where P_{max} is as follows:

$$P_{max} = \frac{v_{SC}^2}{4r_{SC}} \tag{23}$$

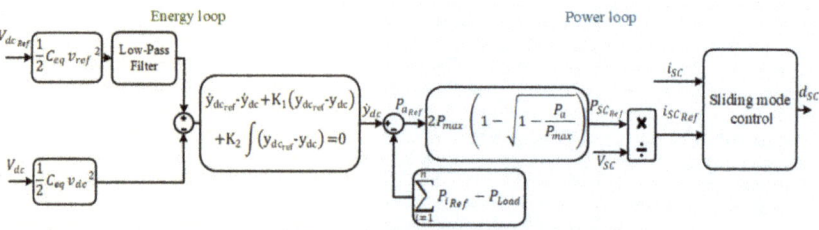

Figure 8. Schematic diagram of the voltage controller.

Using (19)–(23), the power that must be injected by the SC to regulate the DC link voltage is obtained. This control can operate when the inner loop, which controls the power of SC, is fast enough.

To ensure that the electrostatic energy follows its reference, a control method based on the second-order system is used as follows [53]:

$$\dot{y}_{dcref} - \dot{y}_{dc} + K_1(y_{dcref} - y_{dc}) + K_2 \int (y_{dcref} - y_{dc}) = 0 \qquad (24)$$

where K_1 and K_2 are the controlling parameters with $K_1 = 2\zeta\omega_n$, and $K_2 = \omega_n^2$ (where ω_n is the desired cutoff radian frequency of the voltage control loop [53]). $\dot{y}_{dc}$ is obtained from (24). (20) and (24) allow the SC power reference to be obtained. The inner loop goal is to control the power delivered by the SC to track its reference.

Assuming that the dynamic of the SC voltage variations is slow in comparison to the DC bus voltage variations, the SC voltage can be considered as a constant value for the design of the DC bus control. As a result, the reference current of the SC can be obtained by dividing its reference power to its voltage. The sliding mode control method is used in this study to control the current.

A sliding surface is defined as follows [54]:

$$s = i_{SC} - i_{SCref} + k_i \int (i_{SC} - i_{SCref}) \qquad (25)$$

To ensure the zero steady-state error, an integral term is used in this sliding surface [54]. The associated reaching condition is defined as follows:

$$\dot{s}(x) = -\Lambda s(x) \qquad (26)$$

where Λ is a positive constant that determines the speed of attraction to the sliding surface. Using this approach, two poles of the system are $-\Lambda$ and $-k_i$ and independent of the operating point. By differentiating (25) and using (26), the following equation is obtained:

$$\frac{di_{SC}}{dt} + k_i(i_{SC} - i_{SCref}) = -\Lambda s(x) \qquad (27)$$

Using this equation and (16), the equivalent duty cycle of the SC converter can be calculated as follows:

$$d_{SC} = \frac{L_{SC}}{V_{dc}}\left[\frac{V_{dc} - v_{sc} + r_{SC}i_{SC}}{L_{SC}} - k_i(i_{SC} - i_{SCref}) - \lambda\left(i_{SC} - i_{SCref} + k_i \int (i_{SC} - i_{SCref})\right)\right] \qquad (28)$$

3.2. SC Voltage Controller

A similar approach is used to control the voltage of SC. The stored energy in SC is used to controls its voltage. This energy can be calculated as follows:

$$y_{SC} = \frac{1}{2}C_{SC}v_{sc}^2 \qquad (29)$$

where C_{SC} is the capacitance of the SC. Considering that the dynamic of the DC bus voltage loop is widely greater than the dynamic of the SC voltage, the derivation of this energy is as follows:

$$\dot{y}_{SC} = -P_{SC} \approx \sum_{j=1}^{n} P_j - P_{load} \qquad (30)$$

where $\dot{y}_{SC}$ is the power of the SC (P_{SC}) that should be transferred to the SC (negative) or injected by it (positive). The total power of stacks that must be injected in steady-state can be obtained by this equation. A simple proportional gain controller is used to ensure that stored energy in the SC follows its reference:

$$\dot{y}_{SCref} - \dot{y}_{SC} + K_{sc}(y_{SCref} - y_{SC}) = 0 \qquad (31)$$

The proportional gain K_{SC}, which represents the cutoff radian frequency of the SC voltage loop, should respect the low dynamic assumption of the SC voltage. Based on (30), the total power that should be injected by the stacks can be calculated. As seen in Figure 9, a rate limiter is used to respect the dynamical constraint imposed by the fuel cell auxiliaries. V_i and i_i are the voltage and current of the ith stack. d_i is the duty cycle of the boost converter connected to the ith stack.

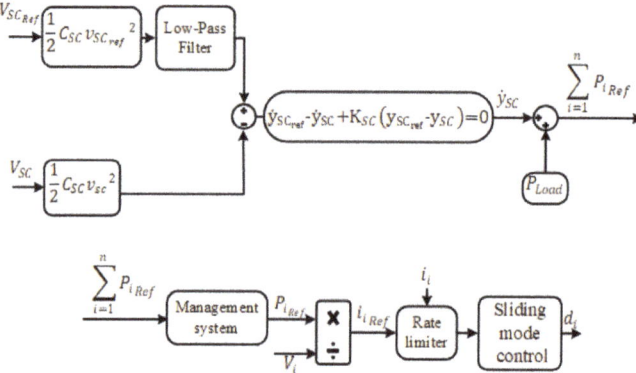

Figure 9. Schematic diagram of the SC voltage controller.

4. Simulation and Experimental Results

4.1. Simulation Results

To evaluate the behavior of the dynamical average model, two simulations are performed when four stacks are connected to four boost converters. The injected power by the first stack is changed from 63 to 100 W at 0.05 s while the first simulation. The injected power of the other stacks is fixed to nominal power (126 W) during this simulation.

The parameters used to obtain simulation results are shown in Table 1. To study the dynamical behavior of the average model, the state-space model of the system based on (1)–(5) is also simulated with the same parameters. The results of the first simulation are shown in Figure 10. As seen in these figures, the average model results are consistent with the results of the state space model. The voltage change of the output capacitor of the boost converters connected to the different stacks is shown in Figure 10. The difference between the two models is due to the linearizing method which was used to calculate (10). Indeed, the transformer current has an exponential form which has been approximated by the first-order polynomial based on the Taylor series. For the proposed average model, the power of the balancing system defined by (10) is a bit overestimated; consequently, the obtained voltage is a bit higher than the switching model. The voltage of the DC bus ($\sum_j V_{C_j}$) is shown in Figure 1b during this simulation. As seen in this figure, the DC voltage is regulated at the reference voltage (48 V). It is notable that the power of the fuel cell is reduced slowly and this undershoot is eliminated when the real low dynamic of the fuel cell is taken into account. Therefore, this simulation was accomplished in the worst condition.

The condition for the second simulation is similar to the first simulation except for the injected power by the second cell. This power is set to 100 W in the second simulation. The voltage changes of different capacitors are shown in Figure 11a. The DC bus voltage is shown in Figure 11b during this simulation. As seen in these figures, the average model is in agreement with the state-space model of the system.

Table 1. Parameters of the proposed equalizer.

Symbol	Unit	Value	Description
C	μF	4700	Electrochemical
A_L	nH/turns^2	12,500	Planar transformer core
N_1	turns	1	Primary winding turns
N_2	turns	4	Secondary winding turns
k	-	0.98	Coupling coefficient
F	kHz	40	Switching frequency of the H-bridge
V_{bat}	V	48	Nominal voltage of the battery
V_C	V	12	Nominal output voltages of boosts
V_d	V	0	Drop voltage of diodes
P_{FC}	W	126	Nominal injected power of different stacks
η	-	1	Efficiency of the equalizer system
ω_f	Rad/s	$2\pi 10^3$	Cut-off radian frequency of the filter
K_p	-	0.1	Proportional gain of the controller
λ	Rad/s	7500	
k_i	Rad/s	7500	
F_s	kHz	29	Switching frequency of the boost converters
K_{SC}	-	0.08	

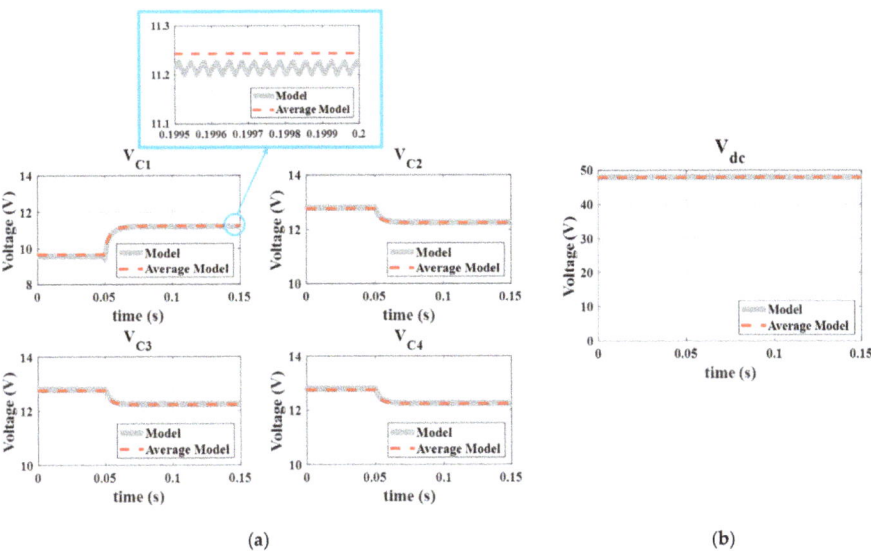

Figure 10. Simulation results in closed-loop when P_1 is increased from 63 to 100 W and the other stacks inject the nominal power: (**a**) Voltage changes of the boost converters output capacitors connected to the stacks; (**b**) DC bus voltage.

4.2. Experimental Results

To verify the validity of the proposed dynamical average model, two experiments are accomplished on a laboratory test bench as seen in Figure 12. Four different power supplies are used to emulate the four stacks of PEMFCs. dSPACE 1005 with the FPGA board is used to receive the information and send the commands. The part number of all the components is summarized in Table 2.

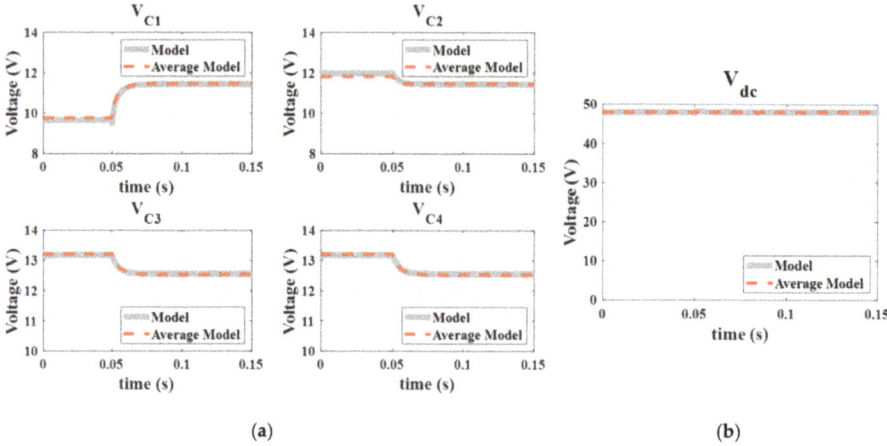

Figure 11. Simulation results in closed-loop when P_1 is changed from 63 to 100 W, P_2 is set to 100 W, and the other stacks inject the nominal power: (**a**) Voltage changes of the boost converters output capacitors connected to the stacks; (**b**) DC bus voltage.

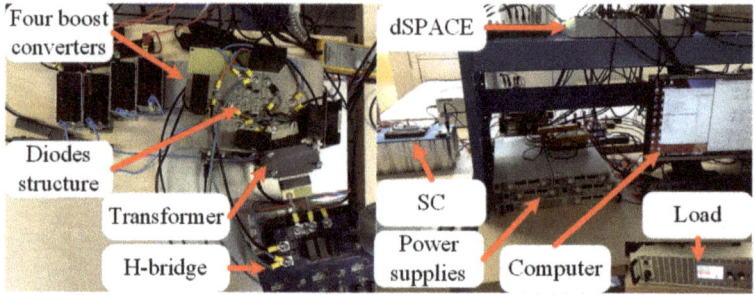

Figure 12. Test bench of the proposed system.

Table 2. Parameters of the used devices in the test bench.

	Unit	Value
FC	Power supply	TDK GENH 750 W
Boost converter	Inductance	(1 mH)
	Switches	IGBT
	Capacitor	Electrochemical (4700 μF)
Equalizer	Diodes	DSS2x121-0045B
	Magnetic core	B66295G Material N87
	H-bridge switches	SiCMOSFET CCS050M12CM2
	Capacitor	film (220 μF)

The same assumption of the first simulation is used to accomplish the first experiment. The experimental results of the output capacitors voltage changes are shown in Figure 13a when the first cell power is increased from 63 to 100 W. As seen in this figure, the voltage of the first capacitor is increased because of the increasing the corresponding stack injected power. The voltage of the other capacitors changed in such a way that the sum of voltage is fixed to 48 V. The average model results are also shown in Figure 13b in the same conditions for a better view. These results are verified by the experimental results. There is always an error between the simulation and experiment results in steady-state. The error between voltages of the capacitors is in a reasonable range and less than 3%. This error is caused by not considering all losses.

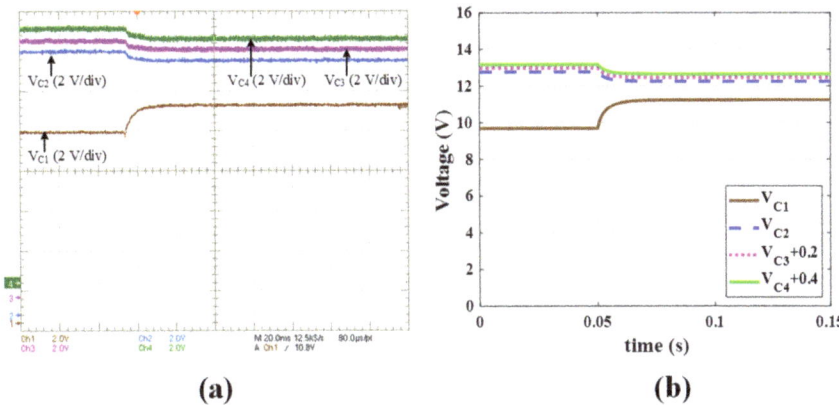

Figure 13. Voltage changes of the boost converters output capacitors connected to the stacks in closed-loop when P_1 is increased from 63 to 100 W, and P_2 is set to the nominal power: (**a**) Experimental results; (**b**) Simulation results of the average model.

The second experiment is accomplished in the same conditions as the second simulation. The experimental results of the voltage changes of output capacitors are shown in Figure 14a. To provide a better view, the average model results are also shown in Figure 14b in the same conditions. As seen in this figure, the experimental results are in agreement with the simulation results. Based on the experimental and simulation results, the proposed model for the management system is valid and it can be used to study and analyze the stability of the system.

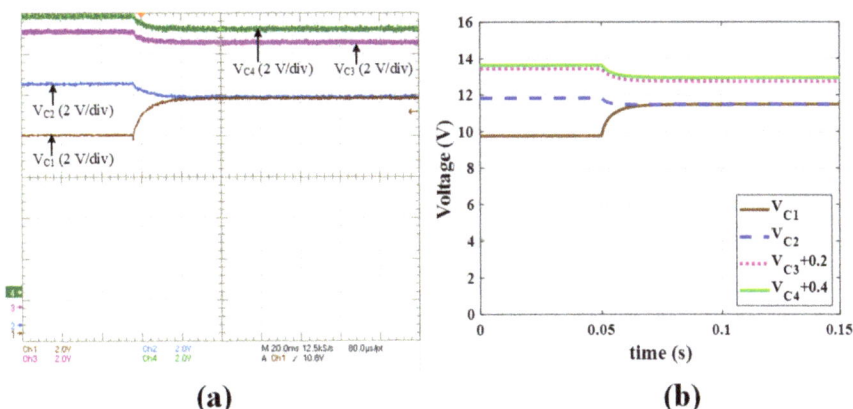

Figure 14. Voltage changes of the boost converters output capacitors in closed-loop when P_1 is changed from 63 to 100 W, and P_2 is set to 100 W: (**a**) Experimental results; (**b**) Simulation results of the average model.

5. Stability Analysis and Energy Management

In this section, the stability of the proposed control method will be investigated. Then, the effectiveness of the proposed system in supplying the load profiles even while curing one of the stacks in two different conditions. Finally, the robustness of the proposed control method will be studied. Since the average model is valid and it can estimate the behavior of the system in closed-loop, this model can be used to analyze the stability of the system. By obtaining the eigenvalues of the Jacobian matrix of the proposed dynamical average model, the local stability around the equilibrium point can be analyzed. Furthermore, this model can be used to size the parameters of the different controllers in such a way that the stability of the system is ensured. The poles of the sliding mode controller (current controller) should be at least less than one-tenth of the switching radian frequency of the boost converters or bidirectional converter connected to the SC. The switching frequency of these converters is equal to 30 kHz. Therefore, the λ and k_i of the sliding mode controller are fixed to 7500 rad/s. As mentioned before, the power loop of the DC bus voltage controller should be faster than the energy loop. As a result, the radian frequency of the energy loop should be slower than one-tenth of the current control loop. Therefore, ω_n equal to 500 rad/s is used to obtain K_1 and K_2. The damping ratio of 0.7 is used to obtain the best behavior. The control of the SC voltage should respect the slow dynamic of the fuel cell. A rate limiter of 4 A/s limits the dynamic of current change to respect the dynamic of the fuel cell. The equalizer system controller should be faster than the DC link voltage controller to ensure the controllability of boost converters even in transient conditions. However, the cut-off frequency of the equalizer controller should be lower than one-tenth of its switching radian frequency. The switching frequency of the H-bridge inverter is equal to 40 kHz. There is a tradeoff between the dynamic of the equalizer controller and the time that the duty cycle of the H-bridge inverter is stuck in the maximum value due to the dynamic saturation. As a result, the cut-off frequency of $2\pi\,10^3$ rad/s is used for the equalizer controller. The dynamic variations of the SC voltage will be neglected in the stability study due to the slow dynamic of the SC voltage and the powers delivered by the fuel cells are supposed to be constant.

To evaluate the stability of the system, the power of the first cell is changed from 0 to 100 W with steps of 10 W. The other cell powers are fixed to 126 W. The other parameters that are used for this simulation is shown in Table 1. Using the proposed dynamical average model, the steady-state voltages of the boost output capacitors are shown in Figure 15. for this simulation. As seen in this figure, the voltage of the first cell is increased by increasing the injected power of the first cell. The voltage of other stacks is decreased to fix the DC bus voltage to 48 V. The eigenvalues of the closed-loop system are depicted in Figure 16. This figure shows the eigenvalues of the Jacobian matrix of the proposed dynamical average model when the stacks inject the nominal power except the first stack. As mentioned above, the injected power of the first stack is changed from zero to 100 W. These eigenvalues with negative real parts prove the stability of the system. As seen in this figure, the different groups of eigenvalues are shown in different circles with different colors and numbers. The purple circles show the eigenvalues that strongly depend on the parameters of the voltage and current controllers. The place of eigenvalues inside the purple circle numbered 1 and 2 can be changed by changing the parameters of the current sliding mode controllers. The multiple eigenvalues in the same place are shown by a circle around the multiplication sign as shown in circle number one. The place of eigenvalues inside purple circle number three and two can be changed by changing the natural radian frequency of the second-order system used to find K_1 and K_2 in (24). The blue circles specify the place of eigenvalues that depend on the parameters of the equalizer system controller. The place of eigenvalues inside blue circle number four can be changed by the operating point and all parameters of equalizer controller consist of the proportional gain and the cut-off frequency used for finding the duty cycle of the H-bridge inverter. The place of eigenvalues inside blue circle number five strongly depends on the operating point and the proportional gain (K_p) of the equalizer system controller.

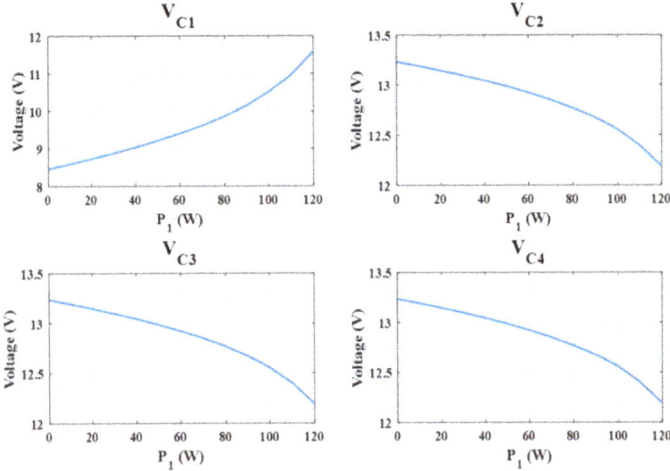

Figure 15. Voltage changes of the boost converters' output capacitor connected to fuel cell stacks in closed-loop by changing the injected power of the first stack while the other stacks inject the nominal power of 126 W.

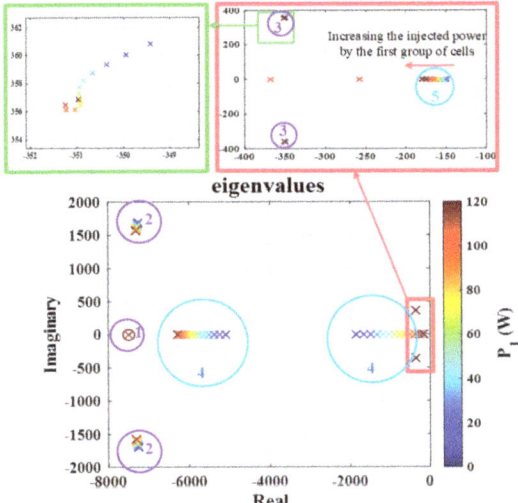

Figure 16. Eigenvalues of the closed-loop system by changing the injected power of the first stack while the other stacks inject the nominal power of 126 W.

To prove the stability of the system when two stacks are in the fault conditions, the power of the second stack is fixed to 50 W that is a little bit lower than the half of nominal power. The injected power of the first stack is increased from 0 to 40W while the injected power by the other stacks is set to 126 W. Using the proposed dynamical average model, the steady-state value of the output capacitor voltages by changing the injected power of the first stack is shown in Figure 17. As seen in this figure, the voltage of the first cell is increased by increasing the injected power of the first cell. The voltage of other stacks is decreased to fix the DC link voltage near 48 V. The eigenvalues of the closed-loop system are shown in Figure 18. This figure shows the eigenvalues of the Jacobian matrix of the proposed dynamical average model when the second stack injects 50 W and the other stacks inject the nominal power except the first stack. As mentioned above, the supplied power by the first stack is changed from zero to 50 W. These eigenvalues with negative real parts prove the stability of the system.

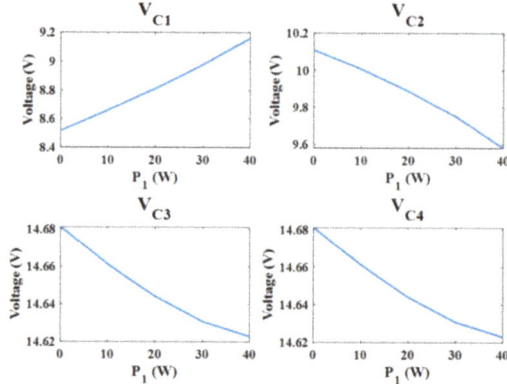

Figure 17. Voltage changes of the boost converters' output capacitor connected to fuel cell stacks in closed-loop by changing the injected power of the first stack while the second stack injects 50 W and the other stacks inject the nominal power of 126 W.

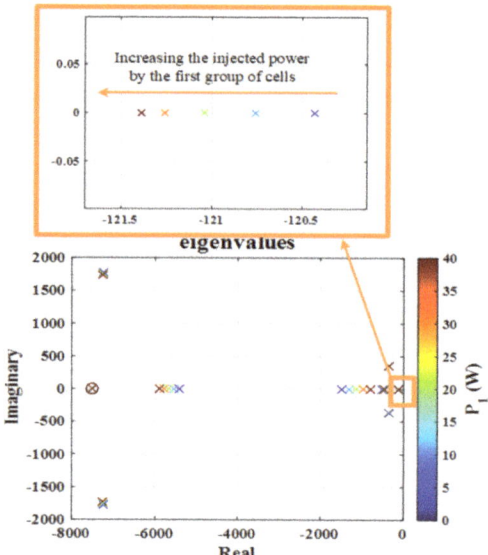

Figure 18. Dominant eigenvalues of the closed-loop system by changing the injected power of the first stack while the second stack injects 50 W and the other stacks inject the nominal power of 126 W.

Regarding the stability of the system, this topology can be used to manage the cells of a stack. To evaluate the effectiveness of the proposed topology in terms of energy management and dynamic performance, three different experiments are performed. Using the test bench shown in Figure 12, these three experiments are performed in the following condition. A step of 400 W for eight seconds is applied to the baseload power in these three experiments. The four stacks are in the normal condition for the first experiment. The load power and the injected power by the first and second stack are shown in Figure 19a,b. The behavior and magnitude of the injected power by two other stacks is very close to the second stack. As seen in this figure, the injected power by the SC is increased rapidly to control the DC link voltage. The injected power by the stacks gradually increased to control the voltage of the SC. Since the load step power is very high, the injected power increased to their maximum power. The difference between the load power and injected power by the stacks is compensated by the SC. Because of the low dynamic of the fuel cells, the injected power by the stacks is gradually decreased and the SC attracts the excess amount of power injected by the stacks whereas the load power decreased to its nominal value. As seen in Figure 19a, the injected power by the SC is equal to zero in steady-state. To provide a better view during the transient conditions of the first experiment, the load power and injected power by the stacks and SC are shown in Figure 19b with the time scale of 4 s/div. The DC link and SC voltages are shown in Figure 19c during this experiment. As seen in this figure, the DC link voltage is well controlled and has a constant value of 48 V. The SC voltage is gradually decreased due to its injected power but then it increased gradually due to the increase of the injected power by the stacks and decrease of load power to its nominal value. To assess the behavior of the DC link voltage controller in transient condition, the SC injected power and the DC link and SC voltages are shown in Figure 19d with the time scale of 4 ms/div. As seen in this figure, the DC link voltage decreases due to the step of load power but the fast increase in the injected power by the SC can compensate the required load power and regulate the DC link voltage. Furthermore, the SC voltage in this figure is in agreement with the SC constant voltage assumption over the transient conditions.

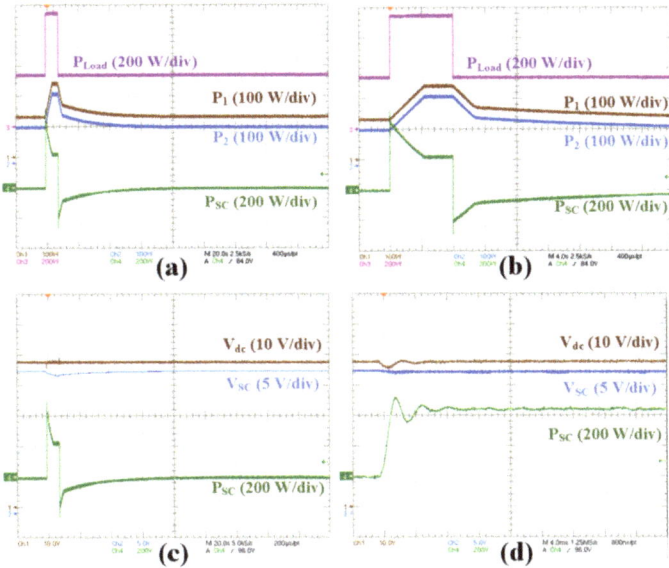

Figure 19. Experimental result of the energy management system when the stacks are in the normal condition and an overload occurs: (**a**) Injected power and load power variations; (**b**) Zoom on powers; (**c**) DC link and SC voltage changes; (**d**) zoom on the part c.

In the second experiment, it is assumed that the first stack is in the drying condition whereas the other stacks are in the normal condition. To produce more water in the first stack, its current should be controlled close to its maximum in the worst condition. As seen in Figure 20a, the injected power by the first stack is controlled close to its maximum (240 W) during this experiment. A zoom is realized on the transient condition of the injected power by the SC and stacks in Figure 20b. As seen in Figure 20c, the DC link voltage is well controlled and it has a constant value of 48 V. The

SC voltage is reduced due to its injected power to supply the overload. The SC voltage gradually increased and approaches to its nominal value of 24 V by increasing the injected power of stacks and decreasing the load power to its nominal value. The voltage of the output capacitors of the boost converters is shown in Figure 20d. Since the step load leads to an increase in the injected power by the stacks, the output capacitor voltages increase. The first stack injects the maximum power during this experiment. Therefore, the first capacitor voltage decreases because of the DC link voltage stabilization by the SC. The output capacitors have an identical voltage when the injected power by the stacks reach to their maximum power. The difference between the voltages at this point is originated from the losses difference in the boost converters. The injected power by the different stacks except the first stack is gradually decreased when the load power decreases to its nominal value. As a result, the difference between the first output capacitor and the other output capacitors increases. Notable that the voltage of the first stack in the drying condition is kept in an acceptable range due to the improved equalizer controller.

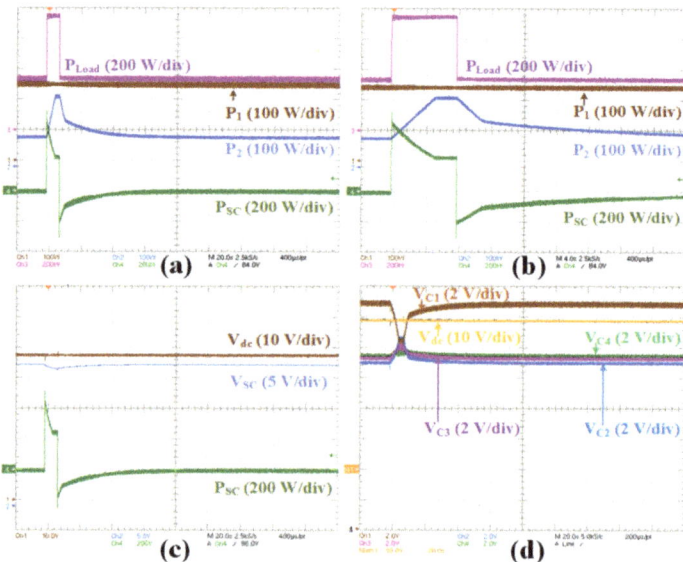

Figure 20. Experimental result of the energy management system when the stacks are in the normal condition except the first stack (drying condition) and an overload occurs: (**a**) Injected power and load power variations; (**b**) Zoom on the power; (**c**) DC link and SC voltage changes; (**d**) Output capacitor voltage changes.

It is assumed that the first stack is in the flooding condition during the third experiment. The other stacks are in normal condition. In this case, the water production inside the cells of the first stack should be reduced. Therefore, the current or power injected by the first stack should be controlled close to zero in the worst condition. The equalizer ensures the controllability of the boost converters in such conditions. As seen in Figure 21a, the injected power by the first stack is controlled at zero Watt during this experiment. More details of the injected power by the SC and stacks are demonstrated in Figure 21b. Most of the overload is supplied by the SC due to the no injected power of the first stack and limits in the maximum injected power by the stacks. As a result, the SC voltage reduces more than the two previous experiments as shown in Figure 21c. The DC link voltage is also shown in this figure and it is controlled at 48 V. The voltage of the output capacitors is shown in Figure 21d during this experiment. Due to the use of the equalizer system, despite the lack of power injection by the first stack in this experiment, the first capacitor has a voltage of 10 V. This experiment confirms the function of the used equalizer system to ensure the controllability of the boost converters.

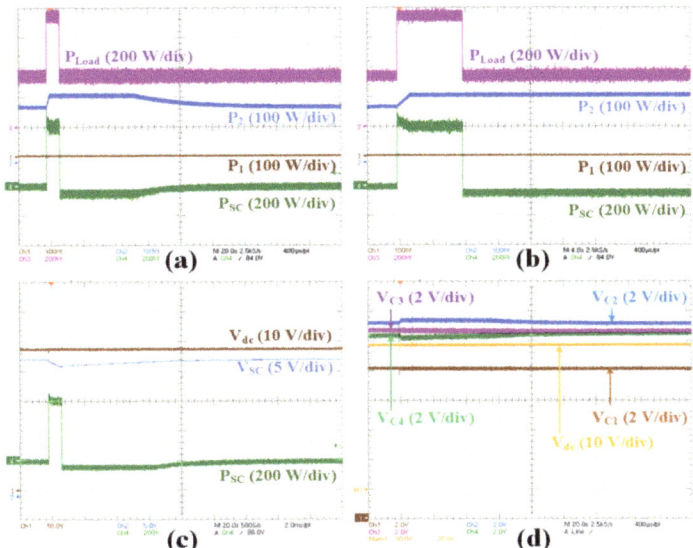

Figure 21. Experimental result of the energy management system when the stacks are in the normal condition except the first stack (flooding condition) and an overload occurs: (**a**) Injected power and load power variations; (**b**) Zoom on the powers; (**c**) DC link and SC voltage changes; (**d**) Output capacitor voltage changes.

The other issue that needs to be addressed is the robustness of the control method. Robustness is the ability of the closed-loop system in being insensitive to perturbations and tolerating the component variation. To assess the robustness of the controller, the eigenvalues of the system are studied under changing some parameters of the system. If the real part of eigenvalues remains negative, it can be proved that it is a robust controller. In this system, the capacitor change can have an important effect on the stability of the system. The capacitance can be changed by aging and temperature effect. The optimum value of the capacitance is also a key factor in reducing the size of the system. In [51], a method was proposed to calculate the minimum value of the equivalent DC-link capacitance. Based on this method and assuming that the maximum permitted voltage drop in DC link is 7 V, the minimum value of each four capacitors in series connection should be 4.7 mF to tolerate a variation of the load power from −400 to 400 W.

To check the robustness of the control method, the capacitance of the first capacitor is changed from 2.35 mF to 9.4 mF. The injected power of the first cell is fixed to zero Watt and the other stacks inject the nominal power. The same parameters as Table 1 are used for the controller. The evolution of the eigenvalues of the system is depicted in Figure 22. This figure shows the eigenvalues of the Jacobian matrix of the proposed dynamical average model when the stacks inject the nominal power except the first stack. The first stack injects no power and the first capacitor value is changed as mentioned earlier. The dominant eigenvalue of the system becomes less negative by further decreasing the capacitance value from 4 mF. The place of multiple eigenvalues that are seen in this figure strongly depends on the control parameters and operating point or other parameters such as the coupling coefficient.

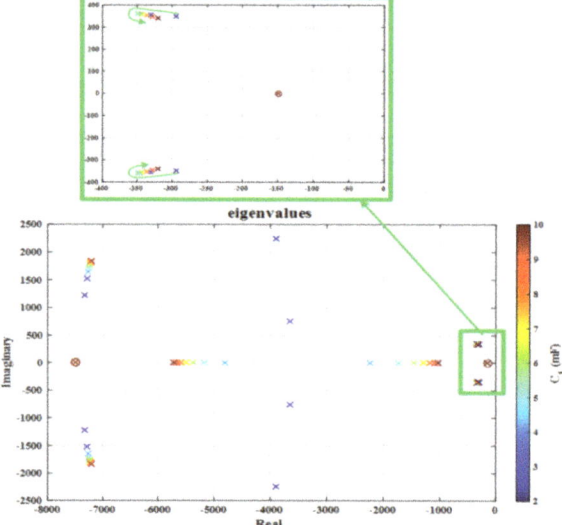

Figure 22. Dominant eigenvalues of the closed-loop system by changing the capacitance value of the first capacitor when the power of the first stack is equal to zero W and the other stacks inject the nominal power of 126 W.

Another parameter that its value can have an impact on stability is the coupling coefficient between the windings of the transformer. To evaluate the impact of this parameter, the power of the first stack is fixed to zero Watt while the other stacks inject the nominal power. The coupling coefficient is changed from 0.7 to 0.97. The parameters of the controller are as same as before. The dominant eigenvalue of the system is depicted in Figure 23. This figure shows the eigenvalues of the Jacobian matrix of the proposed dynamical average model when the stacks inject the nominal power except the first stack. The first stack injects no power and the coupling coefficient value is changed as mentioned earlier.

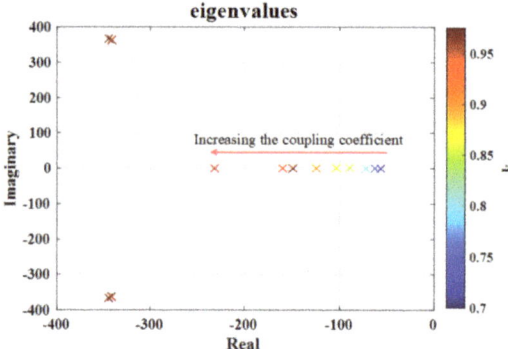

Figure 23. Dominant eigenvalues of the closed-loop system by changing the coupling coefficient when the power of the first stack is equal to zero W and the other stacks inject the nominal power of 126 W.

As seen in the previous results, the real part of the eigenvalues of the system has always a negative value of less than −50. Therefore, the system is stable for a wide range of parameters change. This value depends strongly on the equalizer controller parameters.

6. Conclusions

In this paper, a power electronic system is proposed to enhance the durability of the fuel cells in multi stack architecture. Using this architecture, energy management can be performed even in the fault mode of stack and while the system is trying to cure a stack. To investigate the dynamical properties of the system and to prove its stability, a dynamical average model is proposed taking into account the order reduction induced by the presence of the HF transformer and cross-coupling effects due to the serial connection of output capacitors. The validity of the reduced model is verified through the simulation and experimental results. The stability analysis based on the proposed model proved the asymptotic stability of the system in different conditions. The robustness of this control method was also investigated based on the stability analysis by changing system parameters. The experimental results, which show the behavior of the system in different operating conditions, validated the effectiveness of the proposed topology and its associated energy management functionalities. These results also confirmed that the controllability and the DC-link voltage regulation are always ensured even during drying or flooding conditions concerning one of the stacks in which the power injected by each stack is modified to improve durability.

Author Contributions: Conceptualization, M.B. and S.P.; Data curation, M.B.; Formal analysis, M.B., S.P. and M.Z.; Funding acquisition, S.P.; Investigation, M.B. and S.P.; Methodology, M.B., S.P. and F.M.-T.; Project administration, G.M. and S.P.; Resources, S.P., and M.W.; Software, M.B.; Supervision, J.-P.M., G.M. and S.P.; Validation, M.B. and S.P.; Visualization, M.B.; Writing—original draft, M.B.; Writing—review & editing, M.B., J.-P.M., G.M. and S.P. All authors have read and agreed to the published version of the manuscript.

Funding: This research received no external funding.

Conflicts of Interest: The authors declare no conflict of interest.

References

1. Bizon, N.; Thounthong, P. Energy efficiency and fuel economy of a fuel cell/renewable energy sources hybrid power system with the load-following control of the fueling regulators. *Mathematics* **2020**, *8*, 151. [CrossRef]
2. Zandi, M.; Bahrami, M.; Eslami, S.; Gavagsaz-Ghoachani, R.; Payman, A.; Phattanasak, M.; Nahid-Mobarakeh, B.; Pierfederici, S. Evaluation and comparison of economic policies to increase distributed generation capacity in the Iranian household consumption sector using photovoltaic systems and RET Screen software. *Renew. Energy* **2017**, *107*, 215–222. [CrossRef]
3. Bahrami, M.; Gavagsaz-Ghoachani, R.; Zandi, M.; Phattanasak, M.; Maranzanaa, G.; Nahid-Mobarakeh, B.; Pierfederici, S.; Meibody-Tabar, F. Hybrid maximum power point tracking algorithm with improved dynamic performance. *Renew. Energy* **2019**, *130*, 982–991. [CrossRef]
4. El-fergany, A.A.; Hasanien, H.M.; Agwa, A.M. Semi-empirical PEM fuel cells model using whale optimization algorithm. *Energy Convers. Manag.* **2019**, *201*, 112197. [CrossRef]
5. Parol, M.; Wójtowicz, T.; Księżyk, K.; Wenge, C. Optimum management of power and energy in low voltage microgrids using evolutionary algorithms and energy storage. *Int. J. Electr. Power Energy Syst.* **2020**, *119*, 105886. [CrossRef]
6. Lan, T.; Strunz, K. Modeling of multi-physics transients in PEM fuel cells using equivalent circuits for consistent representation of electric, pneumatic, and thermal quantities. *Int. J. Electr. Power Energy Syst.* **2020**, *119*, 105803. [CrossRef]
7. Yoshida, T.; Kojima, K. Toyota MIRAI fuel cell vehicle and progress toward a future hydrogen society. *Interface Mag.* **2015**, *24*, 45–49. [CrossRef]
8. Matsunaga, M.; Fukushima, T.; Ojima, K. Powertrain system of Honda FCX clarity fuel cell vehicle. *World Electr. Veh. J.* **2009**, *3*, 820–829. [CrossRef]
9. Wang, J. System integration, durability and reliability of fuel cells: Challenges and solutions. *Appl. Energy* **2017**, *189*, 460–479. [CrossRef]
10. Zhang, H.; Li, X.; Liu, X.; Yan, J. Enhancing fuel cell durability for fuel cell plug-in hybrid electric vehicles through strategic power management. *Appl. Energy* **2019**, *241*, 483–490. [CrossRef]
11. Samavatian, V.; Radan, A. A novel low-ripple interleaved buck—Boost converter with high efficiency and low oscillation for fuel-cell applications. *Int. J. Electr. Power Energy Syst.* **2014**, *63*, 446–454. [CrossRef]

12. Maranzana, G.; Didierjean, S.; Dillet, J.; Thomas, A.; Lottin, O. Improved Fuel Cell. U.S. Patent WO/2014/060198, 24 April 2014.
13. Wang, T.; Li, Q.; Yin, L.; Chen, W. Hydrogen consumption minimization method based on the online identification for multi-stack PEMFCs system. *Int. J. Hydrog. Energy* **2019**, *4*, 5074–5081. [CrossRef]
14. Ozpineci, B. Optimum fuel cell utilization with multilevel DC—DC converters. In Proceedings of the 19th Annual IEEE Applied Power Electronics Conference and Exposition, Anaheim, CA, USA, 22–26 February 2004.
15. Ozpineci, B.; Case, A.A. Multiple input converters for fuel cells. In Proceedings of the Conference Record of the 2004 IEEE Industry Applications Conference, Seattle, WA, USA, 3–7 October 2004.
16. Lopes, C.; Kelouwani, S.; Boulon, L.; Agbossou, K.; Marx, N.; Ettihir, K. Neural network modeling strategy applied to a multi-stack PEM fuel cell system. In Proceedings of the 2016 IEEE Transportation Electrification Conference and Expo (ITEC), Dearborn, MI, USA, 27–29 June 2016; pp. 1–7.
17. Ramadan, H.S.; de Bortoli, Q.; Becherif, M.; Claude, F. Multi-stack fuel cell efficiency enhancement based on thermal management. *IET Electr. Syst. Transp.* **2016**, *7*, 65–73.
18. Marx, N.; Cardenas, D.C.T.; Boulon, L.; Gustin, F.; Hissel, D. Degraded mode operation of multi-stack fuel cell systems. *IET Electr. Syst. Transp.* **2016**, *6*, 3–11. [CrossRef]
19. Cardenas, D.C.T.; Hissel, D.; Marx, N.; Boulon, L.; Gustin, F. Degraded mode operation of multi-stack fuel cell systems. In Proceedings of the 2014 IEEE Vehicle Power and Propulsion Conference (VPPC), Coimbra, Portugal, 27–30 October 2014.
20. Garcia, J.E.; Herrera, D.F.; Boulon, L.; Sicard, P.; Hernandez, A. Power sharing for efficiency optimisation into a multi fuel cell system. In Proceedings of the 2014 IEEE 23rd International Symposium on Industrial Electronics (ISIE), Istanbul, Turkey, 1–4 June 2014; pp. 218–223.
21. Candusso, D.; De Bernardinis, A.; Peera, M.-C.; Harel, F.; Francois, X.; Hissel, D.; Coquery, G.; Kauffmann, J.-M. Fuel cell operation under degraded working modes and study of diode by-pass circuit dedicated to multi-stack association. *Energy Convers. Manag.* **2008**, *49*, 880–895. [CrossRef]
22. Lape, N.; Dudfield, C.; Orsillo, A. Environmentally friendly power sources for aerospace applications. *J. Power Sources* **2008**, *181*, 353–362.
23. Miller, A.R.; Hess, K.S.; Barnes, D.L.; Erickson, T.L. System design of a large fuel cell hybrid locomotive. *J. Power Sources* **2007**, *173*, 935–942. [CrossRef]
24. Von Helmolt, R.; Eberle, U. Fuel cell vehicles: Status 2007. *J. Power Sources* **2007**, *165*, 833–843. [CrossRef]
25. Hamad, K.B.; Taha, M.H.; Almaktoof, A.; Kahn, M.T.E. Modelling and analysis of a grid-connected megawatt fuel cell stack. In Proceedings of the 2019 International Conference on the Domestic Use of Energy (DUE), Wellington, South Africa, 25–27 March 2019; pp. 147–155.
26. Cardozo, J.; Marx, N.; Hissel, D. Comparison of multi-stack fuel cell system architectures for residential power generation applications including electrical vehicle charging. In Proceedings of the IEEE Vehicle Power and Propulsion Conference (VPPC), Montreal, QC, Canada, 19–22 October 2015; pp. 2–7.
27. Wang, C.; Nehrir, M.H.; Gao, H. Control of PEM fuel cell distributed generation systems. *IEEE Trans. Energy Convers.* **2006**, *21*, 586–595. [CrossRef]
28. Marx, N.; Boulon, L.; Gustin, F.; Hissel, D.; Agbossou, K. A review of multi-stack and modular fuel cell systems: Interests, application areas and on-going research activities. *Int. J. Hydrogen Energy* **2014**, *39*, 12101–12111. [CrossRef]
29. Arunkumari, T.; Indragandhi, V. An overview of high voltage conversion ratio DC-DC converter configurations used in DC micro-grid architectures. *Renew. Sustain. Energy Rev.* **2017**, *77*, 670–687. [CrossRef]
30. Wu, Q.; Member, S.; Wang, Q.; Xu, J. A high efficiency step-up current-fed push-pull quasi-resonant converter with fewer components for fuel cell application. *IEEE Trans. Ind. Electron.* **2016**, *64*, 6639–6648. [CrossRef]
31. Bahrami, M.; Martin, J.; Maranzana, G.; Pierfederici, S.; Weber, M.; Meibody-Tabar, F.; Zandi, M. Design and modeling of an equalizer for fuel cell energy management systems. *IEEE Trans. Power Electron.* **2019**, *24*, 10925–10935. [CrossRef]
32. Chen, Y.; Liu, X.; Cui, Y.; Zou, J.; Yang, S. A multi winding transformer cell-to-cell active equalization method for lithium-ion batteries with reduced number of driving circuits. *IEEE Trans. Power Electron.* **2016**, *31*, 4916–4929. [CrossRef]

33. Sulaiman, N.; Hannan, M.A.; Mohamed, A.; Ker, P.J.; Majlan, E.H.; Wan Daud, W.R. Optimization of energy management system for fuel-cell hybrid electric vehicles: Issues and recommendations. *Appl. Energy* **2018**, *228*, 2061–2079. [CrossRef]
34. Lü, X.; Wu, Y.; Lian, J.; Zhang, Y.; Chen, C.; Wang, P.; Meng, L. Energy management of hybrid electric vehicles: A review of energy optimization of fuel cell hybrid power system based on genetic algorithm. *Energy Convers. Manag.* **2020**, *205*, 112474. [CrossRef]
35. Marzougui, H.; Kadri, A.; Martin, J.; Amari, M.; Pierfederici, S. Implementation of energy management strategy of hybrid power source for electrical vehicle. *Energy Convers. Manag.* **2019**, *195*, 830–843. [CrossRef]
36. Lü, X.; Qu, Y.; Wang, Y.; Qin, C.; Liu, G. A comprehensive review on hybrid power system for PEMFC-HEV: Issues and strategies. *Energy Convers. Manag.* **2018**, *171*, 1273–1291. [CrossRef]
37. Marx, N.; Hissel, D.; Gustin, F.; Boulon, L.; Agbossou, K. On the sizing and energy management of an hybrid multistack fuel cell e battery system for automotive applications. *Int. J. Hydrog. Energy* **2016**, *42*, 1518–1526. [CrossRef]
38. Lü, X.; Wang, P.; Meng, L.; Chen, C. Energy optimization of logistics transport vehicle driven by fuel cell hybrid power system. *Energy Convers. Manag.* **2019**, *199*, 111887. [CrossRef]
39. Zhan, Y.; Wang, H.; Zhu, J. Modelling and control of hybrid UPS system with backup PEM fuel cell/battery. *Int. J. Electr. Power Energy Syst.* **2012**, *43*, 1322–1331. [CrossRef]
40. Li, T.; Liu, H.; Ding, D. Predictive energy management of fuel cell supercapacitor hybrid construction equipment. *Energy* **2018**, *149*, 718–729. [CrossRef]
41. Bizon, N.; Lopez-Guede, J.M.; Kurt, E.; Thounthong, P.; Gheorghita, A.; Mihai, L.; Iana, G. Hydrogen economy of the fuel cell hybrid power system optimized by air flow control to mitigate the effect of the uncertainty about available renewable power and load dynamics. *Energy Convers. Manag.* **2019**, *179*, 152–165. [CrossRef]
42. Bizon, N.; Cristian, I. Hydrogen saving through optimized control of both fueling flows of the fuel cell hybrid power System under a variable load demand and an unknown renewable power profile. *Energy Convers. Manag.* **2019**, *184*, 1–14. [CrossRef]
43. Phattanasak, M.; Martin, J.; Pierfederici, S.; Davat, B. Predicting the onset of bifurcation and stability study of a hybrid current controller for a boost converter. *Math. Comput. Simul.* **2013**, *91*, 262–273.
44. Gavagsaz-ghoachani, R.; Phattanasak, M.; Zandi, M.; Martin, J.; Pierfederici, S.; Nahid-mobarakeh, B.; Davat, B. Estimation of the bifurcation point of a modulated-hysteresis current-controlled DC—DC boost converter: Stability analysis and experimental verification. *IET Power Electron.* **2015**, *8*, 2195–2203. [CrossRef]
45. Saublet, L.; Gavagsaz-Ghoachani, R.; Martin, J.; Nahid-mobarakeh, B.; Member, S.; Pierfederici, S. Bifurcation analysis and stabilization of DC power systems for electrified transportation systems. *IEEE Trans. Transp. Electrif.* **2016**, *2*, 86–95. [CrossRef]
46. Mira, M.C.; Zhang, Z.; Knott, A.; Andersen, M.A. Analysis, design, modeling, and control of an Interleaved-boost full-bridge three-port converter for hybrid renewable energy systems publication date: Analysis, design, modelling and control of an interleaved-boost full-bridge three-port converter f. *IEEE Trans. Power Electron.* **2017**, *32*, 1138–1155. [CrossRef]
47. Wang, L.; Member, S.; Vo, Q.; Prokhorov, A. V Dynamic stability analysis of a hybrid wave and photovoltaic power generation system integrated into a distribution power grid. *IEEE Trans. Sustain. Energy* **2017**, *8*, 404–413. [CrossRef]
48. Yuhimenko, V.; Member, S.; Geula, G.; Member, S.; Agranovich, G.; Averbukh, M.; Kuperman, A.; Member, S. Average modeling and performance analysis of voltage sensorless active supercapacitor balancer with peak current protection. *IEEE Trans. Power Electron.* **2017**, *32*, 1570–1578. [CrossRef]
49. Qin, H.; Kimball, J.W. Generalized average modeling of dual active. *IEEE Trans. Power Electron.* **2012**, *27*, 2078–2084. [CrossRef]
50. Renaudineau, H.; Martin, J.; Nahid-Mobarakeh, B.; Member, S.; Pierfederici, S. DC—DC converters dynamic modeling with state observer-based parameter estimation. *IEEE Trans. Power Electron.* **2015**, *30*, 3356–3363. [CrossRef]
51. Payman, A.; Pierfederici, S.; Meibody-Tabar, F.; Davat, B. An adapted control strategy to minimize DC—Bus capacitors of a parallel fuel cell/ultracapacitor hybrid system. *IEEE Trans. Power Electron.* **2011**, *26*, 3843–3852. [CrossRef]
52. Payman, A.; Pierfederici, S.; Meibody-Tabar, F. Energy control of supercapacitor/fuel cell hybrid power source. *Energy Convers. Manag.* **2008**, *49*, 1637–1644. [CrossRef]

53. Thounthong, P.; Pierfederici, S.; Davat, B. Analysis of differential flatness-based control for a fuel cell hybrid power source. *IEEE Trans. Energy Convers.* **2010**, *25*, 909–920. [CrossRef]
54. Tabart, Q.; Vechiu, I.; Etxeberria, A.; Bacha, S. Hybrid energy storage system microgrids integration for power quality improvement using four leg three level NPC inverter and second order sliding mode control. *IEEE Trans. Ind. Electron.* **2018**, *65*, 424–435. [CrossRef]

© 2020 by the authors. Licensee MDPI, Basel, Switzerland. This article is an open access article distributed under the terms and conditions of the Creative Commons Attribution (CC BY) license (http://creativecommons.org/licenses/by/4.0/).

Article

Common-Ground-Type Single-Source High Step-Up Cascaded Multilevel Inverter for Transformerless PV Applications

Hossein Khoun Jahan [1,*], Naser Vosoughi Kurdkandi [1], Mehdi Abapour [1], Kazem Zare [1], Seyed Hossein Hosseini [1], Yongheng Yang [2] and Frede Blaabjerg [2,*]

1. Faculty of Electrical and Computer Engineering, University of Tabriz, Tabriz 51368, Iran; naser.vosoughi@yahoo.com (N.V.K.); abapour@tabrizu.ac.ir (M.A.); kazem.zare@tabrizu.ac.ir (K.Z.); hosseini@tabrizu.ac.ir (S.H.H.)
2. Department of Energy Technology, Aalborg University, 9220 Aalborg, Denmark; yoy@et.aau.dk
* Correspondence: hosseinkhounjahan@yahoo.com (H.K.J.); fbl@et.aau.dk (F.B.)

Received: 9 August 2020; Accepted: 23 September 2020; Published: 7 October 2020

Abstract: The cascaded multilevel inverter (CMI) is one type of common inverter in industrial applications. This type of inverter can be synthesized either as a symmetric configuration with several identical H-bridge (HB) cells or as an asymmetric configuration with non-identical HB cells. In photovoltaic (PV) applications with the CMI, the PV modules can be used to replace the isolated dc sources; however, this brings inter-module leakage currents. To tackle the issue, the single-source CMI is preferred. Furthermore, in a grid-tied PV system, the main constraint is the capacitive leakage current. This problem can be addressed by providing a common ground, which is shared by PV modules and the ac grid. This paper thus proposes a topology that fulfills the mentioned requirements and thus, CMI is a promising inverter with wide-ranging industrial uses, such as PV applications. The proposed CMI topology also features high boosting capability, fault current limiting, and a transformerless configuration. To demonstrate the capabilities of this CMI, simulations and experimental results are provided.

Keywords: cascaded multilevel inverter; photovoltaic; leakage current

1. Introduction

Multilevel inverters (MIs) are attractive devices in many industrial applications. These devices can reduce the total harmonic distortion (THD), electromagnetic interference (EMI), dv/dt, switching frequency and voltage stress. One of the most regarded applications of MIs is PV application. The neutral point clamped converter (NPC) and cascaded multilevel inverter (CMI) are two types of multilevel inverters, which are popular in PV applications [1,2]. Between the two topologies, the CMI stands out for its modularity and high magnitude of the output voltage. However, this topology requires several isolated dc sources. This drawback not only calls for a complex control system, but also it gives rise to inter module leakage currents in grid-tied PV applications. The inter-module leakage currents result from differential-mode voltage (DMV) and common mode voltage (CMV) variations. In order to tackle the issue, several topologies are suggested in the literature [3–5]. One solution is using only one dc-source along with some passive components. Single-source CMIs are categorized into three types. (i) Topologies which use low frequency transformers instead of several isolated dc sources. These topologies are referred to as cascaded transformers multilevel inverters (CTMIs) [6–9]. (ii) Topologies which provide the isolated dc sources by adopting a high-frequency link and a single dc-source (HFLMI) [10,11]. (iii) A switched-capacitor (SC) based cascaded multilevel inverter

(SC-CMI) [12,13]. The main advantage of CTMIs is their ability to provide galvanic isolation between the dc source and the load/grid. This is also the case when applying HFLMIs in PV applications, where the leakage current issue is addressed. On the contrary, CTMIs need several bulky and inefficient transformers. Although the transformer size in HFLMIs is reduced due to the use of a high frequency link, many rectifiers are required to convert the isolated high frequency voltages to the desired dc voltages. Thus, the reliability decreases and the cost increase in this topology. Alternatively, SC-CMIs employ several capacitors instead of the isolated dc sources. Therefore, the SC-CMI topologies have a compact size and lower cost. However, these kinds of multilevel inverters lack galvanic isolation.

Moreover, many attempts have been made to use isolated PV arrays as the isolated dc sources in grid-tied CMIs [14]. However, as illustrated in [15], the main constraint of these configurations is the capacitive leakage currents between the H-bridge (HB) cells and grid. Even using an interfacing transformer cannot address the mentioned problem, because inter-module leakage currents appear and circulate between the cascaded HB cells. In ref. [15], the mentioned problem was addressed by equipping each HB cell with additional ac and dc side filters. Apart from limiting various leakage currents, these filters are deemed to eliminate the EMI; however, equipping each cell with several filters increases the volume and cost of the inverter. In ref. [16], several level-double networks (LDN) are used as the auxiliary blocks to enhance the quality of the output voltage. This topology can also offer a common ground between the PV module and the grid, which results in the elimination of the leakage current. Although the suggested topology can eliminate the leakage current in PV applications, balancing of the capacitor voltage in the auxiliary cell is challenging. In another attempt, a two-stage inverter was suggested in [17], which can be regarded as a combination of the H5 and Highly Efficient and Reliable Inverter Concept (HERIC) topologies. When the output voltage is higher than the grid voltage, the inverter operates in the H5 mode; when the dc link voltage decreases, the inverter is switched to the two-stage HERIC mode. This inverter can properly deal with voltage variation. However, it uses a complicated structure and control approach. Moreover, a charge pump circuit was employed to eliminate the leakage in [18]. The topology is simple and compact, but it imposes a non-continuous current to the input side. Notably, in ref. [19] a comprehensive study was conducted to investigate the state-of-the-art inverters for grid-tied PV applications.

In light of the above, a single-source asymmetric CMI is proposed in this paper, which provides a common ground for ac and dc sides. This topology uses capacitors instead of the isolated dc sources in the HB cells. Each capacitor is independently charged through a charging switch. Since there is a common ground for ac and dc sides, the common mode voltage is zero; hence, this topology can totally eliminate the leakage current in grid-tied PV applications. Another merit of the proposed topology is the capability to boost the input dc voltage; this is also an advantage in many applications such as grid-tied transformerless PV and fuel cell systems. In addition to the mentioned features, the three-phase configuration of the proposed topology draws a continuous input current, which makes it feasible in battery, un-interruptible power supply, and PV applications. The proposed topology can exchange reactive power with the load and the grid as well. Furthermore, it can smoothly charge the capacitors, facilitate the protection, and avoid bulky and expensive transformers in the grid-tied mode. As mentioned, the main issue of the conventional CMI in PV applications is the inter-module leakage currents. However, the proposed topology can address this problem properly and effectively. Compared to the transformer-based single-source multilevel inverters, the proposed topology is smaller in size, lower in cost, and higher in efficiency. Additionally, considering that the SC-based single-source MIs mostly suffer from inrush currents, the proposed topology is, however, an inrush-current free CMI, being a promising converter in many industrial applications.

The rest of the paper is organized as follows: In Section 2, the structure and operation principle of the proposed topology are illustrated. In Section 3, the proposed MI is compared with state-of-the-art MI topology. In Section 4 the performance of the proposed topology in off-grid and grid-tied modes is investigated through simulations. Experimental tests are provided in Section 5, where a fifteen-level 0.55 kVA prototype is adopted to demonstrate the off-grid performance of the proposed topology.

Moreover, a seven-level 1.5 kVA prototype is used to extract the grid-tied results. Finally, the overall work is concluded in Section 6.

2. Proposed Topology and Operation Principle

2.1. Conventional CMI in PV Systems

Many solutions are presented in the literature to improve the performance of the CMIs in PV systems. The main problem arises due to the parasitic capacitor in each HB cell that brings inter-module leakage currents [20,21]. These circulating currents cause power loss, EMI, and safety problems [15]. Figure 1a,b shows a grid-tied PV system with a three-cell CMI, and equivalent circuit of the CMV, DMV and leakage currents, respectively.

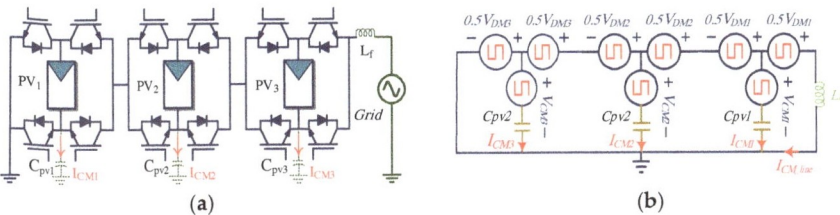

Figure 1. Conventional cascaded multilevel inverter (CMI)-based PV system: (**a**) a three-cell grid-tied CMI; (**b**) equivalent circuit to illustrate the common mode voltage (CMV), differential-mode voltage (DMV) and inter-module currents.

2.2. General Structure of the Proposed Topology

The proposed topology is synthesized with two parts, namely the main and charging parts. The main part is the conventional asymmetric CMI, in which the isolated dc sources are replaced with capacitors ($C_1, C_2, \ldots, C_n$). The charging part is composed of a single dc source (e.g., a PV string, fuel-cell, and batteries), a charging inductor, a freewheeling diode and charging switches ($S_{c1}, S_{c2}, \ldots, S_{cn}$). The general grid-tied configuration of the proposed topology (a configuration with n HB cells) is depicted in Figure 2. Where the main part is colored in black, the charging part is in blue.

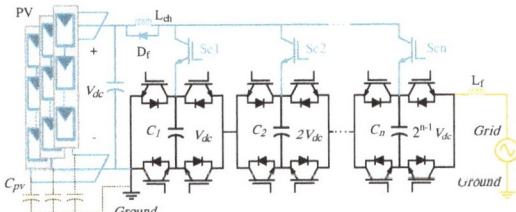

Figure 2. General configuration of the proposed topology in grid-tied PV applications.

As mentioned, the most undesirable phenomenon in an SC-based converter is the inrush currents that emerge in the charging stage of the capacitors. This phenomenon can adversely affect the charging switches and capacitors. In order to limit these currents, a charging inductor (L_{ch}) is connected in series with the dc source, as shown in Figure 2. This inductor can effectively limit the inrush currents. On the contrary, the mentioned inductor can cause voltage spikes and commutation problems in the charging switches. To avoid this and alleviate the EMI, the size of the charging inductor (L_{ch}) can be obtained as

$$L_{ch} = \frac{1}{(4\pi f)^2 C_n} \quad (1)$$

where f and C_n are the output voltage frequency and equivalent capacitance of the capacitors.

It should be noted that a larger inductor can be used to further reduce the inrush currents. However, this increases the cost and volume of the inverter. A large inductor can also cause overvoltage across the capacitors. To avoid this, as shown in Figure 2, a freewheeling diode (D_f) is connected in parallel with the inductor.

In order to illustrate the operation principle of the proposed topology, a fifteen-level configuration, which is depicted in Figure 3, is exemplified. Table 1 shows the switching pattern for each level and different states of the capacitors. It should be mentioned that in Table 1, "on" and "off" states of the switches are indicated by "1" and "0". The capacitors in the proposed topology experience three states namely the charging, discharging, and floating states. In Table 1, "C", "D", and "F" denote the charging, discharging, and floating states of the capacitors. In addition, since the upper switches of the main part (S_{11}, S_{31}, S_{12}, S_{32}, S_{13}, and S_{33}) have complementary states with the lower switches (S_{21}, S_{41}, S_{22}, S_{42}, S_{23}, and S_{43}), only the states of the upper switches are indicated in Table 1 for simplicity.

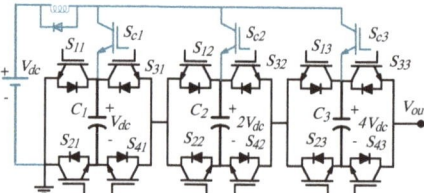

Figure 3. A fifteen-level configuration of the proposed topology.

Table 1. Operation states of components shown in Figure 3.

Levels	Main Switches S_{11}, S_{31}, S_{12} S_{32} S_{13}, S_{33}	Charging Switches S_{c1}, S_{c2}, S_{c2}	Capacitors C_1, C_2, C_3	V_{out}
7	010101	100	C,D,D	$7V_{dc}$
6	000101	100	C,D,D	$6V_{dc}$
5	100101	010	D,C,D	$5V_{dc}$
4	000001	100	C,D,D	$4V_{dc}$
3	100001	010	D,C,D	$3V_{dc}$
2	001001	100	C,D,D	$2V_{dc}$
1	101001	001	D,D,C	$1V_{dc}$
0	000000	100	C,F,F	0
−1	100000	010	D,C,D	$-1V_{dc}$
−2	001000	100	C,D,D	$-2V_{dc}$
−3	101000	001	D,D,C	$-3V_{dc}$
−4	000010	100	C,F,D	$-4V_{dc}$
−5	100010	010	D,C,D	$-5V_{dc}$
−6	001010	100	C,D,D	$-6V_{dc}$
−7	101010	000	D,D,D	$-7V_{dc}$

To clarify, the equivalent circuits of the voltage levels are provided. Due to the page limit, only the positive voltage levels are demonstrated in Figure 4. The negative levels can be found by referring to Table 1. In Figure 4, the charging paths, the capacitors under charge and the load current paths are in red, blue, and dark blue, respectively.

2.3. Three-Phase Configuration

Continuity of the input current in many applications is of high importance. A continuous input current can facilitate the maximum power point tracking (MPPT) process in PV applications and prolong battery life span in storage systems. Referring to Table 1, it can be seen that a single-phase configuration of the proposed topology cannot guarantee a continuous input current because there is no possibility to connect the dc source to any of the capacitors when producing the highest negative voltage level (this is the case for a configuration with any number of voltage-levels). Since the input current is only interrupted in the highest negative voltage level, which is a short interval, this problem will not exist in a three-phase configuration. In such a configuration, when the input current is interrupted in one phase, there are always two paths in the other two phases for the current to flow. Figure 5 depicts the general three-phase configuration of the proposed topology. It is worth mentioning that in the three-phase configuration, the CMV is reduced but not totally eliminated. Thus, in a grid-tied PV application with the three-phase configuration, a limited leakage current is achieved. However, the inter-module leakage currents are totally cancelled out in this configuration.

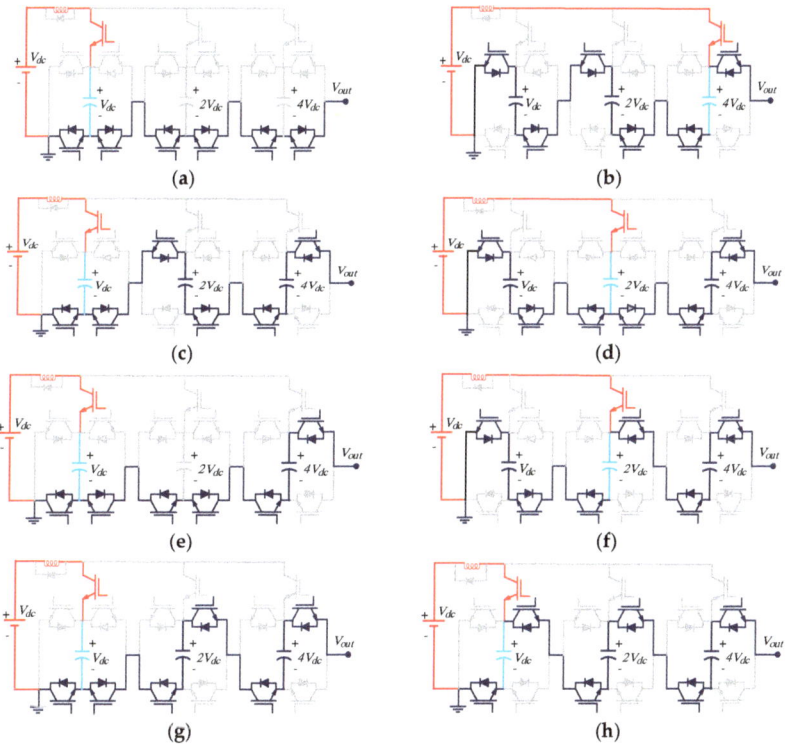

Figure 4. Charging and load current paths: (**a**–**h**) zero to seventh voltage-levels of the topology in Figure 3, respectively, where the PV module is replaced with a dc source for clarity.

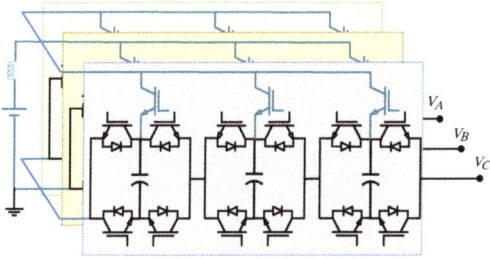

Figure 5. General three-phase configuration of the proposed topology.

2.4. Component Design

Referring to Figure 4 and Table 1, it is seen that during one cycle, the lower the dc voltage a HB cell contains, the longer time it resides in the charging mode. For example, as shown in Table 1, the first HB cell, which contains 1 pu voltage, resides in the charging mode for eight times. The number of being in the charging mode for the second and third HB cells is four and two, respectively. Thus, in an l-level structure, the number of being in the charging state for the nth HB cell is calculated as

$$Nch_n = 2^{\frac{\ln(l+1)}{\ln 2} - n} \tag{2}$$

In respect to this, an HB cell with a higher dc voltage will provides the load current for a longer time than others. Therefore, it experiences the highest voltage ripple. The highest voltage ripple of the nth HB cell (Δv_n) is given as

$$\begin{cases} \Delta v_n = \frac{I_m \Delta t_n}{C_n} \\ \Delta t_n = T(l - Nch_n) \end{cases} \tag{3}$$

where I_m, T, and C_n are the maximum value of the load current, time duration of a cycle, and capacitance of the nth capacitor, respectively. This equation can be used to select a proper capacitor for the nth HB cell.

Considering Figure 2, the equivalent circuit of the capacitor experiencing the highest voltage ripple (C_n) is shown in Figure 6. Taking the parameters indicated in Figure 6 into account, the instantaneous voltage in the nth capacitor and the voltage of the mentioned capacitor at the end of a half cycle are, respectively, given as

$$v_{C_n}(t) = (2^{n-1}) v_{dc} e^{\frac{-t}{RC_n}} \tag{4}$$

$$v_{C_n}(T_d) = (2^{n-1}) v_{dc} e^{\frac{-T_d}{RC_n}} \tag{5}$$

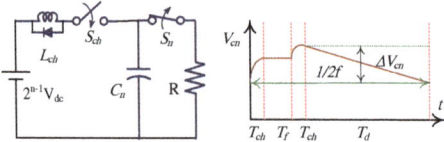

Figure 6. Equivalent circuit and discharging diagram of a capacitor in an HB cell.

The maximum voltage ripple in the nth capacitor can be given as

$$\Delta v_{C_n} = fT_d\left((2^{n-1})v_{dc} - v_{C_n}(T_d)\right) = fT_d(2^{n-1})v_{dc}\left(1 - e^{\frac{-T_d}{RC_n}}\right) \tag{6}$$

As it is an asymmetric topology, the HB cells in the proposed topology include different dc voltage values. Considering v_{dc} as the input voltage, the voltage across the nth cell is given as

$$V_{C_n} = 2^{n-1} v_{dc} \tag{7}$$

The voltage stress on the main and charging switches in the nth HB cell is equal to the voltage of capacitor in that HB cell.

The peak output voltage of an n-cell configuration is given as

$$v_m = v_{dc} \sum_{k=1}^{n} 2^{k-1} \tag{8}$$

The number of switches of an *l*-level configuration of the proposed and conventional asymmetric CMI topologies is, respectively, indicated as

$$N_{sw}^P \frac{5\ln\left(\frac{l+1}{2}\right)}{\ln 2} \tag{9}$$

$$N_{sw}^C \frac{4\ln\left(\frac{l+1}{2}\right)}{\ln 2} \tag{10}$$

This implies that the proposed topology requires one extra switch in each cell (one charging switch for each cell).

The total voltage stresses of the switches in the proposed and the conventional asymmetric CMI topologies are, respectively, indicated as

$$TVS_p = \frac{5v_{dc}}{N_{sw}^P} \sum_{k=1}^{N_{sw}^P/5} 2^{k-1} \tag{11}$$

$$TVS_c = \frac{4v_{dc}}{N_{sw}^C} \sum_{k=1}^{N_{sw}^C/4} 2^{k-1} \tag{12}$$

Implying that the voltage stresses of the switches in both topologies are the same.

3. Benchmarking with Prior-Art Inverters

Several efforts have been done to make the CMI compatible with grid-tied PV applications [22–29]. The main difficulties with the CMI in PV applications are the leakage current and complicated MPPT [14]. Single-source CMIs facilitate the MPPT, but the leakage current problem remains. A transformer can solve the problem, however, transformers are not recommended in grid-tied PV applications due to extra power losses and additional costs. Therefore, as stated previously, the SC-based CMI can fulfill many requirements. The state-of-the-art PV MI topologies are compared with the proposed MI topology in this section to assess its pros and cons. Table 2 lists the main features of the considered MI topologies.

Table 2. Comparison.

Topology	Nsw	Nd	Nc	G	TSV	Coupled Inductor	Leakage Current Limiting
[22]	14	0	2	3	4.67	no	no
[23]	12	-	2	2	5.5	no	no
[24]	10	-	2	0.5	8	yes	yes
[25]	8	3	3	4	5.75	no	no
[26]	12		3	4	5.25	no	no
[27]	9	-	2	2	5.5	no	no
[28]	11	-	3	2	5	no	no
[29]	8	4	4	2	6	no	no
[Proposed]	10	0	2	3	5	no	yes

In table, N_{sw}, N_d, N_c, G, and TSV are the number of switches, diodes, capacitors, voltage gain, and the total standing voltage of the switches. The TSV is calculated as

$$TSV = \frac{\sum_{n=0}^{n=k} V_{sw_n} + \sum_{n=0}^{n=k} V_{sd_n}}{V_{out}} \tag{13}$$

The proposed topology and the topology in [29] can be scaled up to obtain higher voltage gains and levels. However, this is not the case for the other topologies. Since the proposed MI topology is a common-ground-type inverter and the topology in [24] is a mid-point-grounded topology, these two topologies can limit the leakage current in grid-tied PV applications. In this regard, the other topologies listed in Table 2 encounter serious problems. The main disadvantage of the MI topologies in [24] and [25] is that they require a complicated control approach to balance the voltages across the capacitors. The voltage balancing system of these topologies should sense the direction of the ac current and the capacitor voltage magnitude, and then the sensed values are processed though the processor to execute the right switching pattern to balance the voltage of the capacitors. However, this does not happen in the other topologies and the proposed one. Notably, the topologies in [23,25,27,28] suffer from high inrush currents in the charging stage of the capacitor. Owing to the controlled voltage balancing of the capacitors, the inrush current does not appear in the topologies in [24,25]. In the proposed topology and the topology in [29], the inrush current is limited through the charging inductor. As it is seen in Table 2, the proposed inverter has a fairly low TSV, high voltage gain and fewer components.

4. Simulation Results

In order to verify the performance of the proposed topology, both the single-phase and three-phase configurations are simulated under MATLAB/Simulink. The main parts in the considered configurations are assumed to be a fifteen-level CMI. The simulated models are tested under off-grid and grid-tied modes. In the off-grid mode, a general dc source supplies the load through the proposed topology.

4.1. Off-Grid Mode

As illustrated earlier, the three-phase and single-phase configurations only differ in the input current shapes. For this reason, mostly the single-phase configuration is investigated. Table 3 shows the characteristics of the utilized components in the off-grid mode. Figure 7a shows the output and capacitor voltages under no-load condition. A fast Fourier transform (FFT) analysis of the output voltage is depicted in Figure 7b.

Table 3. Components of the off-grid model.

Component	Value	Component	Value
V_{dc}	46 V	L_{ch} (3ϕ)	0.5 mH
Power rating	550 W	C_1, C_2, C_3	3300 µF
L_{ch} (ϕ)	1.8 mH	f_{sw}	5 kHz
Reference voltage		311 V (peak), 50 Hz	

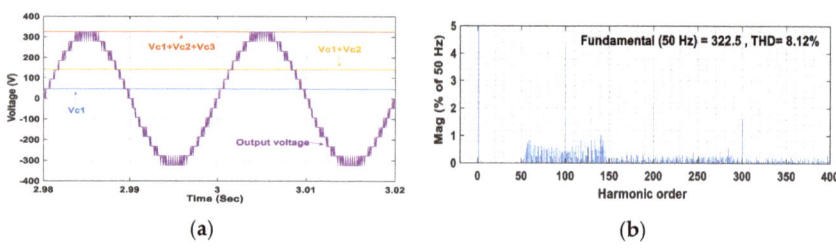

(a)   (b)

Figure 7. Simulation results of the single-phase configuration in the off-grid mode (no-load condition): (**a**) capacitor and output voltages; (**b**) fast Fourier transform (FFT) analysis of the output voltage.

Furthermore, the output voltage, load current, and capacitor voltages, when supplying a purely resistive load of 0.55 kW, are shown in Figure 8a. As seen in Figure 8a, under this condition, the voltage across the capacitors is properly balanced through the charging circuit. Additionally, the capacitor currents along with the input current under the studied loading condition are shown in Figure 8b. It can be seen that the charging unit can properly limit the inrush current of the capacitors. However, the main demerit of the charging process is the discontinuity of the input current due to the absence of a path for the input current when developing the highest negative voltage level.

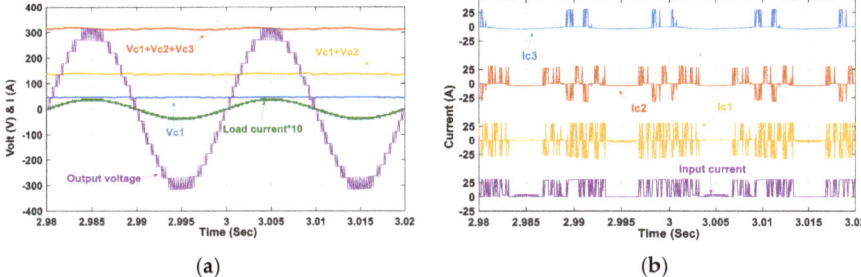

Figure 8. Simulation results of the single-phase configuration in the off-grid mode (under a purely resistive loading condition): (**a**) load current along with the capacitor and output voltages; (**b**) capacitor and input currents.

In order to demonstrate the ability of the proposed topology to provide reactive power, a resistive-inductive load of 0.5 kW + 0.35 kVar is connected. Figure 9a shows the output voltage and load current under the mentioned condition. As shown in Figure 9a, the proposed topology can satisfactorily supply the reactive power. Additionally, the voltage stress and current of the charging switches are shown in Figure 9b. According to Figure 9b, it is known that the charging switch in the last cells can tolerate the highest voltage stress.

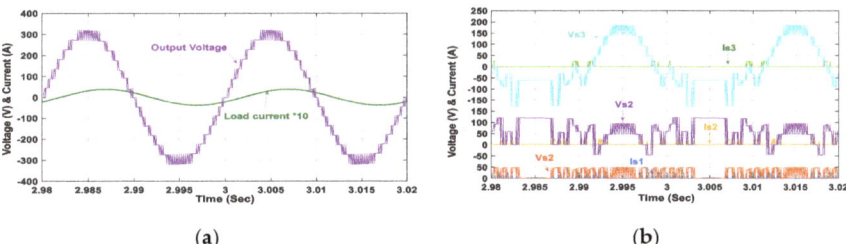

Figure 9. Simulation results of the single-phase configuration in the off-grid mode (under a resistive-inductive loading condition): (**a**) load current and output voltage; (**b**) voltage stress and current of the charging switches.

As mentioned previously, a three-phase configuration of the proposed topology draws a continuous current from the input side. This is proven by considering a three-phase fifteen-level configuration, which supplies a balanced three-phase load under three loading cases (3.8 kW, 4.8 kW + 1.2 kVar, 3 kW + 1.2 kVar). The input current of the phases and the total input current under the mentioned loading condition are shown in Figure 10a. As it is seen in Figure 10a, the input current is a continuous current. Moreover, the output voltages together with the load current under the mentioned condition are shown in Figure 10b,c, respectively. When the freewheeling diode is removed, the capacitors are exposed to overvoltage at the initial instance. Soft starting strategies can be employed to avoid the overvoltage of the capacitors.

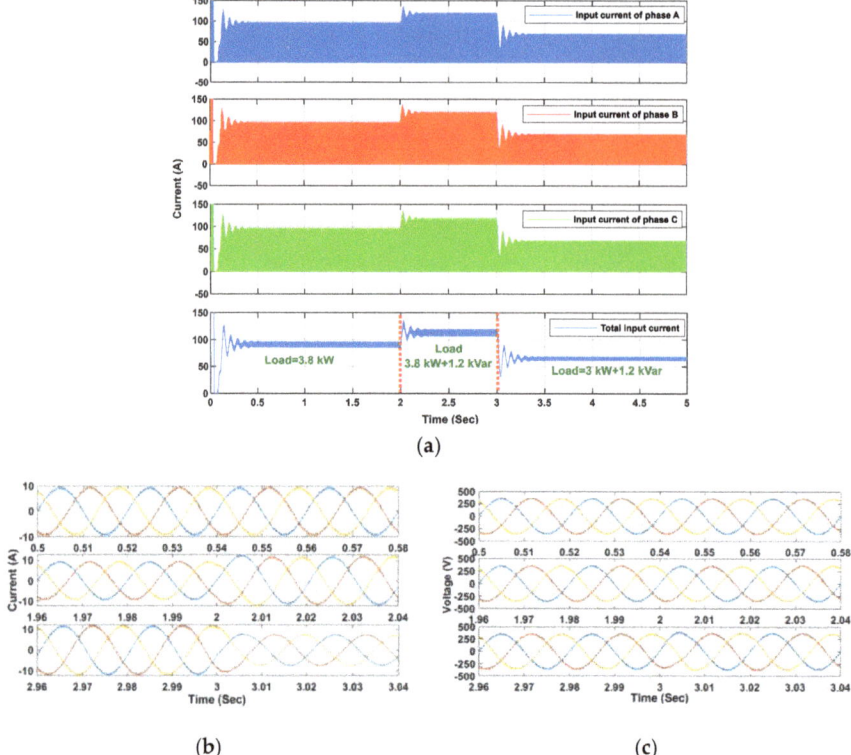

Figure 10. Simulation results of the three-phase configuration in the off-grid mode: (**a**) input current of the phase and total input current; (**b**) load current; (**c**) output voltage.

4.2. Grid-Tied Mode

Similar to the off-grid mode, a fifteen-level configuration of the proposed topology is used to deliver the desired powers to the grid. To this end, the ac components are transferred to the dq0 frame and two proportional-integral (PI) controllers are employed to control the active and reactive powers. Table 4 shows the characteristics of the considered system.

Table 4. Component of the grid-connected model.

Ki	42.3	Grid-side filter	2.8 mH + 30 mΩ
Kp	700	f_{sw}	5 kHz
$V_{g\max}$	320 V	L_{ch}	1.8 mH
f	50 Hz	C_1, C_2, C_3	3300 µF

The simulation results of the single-phase grid-connected model are shown in Figure 11. The desired (reference) and delivered active power to the grid is shown in Figure 11a. The reference of the active power can be obtained by the MPPT system in PV applications. The reference and developed reactive powers are exhibited in Figure 11b. As shown in Figure 11b, the proposed topology has succeeded to provide a bidirectional reactive power flow. The input current is depicted in Figure 11c. The output voltage of the inverter along with the injected current is exhibited in Figure 11d.

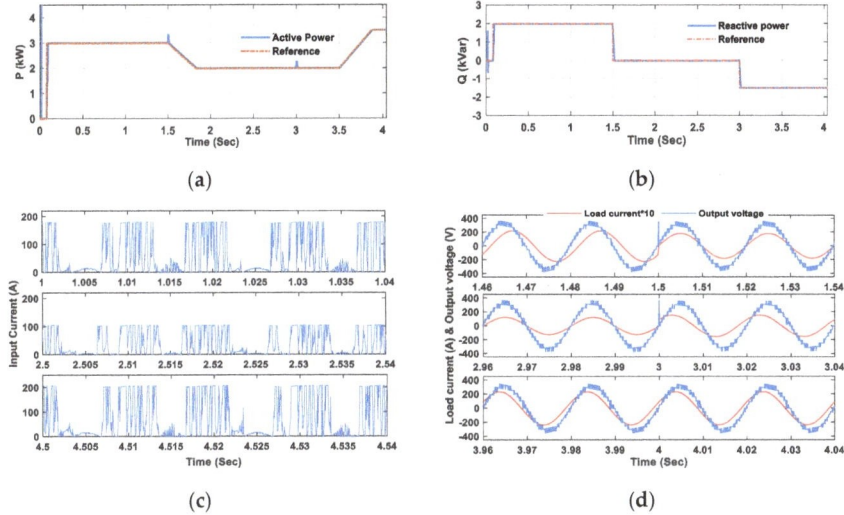

Figure 11. Simulation results of the single-phase grid-connected model: (**a**) the developed and reference of the active power; (**b**) the developed and reference of the active power; (**c**) input current; (**d**) output voltage and load current.

It should be pointed out that one of the significant features of the proposed MI topology is its ability to eliminate the leakage current in grid-tied PV systems without using any additional components.

Furthermore, the simulation results of a grid-connected three-phase model are demonstrated in Figure 12. The injected active and reactive powers to the grid are depicted in Figure 12a,b. As seen in Figure 12, the proposed MI has deservedly developed the desired powers.

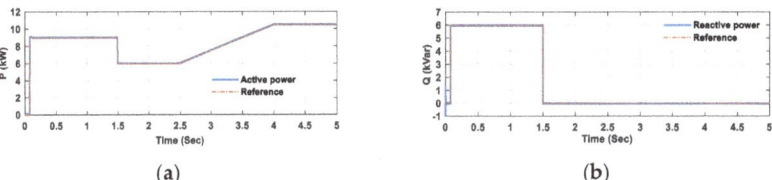

Figure 12. Simulation results of the active and reactive power of the three-phase grid-connected model: (**a**) the injected active power to the grid; (**b**) the injected reactive power to the grid.

As discussed previously, the three-phase configuration of the proposed topology draws a continuous current from the dc-link. Figure 13a shows the input current and proves this. In order to investigate the leakage current, a parasitic capacitor of 200 nF is considered between the negative pole of the dc-side and ac-ground, Figure 13b shows the leakage current. As shown in Figure 13b, the root mean square (RMS) value of the leakage current is in the acceptable range. However, it is possible to reduce this through the proper control and/or switching approaches. It is worth mentioning that since the proposed topology does not use PV modules inside the H-bridge cells, there are no inter-module leakage currents.

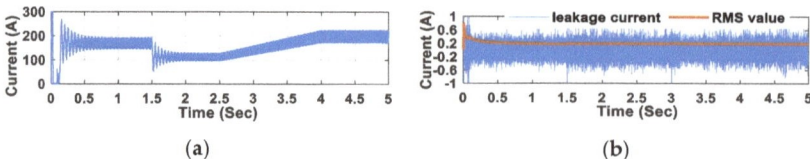

Figure 13. Simulation results of input and leakage current of the three-phase grid-connected model: (**a**) input current; (**b**) leakage current.

Furthermore the output voltage and the injected current are shown in Figure 14. It is to be noted that this paper is not aimed at designing a proper control system. It is possible to obtain a more accurate result through a precise control approach.

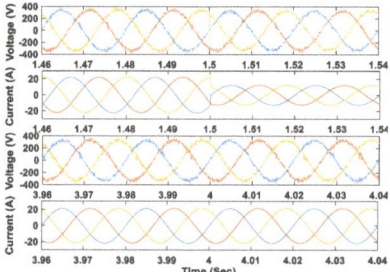

Figure 14. Simulation results of the output voltage and injected current of the three-phase grid-connected model.

5. Experimental Results

5.1. Off-Gird Results

In order to validate the feasibility of the proposed topology, a laboratory-scale prototype is tested. Figure 15 depicts the employed prototype and Table 5 lists the utilized components. It should be noted that the level-shifted SPWM strategy is adopted to compute the switching signals.

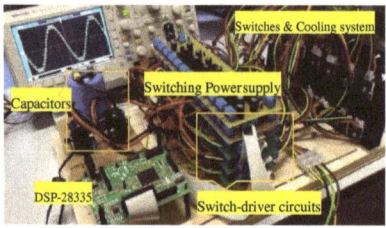

Figure 15. Experimental setup of the proposed topology (15-level).

Table 5. Electrical parameters and component specifications.

Component	Specification	Electrical Parameter	Value
Main Switches	IRFP350	Resistive load	550 W
Charging switches	IRFP460	RL load	650 VA
Opto-coupler	TLP250	$V_{out}(RMS)$	220 v, 50 Hz
Capacitors	3300 µF	V_{dc}	47 V
L_{ch}	2.8 mH	f_{sw}	5 kHz
Diodes	FFPF20UP40S	# of HB cells	3 (15-level)

Figure 16 exhibits the output voltage under the no-load condition and the FFT analysis of the voltage. As can be seen, the harmonics around the fundamental frequency have negligible magnitude, while the harmonics around the multiples of the switching frequency are of high amplitude. Since these harmonics are far away from the fundamental frequency, they can easily be eliminated using small filters.

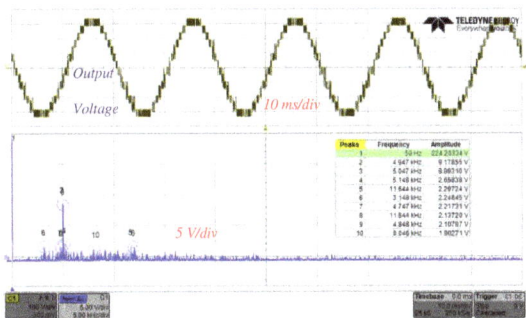

Figure 16. Measured output voltage of the proposed single-phase configuration (15-level) under no-load condition and its FFT analysis.

The output voltage along with the load current, when the prototype supplies a purely resistive load of 550 W is shown in Figure 17a. In order to assess the voltage ripple of the capacitors, the ac components of the capacitor voltages are shown in Figure 17b–d. Additionally, the charging current of the capacitor under 550 W load is shown in Figure 17e. As it is seen, there is no sharp spike on the charging current of the capacitors, which implies that the charging inductor smoothen the charging currents.

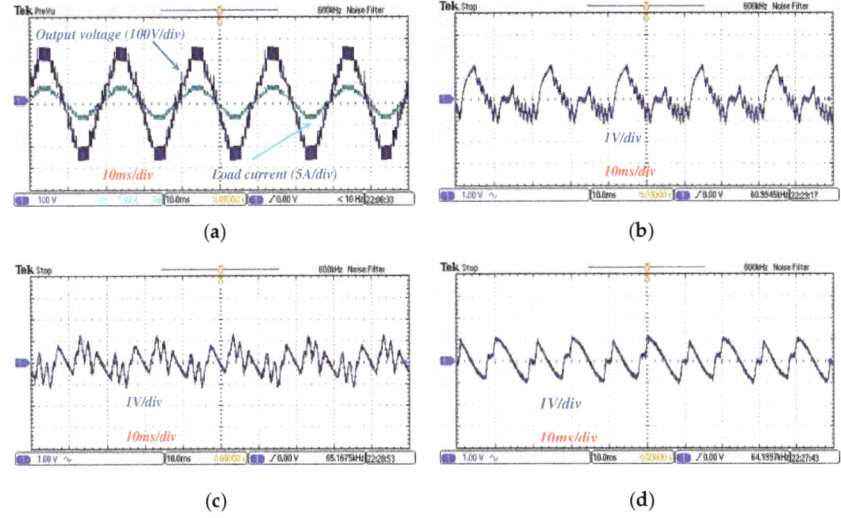

Figure 17. Cont.

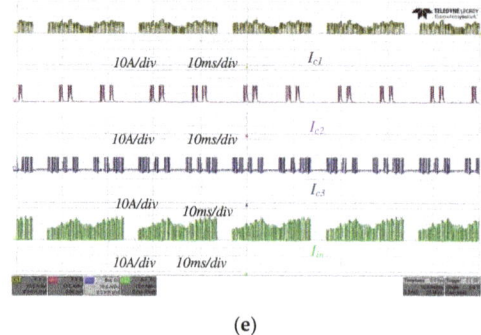

(e)

Figure 17. Experimental results under a purely resistive loading condition: (**a**) output voltage and load current under the purely resistive loading condition; (**b**,**c**), and (**d**) ac components of the capacitor voltages; (**e**) input current and charging current of the capacitors.

In order to prove the capability of the proposed topology to provide reactive power, a resistive-inductive load of 500 W + 350 Var is then considered. Figure 18 exhibits the output voltage and load current under this condition. As can be seen, the proposed topology can supply the reactive power without any constraints.

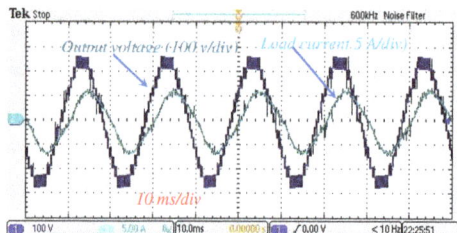

Figure 18. Output voltage and load current under the inductive-resistive loading condition.

5.2. Grid-Tied Results

In order to extract the grid-tied results a seven-level prototype with two cells is employed. The characteristic of the prototype and grid is listed in Table 6. In this test the sample based current control is used to inject the desired active and reactive powers to the grid.

Table 6. Component of the grid-connected model.

Main Switches	FQA14N30	Grid-side filter	1.73 mH
Charging switches	STP30NM30N	Switching frequency (f_{sw})	22 kHz
RMS grid voltage	220 V	L_{ch}	1.6 mH
Grid frequency (f)	50 Hz	C_1, C_2	3300 µF

Three scenarios are considered in grid-tied test. In the first scenario a pure active power of 1.5 kW is injected to the grid. The injected current and grid voltage under this condition are shown in Figure 19a. The FFT analysis of the injected current under the mentioned condition is shown in Figure 19b. Furthermore, the output voltage of the inverter along with the provided current is shown in Figure 19c.

In the second scenario the active power of 1.2 kW and reactive power of 0.9 kVar is injected to the grid and in the third scenario the active power of 1.2 kW is injected to the grid and reactive power of 0.9 kVar absorbed from the grid. The grid voltage together with the injected current to the grid is shown in Figure 20a,b.

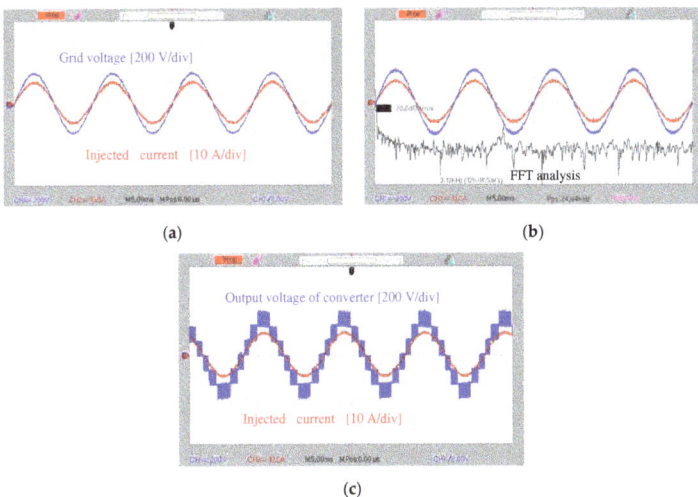

Figure 19. Grid-tied results of the active power: (**a**) grid voltage and injected current; (**b**) FFT analysis of the injected current; (**c**) output voltage and current of the prototype.

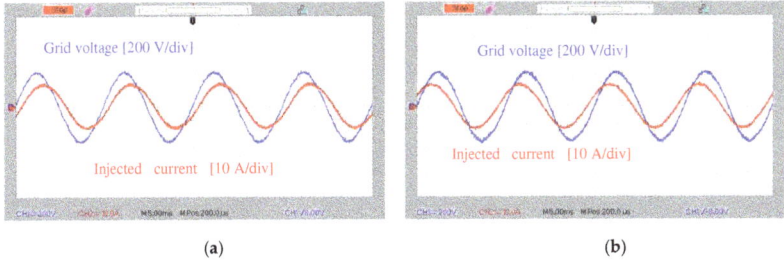

Figure 20. Grid-tied results of the injected current: (**a**) the output voltage and load current when injecting the active and reactive current to the grid; (**b**) grid-voltage and injected current when injecting the active power and absorbing the reactive power.

The seven-level prototype for grid-tied application is exhibited in Figure 21.

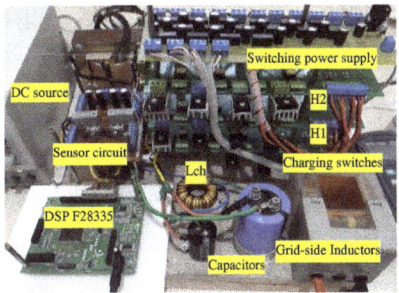

Figure 21. Seven-level setup for grid-tied application.

6. Conclusions

In this paper, a single-source high step-up asymmetric power converter topology is proposed. The proposed topology offers several advantages in many industrial applications such as PV, fuel cell, etc. It is synthesized with two parts, namely, the main and charging parts. The main part is the same as the conventional asymmetric CMI with certain capacitors instead of the isolated dc sources. The charging part, however, consists of charging switches, a charging inductor, a freewheeling diode, and one dc source. The main feature of the proposed topology is to provide a common-ground for ac and dc sides, which eliminates the leakage current in grid-tied PV applications. It also has the ability to boost the input voltage. Thus, in the grid-tied PV applications, bulky and expensive transformers can be avoided. Moreover, it uses only one dc source, at the expense of using many switches, employing many switches can be considered as the main drawback of the proposed inverter. Simulations and experiments were performed to verify the effectiveness of the proposed topology. Through simulations, the performances of the suggested topology with single- and three-phase configuration, in both off-grid and grid-tied conditions, were assessed. In the experimental tests, the performance of the proposed topology was studied in the presence of a 550-VA load. Moreover, using a 1.5 kVA seven-level prototype the grid-tied results were provided. Both results have demonstrated the feasibility of the proposed multilevel inverter in terms of the ability to develop a boosted voltage of high quality using only one dc source.

Author Contributions: The concept, theoretical analysis, and preparation of the manuscript were with H.K.J. N.V.K. contributed to extract the grid-tied experimental results. M.A., K.Z. and S.H.H. were supervisors and helped to investigate and assess the performance of the proposed inverter topology. Y.Y. and F.B. supervised the work, helped to extract experimental results, edited the manuscript, and provided financial supports. All authors have read and agreed to the published version of the manuscript.

Funding: This work was supported under the research project—Reliable Power Electronic based Power Systems (REPEPS) by The Velux Foundations under Award No.: 00016591.

Conflicts of Interest: The authors declare no conflict of interest.

References

1. Wu, F.; Li, X.; Feng, F.; Gooi, H.B. Modified cascaded multilevel grid-connected inverter to enhance European efficiency and several extended topologies. *IEEE Trans. Ind. Inform.* **2015**, *11*, 1358–1365. [CrossRef]
2. Sepahvand, H.; Liao, J.; Ferdowsi, M.; Corzine, K.A. Capacitor voltage regulation in single-DC-source cascaded H-bridge multilevel converters using phase-shift modulation. *IEEE Trans. Ind. Electron.* **2013**, *60*, 3619–3626. [CrossRef]
3. Banaei, M.R.; Jahan, H.K.; Salary, E. Single-source cascaded transformers multilevel inverter with reduced number of switches. *IET Power Electron.* **2012**, *5*, 1748–1753. [CrossRef]
4. Vazquez, S.; Leon, J.I.; Franquelo, L.G.; Padilla, J.J.; Carrasco, J.M. DC-voltage-ratio control strategy for multilevel cascaded converters fed with a single DC source. *IEEE Trans. Ind. Electron.* **2009**, *56*, 2513–2521. [CrossRef]
5. Taghvaie, A.; Adabi, J.; Rezanejad, M. A multilevel inverter structure based on a combination of switched-capacitors and DC sources. *IEEE Trans. Ind. Inform.* **2017**, *13*, 2162–2171. [CrossRef]
6. Jahan, H.K.; Zare, K.; Abapour, M. Verification of a low components nine-level cascaded-transformer multilevel inverter in grid-tied mode. *IEEE J. Emerg. Sele. Topi. Power Electron.* **2018**, *6*, 429–440. [CrossRef]
7. Song, S.G.; Kang, F.S.; Park, S.J. Cascaded multilevel inverter employing three-phase transformers and single DC input. *IEEE Trans. Ind. Electron.* **2009**, *56*, 2005–2014. [CrossRef]
8. Jahan, H.K.; Naseri, M.; Haji-Esmaeili, M.M.; Abapour, M.; Zare, K. Low component merged cells cascaded-transformer multilevel inverter featuring an enhanced reliability. *IET Power Electron.* **2017**, *10*, 855–862. [CrossRef]
9. Panda, A.K.; Suresh, Y. Performance of cascaded multilevel inverter by employing single and three-phase transformers. *IET Power Electron.* **2012**, *5*, 1694–1705. [CrossRef]
10. Pereda, J.; Dixon, J. High-frequency link: A solution for using only one dc source in asymmetric cascaded multilevel inverters. *IEEE Trans. Ind. Electron.* **2011**, *58*, 3884–3892. [CrossRef]
11. Wang, L.; Zhang, D.; Wang, Y.; Wu, B.; Athab, H.S. Power and voltage balance control of a novel three-phase solid-state transformer using multilevel cascaded H-bridge inverters for microgrid applications. *IEEE Trans. Power Electron.* **2016**, *31*, 3289–3301. [CrossRef]

12. Barzegarkhoo, R.; Kojabadi, H.M.; Zamiry, E.; Vosough, N.; Chang, L. Generalized structure for a single phase switched-capacitor multilevel inverter using a new multiple DC link producer with reduced number of switches. *IEEE Trans. Power Electron.* **2016**, *31*, 5604–5617. [CrossRef]
13. Du, Z.; Ozpineci, B.; Tolbert, L.M.; Chiasson, J.N. DC–AC Cascaded H-bridge multilevel boost inverter with no inductors for electric/hybrid electric vehicle applications. *IEEE Trans. Ind. Appl.* **2009**, *45*, 963–970. [CrossRef]
14. Villanueva, E.; Correa, P.; Rodriguez, J.; Pacas, M. Control of a single-phase cascaded H-bridge multilevel inverter for grid-connected photovoltaic systems. *IEEE Trans. Ind. Electron.* **2009**, *56*, 4399–4406. [CrossRef]
15. Zhou, Y.; Li, H. Analysis and suppression of leakage current in cascaded-multilevel-inverter-based PV systems. *IEEE Trans. Power Electron.* **2014**, *29*, 5265–5277. [CrossRef]
16. Kadam, A.; Shukla, A. A multilevel transformerless inverter employing ground connection between PV negative terminal and grid neutral point. *IEEE Trans. Ind. Electron.* **2017**, *64*, 8897–8907. [CrossRef]
17. Siwakoti, Y.P.; Blaabjerg, F. A single-phase transformerless inverter with charge pump circuit concept for grid-tied PV applications. *IEEE Trans. Ind. Electron.* **2018**, *65*, 2100–2111. [CrossRef]
18. Anurag, A.; Deshmukh, N.; Maguluri, A.; Anand, S. Integrated DC–DC Converter Based Grid-Connected Transformerless Photovoltaic Inverter with Extended Input Voltage Range. *IEEE Trans. Power Electron.* **2018**, *33*, 8322–8330. [CrossRef]
19. Khan, M.N.H.; Forouzesh, M.; Siwakoti, Y.P.; Li, L.; Kerekes, T.; Blaabjerg, F. Transformerless Inverter Topologies for Single-Phase Photovoltaic Systems: A Comparative Review. *IEEE J. Emerg. Sele. Topi. Power Electron.* **2019**, *65*, 805–835. [CrossRef]
20. Kumar, V.V.S.P.; Fernandes, B.G. Minimization of inter-module leakage current in cascaded H-bridge multilevel inverters for grid connected solar PV applications. *Proc. IEEE Appl. Power Electron. Conf. Expo.* **2016**, 2673–2678. [CrossRef]
21. Wang, F.; Li, Z.; Do, H.T.; Zhang, D. A modified phase disposition pulse width modulation to suppress the leakage current for the transformerless cascaded H-bridge inverters. *IEEE Trans. Ind. Electron.* **2018**, *65*, 1281–1289. [CrossRef]
22. Sun, X.; Wang, B.; Zhou, Y.; Wang, W.; Du, H.; Lu, Z. A Single DC Source Cascaded Seven-Level Inverter Integrating Switched-Capacitor Techniques. *IEEE Trans. Ind. Electron.* **2016**, *63*, 7184–7194. [CrossRef]
23. Lee, S.S. Single-Stage Switched-Capacitor Module (S3CM) Topology for Cascaded Multilevel Inverter. *IEEE Trans. Power Electron.* **2018**, *33*, 8204–8207. [CrossRef]
24. Phanikumar, C.; Roy, J.; Agarwal, V. A Hybrid Nine-Level, 1-φ Grid Connected Multilevel Inverter with Low Switch Count and Innovative Voltage Regulation Techniques across Auxiliary Capacitor. *IEEE Trans. Power Electron.* **2019**, *34*, 2159–2170. [CrossRef]
25. Liu, J.; Lin, W.; Wu, J.; Zeng, J. A Novel Nine-Level Quadruple Boost Inverter with Inductive-Load Ability. *IEEE Trans. Power Electron.* **2019**, *34*, 4014–4018. [CrossRef]
26. Sandeep, N.; Yaragatti, U.R. Operation and Control of an Improved Hybrid Nine-Level Inverter. *IEEE Trans. Ind. Appl.* **2017**, *53*, 5676–5686. [CrossRef]
27. Siddique, M.D.; Mekhilef, S.; Shah, N.M.; Sandeep, N.; Ali, J.S.M.; Iqbal, A.; Ahmed, M.; Ghoneim, S.S.; Al-Harthi, M.M.; Alamri, B.; et al. A Single DC Source Nine-Level Switched-Capacitor Boost Inverter Topology with Reduced Switch Count. *IEEE Access* **2020**, *8*, 5840–5851. [CrossRef]
28. Siddique, M.D.; Mekhilef, S.; Shah, N.M.; Ali, J.S.M.; Meraj, M.; Iqbal, A.; Al-Hitmi, M.A. A New Single Phase Single Switched-Capacitor Based Nine-Level Boost Inverter Topology with Reduced Switch Count and Voltage Stress. *IEEE Access* **2019**, *7*, 174178–174188. [CrossRef]
29. Jahan, H.K.; Abapour, M.; Zare, K.; Hosseini, S.H.; Blaabjerg, F.; Yang, Y. A Multilevel Inverter with Minimized Components Featuring Self-balancing and Boosting Capabilities for PV Applications. *IEEE J. Emerg. Sel. Top. Power Electron.* **2019**. [CrossRef]

© 2020 by the authors. Licensee MDPI, Basel, Switzerland. This article is an open access article distributed under the terms and conditions of the Creative Commons Attribution (CC BY) license (http://creativecommons.org/licenses/by/4.0/).

Article

Hybrid Gravitational–Firefly Algorithm-Based Load Frequency Control for Hydrothermal Two-Area System

Deepak Kumar Gupta [1], Ankit Kumar Soni [1], Amitkumar V. Jha [2], Sunil Kumar Mishra [2], Bhargav Appasani [2], Avireni Srinivasulu [3], Nicu Bizon [4,5,6,*] and Phatiphat Thounthong [7,8]

1. School of Electrical Engineering, Kalinga Institute of Industrial Technology, Bhubaneswar 751024, India; deepak.guptafel@kiit.ac.in (D.K.G.); ankit.sonifel@kiit.ac.in (A.K.S.)
2. School of Electronics Engineering, Kalinga Institute of Industrial Technology, Bhubaneswar 751024, India; amit.jhafet@kiit.ac.in (A.V.J.); sunil.mishrafet@kiit.ac.in (S.K.M.); bhargav.appasanifet@kiit.ac.in (B.A.)
3. Department of Electronics Engineering, JECRC University, Jaipur 303905, India; avireni@jecrcu.edu.in
4. Faculty of Electronics, Communication and Computers, University of Pitesti, 110040 Pitesti, Romania
5. ICSI Energy, National Research and Development Institute for Cryogenic and Isotopic Technologies, 240050 Ramnicu Valcea, Romania
6. Doctoral School, Polytehnic University of Bucharest, 313 Splaiul Independentei, 060042 Bucharest, Romania
7. Renewable Energy Research Centre (RERC), Department of Teacher Training in Electrical Engineering, Faculty of Technical Education, King Mongkut's University of Technology North Bangkok, 1518, Pracharat 1 Road, Bangsue, Bangkok 10600, Thailand; phatiphat.t@fte.kmutnb.ac.th
8. Groupe de Recherche en Energie Electrique de Nancy (GREEN), Université de Lorraine, F-54000 Nancy, France
* Correspondence: nicu.bizon@upit.ro

Citation: Gupta, D.K.; Soni, A.K.; Jha, A.V.; Mishra, S.K.; Appasani, B.; Srinivasulu, A.; Bizon, N.; Thounthong, P. Hybrid Gravitational–Firefly Algorithm-Based Load Frequency Control for Hydrothermal Two-Area System. *Mathematics* **2021**, *9*, 712. https://doi.org/10.3390/math9070712

Academic Editor: Aleksandr Rakhmangulov

Received: 8 February 2021
Accepted: 22 March 2021
Published: 25 March 2021

Publisher's Note: MDPI stays neutral with regard to jurisdictional claims in published maps and institutional affiliations.

Copyright: © 2021 by the authors. Licensee MDPI, Basel, Switzerland. This article is an open access article distributed under the terms and conditions of the Creative Commons Attribution (CC BY) license (https://creativecommons.org/licenses/by/4.0/).

Abstract: The load frequency control (LFC) and tie-line power are the key deciding factors to evaluate the performance of a multiarea power system. In this paper, the performance analysis of a two-area power system is presented. This analysis is based on two performance metrics: LFC and tie-line power. The power system consists of a thermal plant generation system and a hydro plant generation system. The performance is evaluated by designing a proportional plus integral (PI) controller. The hybrid gravitational search with firefly algorithm (hGFA) has been devised to achieve proper tuning of the controller parameter. The designed algorithm involves integral time absolute error (ITAE) as an objective function. For two-area hydrothermal power systems, the load frequency and tie-line power are correlated with the system generation capacity and the load. Any deviation in the generation and in the load capacity causes variations in the load frequencies, as well as in the tie-line power. Variations from the nominal value may hamper the operation of the power system with adverse consequences. Hence, performance of the hydrothermal power system is analyzed using the simulations based on the step load change. To elucidate the efficacy of the hGFA, the performance is compared with some of the well-known optimization techniques, namely, particle swarm optimization (PSO), genetic algorithm (GA), gravitational search algorithm (GSA) and the firefly algorithm (FA).

Keywords: load frequency control; automatic generation control; controllers; optimization techniques; multisource power system; interconnected power system; hybrid gravitational with fire fly algorithm; gravitational search algorithm; firefly algorithm

1. Introduction

A power network generally comprises several areas or power systems, interconnected through tie-lines. Distribution systems, transmission lines, and generation systems that may also include renewable energy sources are some of the prime constituents of the power network [1]. The real-time integration of these components and their operation in the dynamic environment cause differences in the active and reactive power demands. The variations in these quantities produce undesired oscillations in the system. These oscillations have to be damped, else they may adversely affect the operation of the power system, and may even lead to a power blackout. Many approaches have been adopted in the literature in the domain of power grid control and stability [2,3]. Y. Li et al. addressed the issues of scalability, privacy, and reliability in a multienergy system [4]. For a power system comprising multiple areas, the issue of optimal generation and distribution has been addressed in [5]. The power management of interconnected single-phase/three-phase microgrids for enhancing voltage quality is considered by J. Zhou et al. in [6]. Researchers have also modernized the power system using smart grid technologies with the objective to make the power system reliable, resilient, secure, and stable [7–10].

For an interconnected power system with multiple sources, the automatic generation control (AGC) methodology has been devised to limit the oscillations produced in the power system due to the mismatch in demand and supply. However, to ensure the control and stable operation of the power system, it is desirable that the oscillations lie within the acceptable range. Further, these oscillations must be controlled within a minimum time to stabilize the system. The main motivation of the AGC is to improve the performance of the interconnected power system by considering several performance metrics, such as load frequency and tie-line power. These performance metrics are highly correlated with generation capacity at the generation side, demand at the consumer side, and total losses at the transmission side. Any mismatch in these may result in deviations in load frequency, as well as in the tie-line power flow. This may lead the system to an unstable state, with severe consequences. Thus, system load frequencies and tie-line power (TLP) must be within the nominal range to realize a stable system. This is generally achieved using the load frequency control (LFC) method. Further, the power systems are stabilized by controlling the speed of the generators (for load frequency) and the TLP based on the area control system. The area control system has basically two objectives: to cater to the demands of its own customers and to respond to the demands of other control areas. In the context of area control systems, area control error (ACE) comprises load frequencies (LF) and tie-line power (TLP).

Automatic generation control has the following major responsibilities [11–13]:
a. To control the system load frequency.
b. To control the tie-line power of the interconnected area.
c. To ensure the economical operation of the power system, including the generation system.

Extensive research in the literature shows an attempt to bridge the gap in the modeling and analysis of the hybrid hydrothermal power system (HTPS) through a linearized approach. The linearized model of the power system is easy for performance evaluation. This has further motivated researchers to present the linearized model of the interconnected power system for AGC analysis. The power system with more than one interconnected area is referred to as a multiarea power system (MAPS). Many aspects of the AGC analysis in the case of interconnected power systems have been thoroughly discussed in [14,15] with several case studies. For the design of the control parameters of proportional-integral-derivative (PID) and proportional plus integral (PI) controllers, the maximum peak resonance method has been reported in [16]. Continuous and discrete mode analysis with a generation rate constraint for interconnected HTPS has been reported in [17]. Problems with load frequency control for hybrid hydrothermal have also been addressed in [18]. Further, Jha et al. have considered the PI controller for the load frequency control of hybrid hydrothermal systems [19]. The use of several artificial optimization techniques were illustrated with case studies in [20] for the LFC in various system operating conditions. A comparative analysis of several soft computing techniques for the LFC have been studied in [21] by Gupta et al. Further, for a power system with interconnected areas, load frequency control analysis has been carried out in [22] using a hybrid adaptive gravitational search and pattern search algorithm. Recently, Gupta et al. reported novel hybrid optimization techniques for addressing the issues of LFC in MAPS comprising multiple sources [23]. Koley et al. [24] presented the issue of LFC by considering a power system involving hybrid power plants such as thermal, wind, and photovoltaic generation stations. The exemplary work of Khadange et al. introduces the hybrid guided gravitational search with pattern search (hGGSA-PS) optimization technique for MAPS in order to analyze the LFC [25]. AGC, using a coordinated design for two-area systems (TAS), was proposed in [26] by Khezri et al. The scope of some of the advanced controllers to achieve AGC for multiarea systems was considered by Gondaliya et al. in [27]. A new approach, referred to as secondary LFC, has been introduced in [28] for a power system with a multigrid configuration. A new fractional order PI controller was proposed by Celik et al. in [29]. Various optimization techniques for controlling the controller parameter for the LFC of multiarea power systems have been proposed in the literature [30]. For example, a wind-driven optimization algorithm has been proposed by Haes et al. in [31]. A social spider optimization technique was presented in [32]. Further, Nilkmanesh et al. have proposed a multiobjective uniform-diversity genetic algorithm (MUGA) for MAPS in [33]. The slap-swarm optimization technique was discussed for the LFC of MAPS by Sahu et al. in [34].

The overall objective of the design is to discuss the LFC of the interconnected MAPS. The novelty and contribution of this study can be highlighted as follows:

- A new hybrid gravitational–firefly algorithm (hGFA), based on the gravitational search algorithm (GSA) and the firefly algorithm (FA), is proposed.
- A new methodology for adjusting the PI parameters to improve hGFA performance is proposed by the specific design of the hGFA for the two-area interconnected HTPS.
- Furthermore, the overall performance of the hGFA is compared with other well-known optimization techniques, such as the genetic algorithm (GA), particle swarm optimization (PSO), GSA, and FA using the ITAE as an objective function in different case studies. Furthermore, it is noted that, for the same computation time, the overshoot and settling time values of the load frequency, as well as tie-line power, are reduced significantly in each case study with the proposed algorithm.

Thus, in summary, this paper presents a hybrid algorithm to tune the PI controller for the optimum LFC of an interconnected MAPS. This hybrid technique merges two well-known optimization techniques, the GSA and the FA. Effectiveness of the proposed controller tuned using this hybrid intelligent optimization technique is compared with controllers tuned using other well-known optimization techniques, such as the GA, the PSO, the GSA, and the FA. The proposed algorithm works well for different operating conditions (such as changes in load), which shows its robustness. The dynamic response of all the state variables has been improved in terms of settling time and overshoot. It is observed that the proposed controller outperformed the other techniques (GA, PSO, GSA, and FA) in terms of performance, stability, and robustness.

The remainder of this paper is presented in five sections. In Section 2, a brief introduction about the proportional integral controller is presented. The test system of hybrid HTPS is also presented in this section. Further, to analyze the test system, it is modeled on the basis of the state space approach. The preliminaries of the optimization techniques along with the design of hGFA for a hydrothermal power system under consideration is presented in Section 3. Section 4 covers the design methodology and the simulation of the HTPS. The analysis of the simulation results and discussion is presented in Section 5. Finally, Section 6 deals with the conclusion of the research work. Some of the system variables are summarized in the Appendix A.

2. State Space Modeling of a Hydrothermal System

This section deals with the PI controller and the modeling of the two-area hydrothermal system using a state space approach.

2.1. The Proportional Plus Integral Controller

We have considered the PI controller for a LFC analysis of the hydrothermal power system. If $e(t)$ represents the error at the input of the PI controller, then its output $u(t)$ can be represented by Equation (1) as given below:

$$u(t) = K_{pr}e(t) + K_i \int e(t)dt \quad (1)$$

where K_{pr}, and K_i, are the controller parameters. In this paper, we consider the objective function as an integral time absolute error (ITAE) [23]. The ITAE cost function is as given by Equation (2) below.

$$ITAE = \int_0^\infty t \times |e(t)|dt \quad (2)$$

2.2. Test Model for the Two-Area Hydrothermal System

In order to obtain a performance analysis of the hydrothermal power system based on hGFA optimization techniques, the test system as illustrated in Figure 1 is considered. The different blocks of the test system are modeled using the standard linearized method and are shown by their respective transfer functions. Here, the two-area power system is considered such that one area consists of a hydropower plant and the other area has a thermal power plant for power generation. Each plant has a PI controller, shown by the transfer function to achieve automatic gain control of the power system.

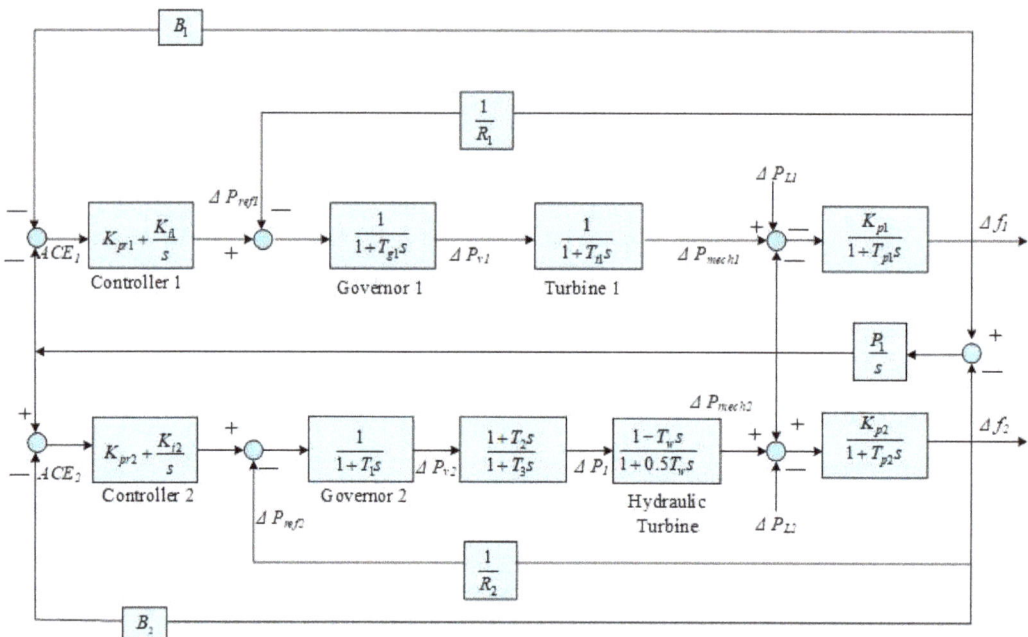

Figure 1. Test system model using transfer functions.

The model of the test system is obtained using the state space approach detailed in [11]. The state space modeling of the test system and its implementation are discussed in the next subsection.

2.3. State Space Modeling

As shown in Figure 1, the test system comprises two interconnected areas resulting in a hydrothermal power system. Each area consists of one PI controller to achieve AGC based on LF, as well as TLP for effective operation of the power system. With the given model and its transfer functions, the test system can be modeled through multidimensional state space analysis. Using a state space approach corresponding to Equations (3) and (4), the transfer function of the overall hydrothermal power system can be obtained.

$$\dot{x} = Ax + Bu \tag{3}$$

$$y = Cx + Du \tag{4}$$

Here, x, y, and u, denotes the state vector, output vector, and input vector, respectively. The different matrices with the real constant model-dependent values of the state space model are represented by variables A, B, C, and D.

On analysis of the state space model, the following equations follow:

$$x = \left[\Delta f_1, \Delta P_{mech1}, \Delta P_{v1}, \Delta f_2, \Delta P_{mech2}, \Delta P_1, \Delta P_{v2}, \Delta P_{12}, \Delta P_{ref1}, \Delta P_{ref2}\right]^T \tag{5}$$

$$u = [\Delta P_{L1}, \Delta P_{L2}]^T \tag{6}$$

$$y = [\Delta f_1, \Delta f_2, \Delta P_{12}]^T \tag{7}$$

These variables are part of the test system's state space model, which characterizes the entire hydrothermal power system as described in [11]. The state space matrices to construct the transfer function is determined by analyzing the differential equations. These are illustrated below.

$$\dot{x}_1 = -\frac{1}{T_{p1}}x_1 + \frac{K_{p1}}{T_{p1}}x_2 - \frac{K_{p1}}{T_{p1}}x_8 - \frac{K_{p1}}{T_{p1}}u_1 \tag{8}$$

$$\dot{x}_2 = -\frac{1}{T_{t1}}x_2 + \frac{1}{T_{t1}}x_3 \tag{9}$$

$$\dot{x}_3 = -\frac{1}{R_1 T_{g1}}x_1 - \frac{1}{T_{g1}}x_3 + \frac{1}{T_{g1}}x_9 \tag{10}$$

$$\dot{x}_4 = -\frac{1}{T_{p2}}x_4 + \frac{K_{p2}}{T_{p2}}x_5 + \frac{K_{p2}}{T_{p2}}x_8 - \frac{K_{p2}}{T_{p2}}u_2 \tag{11}$$

$$\dot{x}_5 = -\frac{2T_2}{R_2 T_1 T_3}x_4 - \frac{2}{T_w}x_5 + \left(\frac{2}{T_w} + \frac{2}{T_3}\right)x_6 + \left(\frac{2T_2}{T_1 T_3} - \frac{2}{T_3}\right)x_7 - \frac{2T_2}{T_1 T_3}x_{10} \tag{12}$$

$$\dot{x}_6 = -\frac{T_2}{R_2 T_1 T_3}x_4 - \frac{1}{T_3}x_6 + \left(\frac{1}{T_3} - \frac{T_2}{T_1 T_3}\right)x_7 + \frac{T_2}{T_1 T_3}x_{10} \tag{13}$$

$$\dot{x}_7 = -\frac{1}{R_2 T_1}x_4 - \frac{1}{T_1}x_7 + \frac{1}{T_1}x_{10} \tag{14}$$

$$\dot{x}_8 = T_s x_1 - T_s x_4 \tag{15}$$

$$\dot{x}_9 = \left(\frac{B_1 K_{pr1}}{T_{p1}} - K_{pr1}T_s - K_{i1}B_1\right)x_1 - \frac{B_1 K_{pr1}K_{p1}}{T_{p1}}x_2 + K_{pr1}T_s x_4 + \left(\frac{B_1 K_{pr1}K_{p1}}{T_{p1}} - K_{i1}\right)x_8 + \frac{B_1 K_{pr1}K_{p1}}{T_{p1}}u_1 \tag{16}$$

$$\dot{x}_{10} = K_{pr2}T_s x_1 + \left(\frac{B_2 K_{pr2}}{T_{p2}} - K_{pr2}T_s - K_{i2}B_2\right)x_4 - \frac{B_2 K_{pr2}K_{p2}}{T_{p2}}x_5 + \left(-\frac{B_2 K_{pr2}K_{p2}}{T_{p2}} + K_{i2}\right)x_8 + \frac{B_2 K_{pr2}K_{p2}}{T_{p2}}u_2 \tag{17}$$

Finally, the state matrices are obtained by incorporating the above differential equations. These are as shown below:

$$A = \begin{bmatrix} -\frac{1}{T_{P1}} & \frac{K_{P1}}{T_{P1}} & 0 & 0 & 0 & 0 & 0 & -\frac{K_{P1}}{T_{P1}} & 0 & 0 \\ 0 & -\frac{1}{T_{t1}} & \frac{1}{T_{t1}} & 0 & 0 & 0 & 0 & 0 & 0 & 0 \\ -\frac{1}{R_1 T_{g1}} & 0 & 0 & \frac{1}{T_{g1}} & 0 & 0 & 0 & 0 & \frac{1}{T_{g1}} & 0 \\ 0 & 0 & 0 & -\frac{1}{T_{P2}} & \frac{K_{P2}}{T_{P2}} & 0 & 0 & \frac{K_{P2}}{T_{P2}} & 0 & 0 \\ 0 & 0 & 0 & \frac{2T_2}{R_2 T_1 T_2} & -\frac{2}{T_w} & (\frac{2}{T_w} + \frac{2}{T_3}) & (\frac{2T_2}{T_1 T_3} - \frac{2}{T_3}) & 0 & 0 & -\frac{2T_2}{T_1 T_3} \\ 0 & 0 & 0 & -\frac{T_2}{R_2 T_1 T_2} & 0 & -\frac{1}{T_3} & (\frac{1}{T_3} - \frac{T_2}{T_1 T_3}) & 0 & 0 & \frac{T_2}{T_1 T_3} \\ 0 & 0 & 0 & -\frac{1}{R_2 T_1} & 0 & 0 & -\frac{1}{T_1} & 0 & 0 & \frac{1}{T_1} \\ T_s & 0 & 0 & -T & 0 & 0 & 0 & 0 & 0 & 0 \\ A_{9,1} & -\frac{B_1 K_{pr1} K_{P1}}{T_{P1}} & 0 & K_{pr1} T_s & 0 & 0 & 0 & (\frac{B_1 K_{pr1} K_{P1}}{T_{P1}} - K_{i1}) & 0 & 0 \\ K_{pr2} T_s & 0 & 0 & A_{10,4} & -\frac{B_2 K_{pr2} K_{P2}}{T_{P2}} & 0 & 0 & (-\frac{B_2 K_{pr2} K_{P2}}{T_{P2}} + K_{i2}) & 0 & 0 \end{bmatrix}$$

where,

$$A_{9,1} = (\frac{B_1 K_{pr1}}{T_{P1}} - K_{pr1} T_s - K_{i1} B_1), \quad A_{10,4} = (\frac{B_2 K_{pr2}}{T_{P2}} - K_{pr2} T_s - K_{i2} B_2)$$

$$B = \begin{bmatrix} -\frac{K_{P1}}{T_{P1}} & 0 & 0 & 0 & 0 & 0 & 0 & \frac{B_1 K_{pr1} K_{P1}}{T_{P1}} & 0 \\ 0 & 0 & 0 & -\frac{K_{P2}}{T_{P2}} & 0 & 0 & 0 & 0 & \frac{B_2 K_{pr2} K_{P2}}{T_{P2}} \end{bmatrix}^T$$

$$C = \begin{bmatrix} 1 & 0 & 0 & 0 & 0 & 0 & 0 & 0 & 0 \\ 0 & 0 & 0 & 1 & 0 & 0 & 0 & 0 & 0 \\ 0 & 0 & 0 & 0 & 0 & 0 & 1 & 0 & 0 \end{bmatrix} \quad D = 0$$

The transfer function (T.F) using matrices A, B, C, and D can be obtained using the relation:

$$T.F = C[SI - A]^{-1} B + D$$

Without loss of generality, the transfer function will be a matrix of the dimension (3×2). The elements of the transfer function are as follows.

$$\Delta f_1 = \frac{\Delta f_1(s)}{\Delta P_{L1}(s)} \Delta P_{L1} + \frac{\Delta f_1(s)}{\Delta P_{L2}(s)} \Delta P_{L2} \tag{18}$$

$$\Delta f_2 = \frac{\Delta f_2(s)}{\Delta P_{L1}(s)} \Delta P_{L1} + \frac{\Delta f_2(s)}{\Delta P_{L2}(s)} \Delta P_{L2} \tag{19}$$

$$\Delta P_{12} = \frac{\Delta P_{12}(s)}{\Delta P_{L1}(s)} \Delta P_{L1} + \frac{\Delta P_{12}(s)}{\Delta P_{L2}(s)} \Delta P_{L2} \tag{20}$$

3. Hybrid Gravitational–Firefly Algorithm

The hybrid gravitational–firefly algorithm uses two well-known optimization techniques, the GSA and the FA [35–38]. Newton's gravitational law is the working principle for the GSA. In this, every object is treated as a candidate solution. The masses of each of these variables or the candidate are used to evaluate their performance depending on the value of the selected objective function [21]. On the other hand, the working of the FA is similar to the flashing behavior of fireflies, which they use to achieve communication amongst themselves. The hybrid gravitational–firefly algorithm takes the properties of both these algorithms and updates the values of its objects (candidate solutions) using the updated equations.

3.1. GSA

The position (x) of each agent (i) out of N agents is represented by x_i. If we assume m dimensional space, then the position of each agent x_i^d is given using the following equation:

$$x_i = (x_i^1, \ldots x_i^d, \ldots x_i^m) \forall i = \{1, 2, \ldots, N\} \tag{21}$$

The mass 'j' applies force on mass 'i', which is given as,

$$F_{ij}^d(t) = G(t) \times \frac{\left(x_j^d(t) - x_i^d(t)\right)}{R_{ij}(t) + \varepsilon} \left(M_{pi}(t) * M_{aj}(t)\right) \tag{22}$$

where $G(t)$ denotes the gravitational constant, $R_{ij}(t)$ denotes the Euclidian distance between agent 'i' and agent 'j', and $M_{pi}(t)$ and $M_{ai}(t)$ represent the passive gravitational and active gravitational masses of the agent 'i' and the agent 'j', respectively. The Euclidian and the gravitational constants are given as below.

$$\left.\begin{array}{l} R_{ij}(t) = ||x_i(t), x_j(t)||_2 \\ G(t) = G(G_0, t) \end{array}\right\} \tag{23}$$

If $M_{ii}(t)$ indicates the inertial mass of the ith agent, the total force and the total acceleration acting on an agent 'i' due to other agents in the d-dimensional space is given as:

$$F_i^d(t) = \sum_{j=1, j \neq i}^{N} rand_j F_{ij}^d(t), \; a_i^d(t) = \frac{F_i^d(t)}{M_{ii}(t)} \tag{24}$$

In each round, Equations (25) and (26) are used to update the velocity as well as the mass of each agent i.

$$m_i(t) = \frac{\text{fit}_i(t) - \text{worst}(t)}{\text{best}(t) - \text{worst}(t)} \tag{25}$$

$$M_i(t) = \frac{m_i(t)}{\sum_{j=1}^{N} m_j(t)} \tag{26}$$

3.2. Firefly Algorithm

The Euclidian distance between the firefly 'i' and firefly 'j' for given position x_i and x_j can be evaluated using the equation given below,

$$r_{ij} = \sqrt{\sum_{k=1}^{d} (x_{i,k} - x_{j,k})^2} \tag{27}$$

where k indicates the kth element of the spatial coordinates.

The attractiveness parameter of every firefly basically can be given by the following equation:

$$\beta = \beta_0 e^{-\gamma r^2} \tag{28}$$

where γ represents the coefficient of absorption, which is used to control the light intensity. We can define the movement of each firefly as follows.

$$v_i^d(t+1) = rand_i \times v_i^d(t) + a_i^d(t) \tag{29}$$

$$x_i^d = x_i^d + \beta_0 e^{-\gamma r_{ij}^2}(x_j - x_i) + v_i^d(t+1) + \alpha \varepsilon \tag{30}$$

Here, x_i is the instantaneous position of an object and $\alpha \varepsilon$ indicates the random behavior of a firefly if no brighter firefly is detected.

4. Methodology and Simulation Results

This section presents a methodology to design and configure two-area hydrothermal power systems using simulation to evaluate the performance of the proposed optimization technique.

4.1. Simulation Methodology

To evalute the efficacy of the proposed hybrid gravitational–firefly algorithm, the test system shown in Figure 1 is designed. The test system comprises a two-area power system, one with a thermal power plant and the other with a hydropower plant. To design and configure the two-area power system, MATLAB Simulink is used. Further, the proposed algorithm is coded in a MATLAB script file and then interfaced to the Simulink model for testing its efficacy on the test system.

The various simulation configuration parameters are as follows: MATLAB (R2016a) software is used along with the Simulink tool. The system used for simulation has an i5-6200 CPU@ processor running at 2.30 GHz frequency and having 8 GB RAM. The proposed algorithm and other optimization techniques are written as MATLAB scripts, which are interfaced to the test power system through the Simulink. A few of the other key parameters are included in the Appendix A section.

4.2. Simulation Results

The simulation was performed to evaluate the load frequencies and tie-line parameters of the test system. The simulations were carried out by tuning the parameters of the PI controller using the novel hybrid gravitational search with firefly algorithm as an optimization technique. To understand the efficacy of this algorithm, the results are compared with PSO, GA, GSA, and FA, which can be used as a benchmark for performance evaluation. In all the optimization techniques, the cost function is ITAE and remains unaltered.

The performance of the proposed algorithm on hydrothermal power systems is evaluated by considering two case studies, which are discussed below.

4.2.1. Case Study-I

To observe performance under a step load change, the load in the area having the thermal power plant (area-1) is incremented up to 20%. The load in the area having the hydropower plant (area-2) is unchanged. The test system is simulated using different optimization techniques, i.e., PSO, GA, GSA, FA, and hGFA. ITAE is used as the objective function. The parameters of the PI controller tuned using these optimization techniques are summarized in Table 1.

Table 1. Optimized parameters of PI controller for case study-I.

O.T.	K_{pr1}	K_{pr2}	K_{i1}	K_{i2}
GA	0.068627	1.9095	0.33831	0.033451
PSO	0.048043	1.2241	0.75369	0.001
GSA	1.4718	9.1185	0.4606	0.0006
FA	0.6366	5.5908	0.8408	0.0343
hGFA	0.0015	3.3537	0.8210	0.0600

The performance of the test system based on LF as well as TLP is analyzed by using the optimized parameters shown in Table 1. The simulation results corresponding to the load frequency in area-1 (Δf_1), the load frequency in area-2 (Δf_2), and the tie-line power flow of the test system are shown in Figures 2–4, respectively.

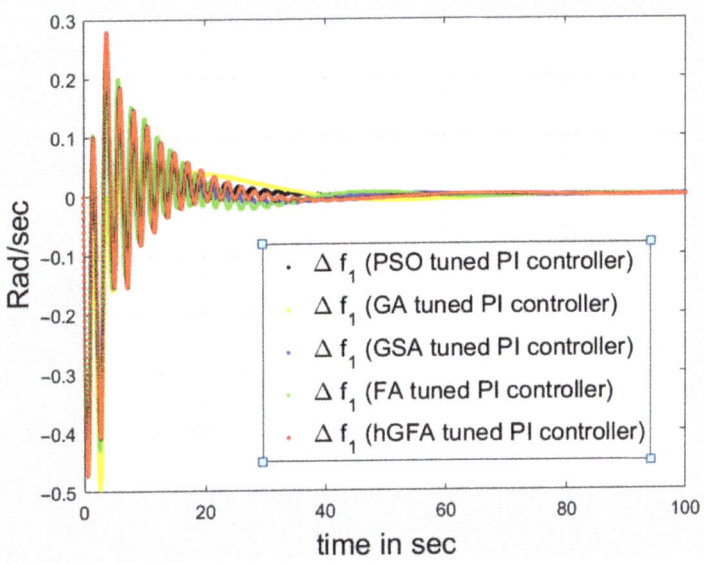

Figure 2. Perturbation in load frequency response in area-1 of HTPS for case study-I.

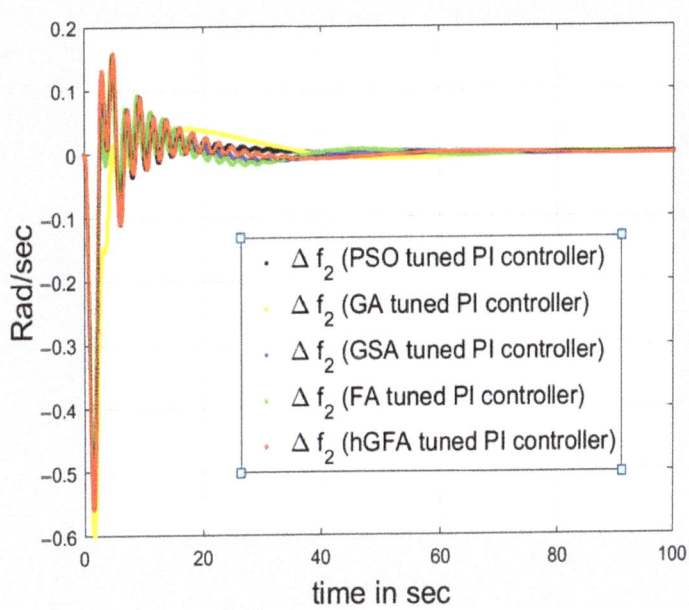

Figure 3. Perturbation in load frequency response in area-2 of HTPS for case study-I.

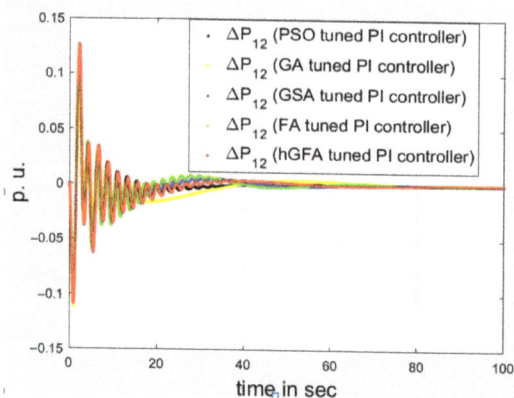

Figure 4. Perturbation in the TLP flow response of HTPS for case study-I.

4.2.2. Case Study-II

To observe performance under a step load change, the load in the area with a thermal power plant (area-1) is incremented up to 5% and the load in area-2 is subjected to a 1% change. The tuned parameters of the PI controller using different optimization techniques are summarized in Table 2.

Table 2. Optimized parameters of the PI controller for case study-II.

O.T.	K_{pr1}	K_{pr2}	K_{i1}	K_{i2}
GA	0.033451	0.092078	0.7018	0.01
PSO	0.0010	1.5064	0.2833	0.0128
GSA	0.4273	1.9126	0.7865	0.0104
FA	1.2880	1.5767	0.5235	0.0333
hGFA	0.0010	1.5534	0.7537	0.0363

The simulation results corresponding to the load frequency in area-1 (Δf_1), the load frequency in area-2 (Δf_2), and the tie-line power flow of the test system are shown in Figures 5–7, respectively.

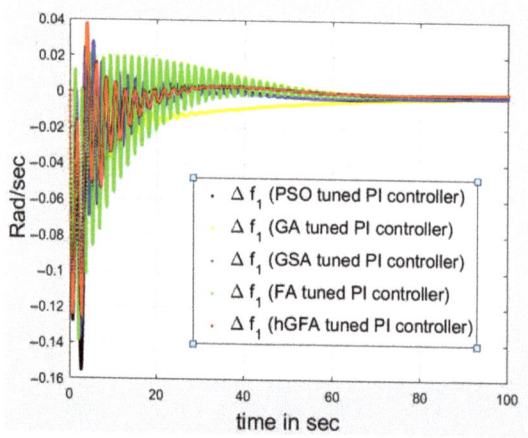

Figure 5. Perturbation in load frequency response in area-1 of HTPS for case study-II.

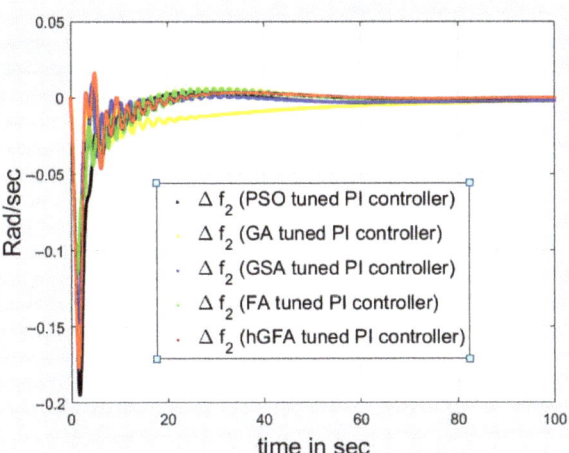

Figure 6. Perturbation in load frequency response in area-2 of HTPS for case study-II.

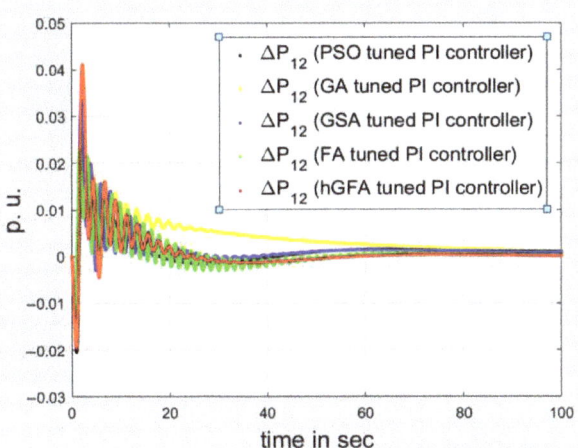

Figure 7. Perturbation in the TLP flow response of HTPS for case study-II.

5. Analysis and Discussion

The simulated results of the test system are analyzed in this section. To perform a comparative analysis, the results achieved using the hGFA are compared with those obtained using the PSO, the GA, the GSA, and the FA. The comparisons are in terms of the overshoot/undershoot, and settling time for the load frequency response and the TLP.

5.1. Case Study-I

For the first case study, for the perturbation in step load change, the load frequency responses in area-1 and area-2 have been shown in Figures 2 and 3, respectively. Figure 4 represents the TLP for the interconnected power system belonging to case study-I. It can be inferred from these responses that the hGFA optimization technique outperforms other optimization techniques. The overshoot/undershoot, and settling time of these results are summarized in Table 3.

Table 3. Comparison of optimization techniques based on overshoot and settling time for case study-I.

Optimization Techniques	System Variables	Case Study	
		Overshoot	Settling Time (s)
Genetic Algorithm (GA)	Δf_1	−0.5	40
	Δf_2	−0.6	41
	ΔP_{12}	−0.12	45
Particle Swarm Optimization (PSO)	Δf_1	−0.47	40
	Δf_2	−0.55	38
	ΔP_{12}	−0.12	39
Gravitational Search Algorithm (GSA)	Δf_1	−0.46	40
	Δf_2	−0.52	40
	ΔP_{12}	−0.12	43
Firefly Algorithm (FA)	Δf_1	−0.47	45
	Δf_2	−0.51	43
	ΔP_{12}	−0.11	42
Hybrid Gravitational–Firefly Algorithm (hGFA)	Δf_1	−0.46	35
	Δf_2	−0.50	33
	ΔP_{12}	−0.11	35

From Table 3, it can be seen that the overshoot/undershoot for the system variable Δf_1 is −0.46, −0.47, −0.5, −0.46, and −0.47 for the hGFA, the PSO, the GA, the GSA, and the FA, respectively. Further, it can also be seen that the overshoot/undershoot for the system variable Δf_2 is −0.50, −0.55, −0.6, −0.52, and −0.51, corresponding to the hGFA, the PSO, the GA, the GSA, and the FA, respectively. Furthermore, the system variable ΔP_{12} is −0.11, −0.12, −0.12, −0.12, and −0.11, corresponding to the hGFA, the PSO, the GA, the GSA, and the FA, respectively. Thus, it can be inferred that the hGFA outperforms other well-known optimization techniques such as PSO, GA, GSA, and FA in terms of overshoot/undershoot for all the system variables of the HTPS. Similarly, it can be inferred by analyzing the results in Table 3 that the hGFA (with T_s = 35 s, 33 s, 35 s, respectively, for Δf_1, Δf_2, ΔP_{12}) performs better in terms of settling time (T_s) compared to its competitor optimization techniques such as PSO (with T_s = 40 s, 38 s, 39 s, respectively, for Δf_1, Δf_2, ΔP_{12}), GA (with T_s = 40 s, 41 s, 45 s, respectively, for Δf_1, Δf_2, ΔP_{12}), GSA (with T_s = 40 s, 40 s, 43 s, respectively, for Δf_1, Δf_2, ΔP_{12}), and FA (with T_s = 45 s, 43 s, 42 s, respectively, for Δf_1, Δf_2, ΔP_{12}). Hence, it can be concluded that, for case study-I, the performance of the proposed hybrid optimization technique is better than other optimization techniques.

5.2. Case Study-II

From the perturbation in responses of frequencies in both the area as well as tie-line power, illustrated in Figures 5–7, it can be inferred that the hGFA optimization technique performs better compared to other techniques. The overshoot/undershoot and settling time for different system variables are summarized in Table 4.

From Table 4, it can be seen that the overshoot/undershoot for the system variable Δf_1 is −0.12, −0.158, −0.13, −0.15, and −0.14, corresponding to hGFA, PSO, GA, GSA, and FA, respectively. Further, it can also be seen that the overshoot/undershoot for the system variable Δf_2 is −0.17, −0.2, −0.17, −0.15, and −0.12, corresponding to hGFA, PSO, GA, GSA, and FA, respectively. The system variable ΔP_{12} is −0.02, −0.021, −0.02, −0.018, and −0.014, corresponding to hGFA, PSO, GA, GSA, and FA, respectively. Similarly, by analyzing Table 4, it can be seen that the hGFA (with T_s = 35 s, 30 s, 30 s, respectively, for Δf_1, Δf_2, ΔP_{12}) performs better in terms of settling time (T_s) compared to its competitor optimization techniques such as PSO (with T_s = 70 s, 60 s, 45 s, respectively, for Δf_1, Δf_2, ΔP_{12}), GA (with T_s = 75 s, 70 s, 70 s, respectively, for Δf_1, Δf_2, ΔP_{12}), GSA (with T_s = 65 s, 65 s, 70 s, respectively, for Δf_1, Δf_2, ΔP_{12}), and FA (with T_s = 70 s, 60 s, 70 s, respectively, for Δf_1, Δf_2, ΔP_{12}). These algorithms offer similar overshoot/undershoot characteristics, but the settling time characteristics vary. The proposed algorithm reduces the settling time considerably, by approximately 50% when compared to other techniques.

Table 4. Comparison of optimization techniques based on overshoot/undershoot and settling time for case study-II.

Optimization Techniques	System Variables	Case Study	
		Overshoot	Settling Time (s)
Genetic Algorithm (GA)	Δf_1	−0.13	75
	Δf_2	−0.17	70
	ΔP_{12}	−0.02	70
Particle Swarm Optimization (PSO)	Δf_1	−0.158	70
	Δf_2	−0.2	60
	ΔP_{12}	−0.021	45
Gravitational Search Algorithm (GSA)	Δf_1	−0.15	65
	Δf_2	−0.15	65
	ΔP_{12}	−0.018	70
Firefly Algorithm (FA)	Δf_1	−0.14	70
	Δf_2	−0.12	60
	ΔP_{12}	−0.014	70
Hybrid Gravitational–Firefly Algorithm (hGFA)	Δf_1	−0.12	35
	Δf_2	−0.17	30
	ΔP_{12}	−0.02	30

6. Conclusions

The TAS consideration of hydrothermal systems is considered in this paper for LFC as well as TLP analysis. Each area has been considered with a proportional integral controller for analysis. The parameters of each PI controller are tuned using several optimization techniques with ITAE as an objective function. From the in-depth simulation presented in Section 4, followed by the analysis in Section 5, it has been observed that the proposed hGFA performs better than the other optimization techniques. Further, the hGFA improves the response of the interconnected power system, resulting in low overshoot and reduced settling time for LF as well as TLP flow. The main findings of this study can be summarized as follows:

- The novel optimization technique, hGFA, based on the GSA and the FA, is proposed in this paper to address the LFC and tie-line power flow issues of the two-area HTPS.
- The performance of the proposed algorithm is compared with other well-known optimization techniques such as PSO, GA, GSA, and FA using ITAE as the objective function.
- For almost the same overshoot, the settling time value of the LF as well as the TLP is lowered by almost 15% in case study-I, and it is lowered by almost 50% in case study-II using the proposed hGFA, as compared to PSO, GA, GSA, and FA.
- The computation time is almost the same for all algorithms analyzed in this study, so the complexity of the proposed algorithm remains comparable to the basic algorithms and is used as a benchmark for evaluating performance.

Hence, it can be inferred that the hGFA outperforms the other optimization techniques (PSO, GA, GSA, and FA). Therefore, it can be successfully applied for AGC of a hydrothermal power system.

Author Contributions: Conceptualization, D.K.G., A.K.S. and B.A.; methodology, D.K.G. and A.K.S.; software, D.K.G.; validation, A.V.J.; investigation, D.K.G.; resources, B.A.; data curation, A.V.J.; writing—original draft preparation, A.V.J. and A.S.; supervision, A.S. and N.B.; project administration, A.S.; formal analysis: N.B.; funding acquisition: N.B. and P.T.; visualization: P.T.; writing—review and editing: N.B., P.T.; figure and table, S.K.M. All authors have read and agreed to the published version of the manuscript.

Funding: This work was partially supported by the International Research Partnerships: Electrical Engineering Thai–French Research Center (EE-TFRC) between King Mongkut's University of Technology North Bangkok and the University of Lorraine under grant KMUTNB−BasicR−64−17.

Institutional Review Board Statement: Not applicable.

Informed Consent Statement: Not applicable.

Data Availability Statement: Not applicable.

Conflicts of Interest: The authors declare no conflict of interest.

Appendix A

Some of the standards values of the system variables are as follows:

$T_{p1} = T_{p2} = 20$ s, $T_g = 0.08$ s, $T_t = 0.3$ s, $B_1 = B_2 = 0.4249$, $T_{12} = 0.0866$, $T_1 = 48.709$, $T_2 = 0.51308$, $T_3 = 10$, $T_w = 1$, $R_1 = R_2 = 2.4$ Hz/p.u. MW, $P_{r1} = P_{r2} = 1200$ MW, $K_{p1} = K_{p2} = 0.120$ Hz/p.u. MW, $D_1 = D_2 = 0.00833$ p.u. MW/Hz.

Hybrid gravitational firefly algorithm parameters: number of objects = 70, number of iterations = 15, $\alpha = 20$, $G_0 = 100$, $\beta_0 = 0.2$, and $\gamma = 1$.

GSA parameters: number of populations = 70, number of iterations = 15, $G_0 = 100$, and $\alpha = 20$.

FA parameters: number of fireflies = 70, number of iterations = 15, $\beta_0 = 0.2$, $\gamma = 1$, and $\alpha = 0.5$.

References

1. Tabatabaei, N.M.; Kabalci, E.; Bizon, N. *Microgrid Architectures, Control and Protection Methods*, 1st ed.; Springer: London, UK, 2019.
2. Thounthong, P.; Mungporn, P.; Pierfederici, S.; Guilbert, D.; Bizon, N. Adaptive Control of Fuel Cell Converter Based on a New Hamiltonian Energy Function for Stabilizing the DC Bus in DC Microgrid Applications. *Mathematics* **2020**, *8*, 2035. [CrossRef]
3. Tabatabaei, N.M.; Ravadanegh, S.N.; Bizon, N. *Power Systems Resiliency: Modeling, Analysis and Practice*; Springer: London, UK, 2018.
4. Li, Y.; Zhang, H.; Liang, X.; Huang, B. Event-Triggered-Based Distributed Cooperative Energy Management for Multienergy Systems. *IEEE Trans. Industr. Inform.* **2019**, *15*, 2008–2022. [CrossRef]
5. Yushuai, L.; Gao, W.; Gao, W.; Zhang, H.; Zhou, J. A Distributed Double-Newton Descent Algorithm for Cooperative Energy Management of Multiple Energy Bodies in Energy Internet. *IEEE Trans. Industr. Inform.* **2020**. [CrossRef]
6. Zhou, J.; Xu, Y.; Sun, H.; Li, Y.; Chow, M. Distributed Power Management for Networked AC–DC Microgrids with Unbalanced Microgrids. *IEEE Trans. Industr. Inform.* **2020**, *16*, 1655–1667. [CrossRef]
7. Jha, A.V.; Ghazali, A.N.; Appasani, B.; Ravariu, C.; Srinivasulu, A. Reliability Analysis of Smart Grid Networks Iincorporating Hardware Failures and Packet Loss. *Rev. Roum. Sci. Tech. El.* **2021**, *65*, 245–252.
8. Mishra, S.K.; Bhargav, A.; Jha, A.V.; Garrido, I.; Garrido, A.J. Centralized Airflow Control to Reduce Output Power Variation in a Complex OWC Ocean Energy Network. *Complexity* **2020**, *2020*, 2625301. [CrossRef]
9. Jha, A.V.; Ghazali, A.N.; Appasani, B.; Mohanta, D.K. Risk Identification and Risk Assessment of Communication Networks in Smart Grid Cyber-Physical Systems. In *Security in Cyber-Physical Systems*; Studies in Systems, Decision and Control; Awad, A.I., Furnell, S., Paprzycki, M., Sharma, S.K., Eds.; Springer: Cham, Switzerland, 2021; Volume 339. [CrossRef]
10. Li, Y.; Gao, W.; Yan, W.; Huang, S.; Wang, R.; Gevorgian, V.; Gao, D.W. Data-Driven Optimal Control Strategy for Virtual Synchronous Generator via Deep Reinforcement Learning Approach. *J. Mod. Power Syst. Clean Energy* **2021**, *9*, 27–36. [CrossRef]
11. Saadat, H. *Power System Analysis*, 2nd ed.; McGraw-Hill: New York, NY, USA, 2009.
12. Kundur, P. *Power System Stability and Control*, 1st ed.; McGraw-Hill: New York, NY, USA, 1994.
13. Wood, A.J.; Woolenberg, B.F. *Power Generation Operation and Control*, 3rd ed.; John Wiley and Sons: Hoboken, NJ, USA, 1984.
14. Nanda, J.; Kaul, B.L. Automatic generation control of an interconnected power system. *Proc. IEEE* **1978**, *125*, 385–390. [CrossRef]
15. Working Group Prime Mover and Energy Supply. Hydraulic turbine and turbine control models for system dynamic studies. *IEEE Trans. Power Syst.* **1992**, *7*, 167–179. [CrossRef]
16. Khodabakhshian, A.; Golbon, N. Unified PID design for load frequency control. In Proceedings of the International Conference on Control Applications, Taipei, Taiwan, 2–4 September 2004; IEEE: New York, NY, USA, 2004; pp. 1627–1632. [CrossRef]
17. Nanda, J.; Kothari, M.L.; Satsangi, P.S. Automatic generation control of an interconnected hydrothermal system in continuous and discrete modes considering generation rate constraints. *Proc. Inst. Elect. Eng.* **1983**, *130*, 17–27. [CrossRef]

18. Ramakrishna, K.S.S.; Bhatti, T.S. Load Frequency Control of Interconnected Hydro-Thermal Power Systems. In Proceedings of the International Conference on Energy and Environment, New Delhi, India, Month–August 2006; Available online: https://scholar.google.com/scholar?hl=en&as_sdt=0%2C5&q=Load+frequency+control+of+interconnected+hydro-thermal+power+systems&btnG= (accessed on 7 February 2021).
19. Jha, A.V.; Gupta, D.K.; Appasani, B. The PI Controllers and its optimal tuning for Load Frequency Control (LFC) of Hybrid Hydro-thermal Power Systems. In Proceedings of the International Conference on Communication and Electronics Systems (ICCES), Coimbatore, India, 17–19 July 2019; IEEE: New York, NY, USA, 2019; pp. 1866–1870. [CrossRef]
20. Arora, K.; Kumar, A.; Kamboj, V.K.; Prashar, D.; Shrestha, B.; Joshi, G.P. Impact of Renewable Energy Sources into Multi Area Multi-Source Load Frequency Control of Interrelated Power System. *Mathematics* **2021**, *9*, 186. [CrossRef]
21. Gupta, D.K.; Naresh, R.; Jha, A.V. Automatic Generation Control for Hybrid Hydro-Thermal System using Soft Computing Techniques. In *2018 5th IEEE Uttar Pradesh Section International Conference on Electrical, Electronics and Computer Engineering (UPCON 2018), Proceedings of the 5th IEEE Uttar Pradesh Section International Conference on Electrical, Electronics and Computer Engineering (UPCON), Gorakhpur, India, 2–4 September 2018*; IEEE: New York, NY, USA, 2018; pp. 1–6. [CrossRef]
22. Prakash, S.; Sinha, S.K. Application of artificial intelligent and load frequency control of interconnected power system. *Int. J. Eng. Sci. Technol.* **2011**, *3*, 377–384. [CrossRef]
23. Gupta, D.K.; Jha, A.V.; Appasani, B.; Srinivasulu, A.; Bizon, N.; Thounthong, P. Load Frequency Control Using Hybrid Intelligent Optimization Technique for Multi-Source Power Systems. *Energies* **2021**, *14*, 1581. [CrossRef]
24. Koley, I.; Bhowmik, P.S.; Datta, A. Load frequency control in a hybrid thermal-wind-photovoltaic power generation system. In Proceedings of the 4th International Conference on Power, Control & Embedded Systems (ICPCES), Allahabad, India, 9–11 March 2017; IEEE: New York, NY, USA, 2017; pp. 1–5. [CrossRef]
25. Khadanga, R.K.; Kumar, A. Hybrid adaptive 'gbest'-guided gravitational search and pattern search algorithm for automatic generation control of multi-area power system. *IET Gener. Transm. Distrib.* **2016**, *11*, 3257–3267. [CrossRef]
26. Khezri, R.; Oshnoei, A.; Oshnoei, S.; Bevrani, H.; Muyeen, S.M. An intelligent coordinator design for GCSC and AGC in a two-area hybrid power system. *Appl. Soft Comput.* **2019**, *76*, 491–504. [CrossRef]
27. Gondaliya, S.; Arora, K. Automatic generation control of multi area power plants with the help of advanced controller. *Int. J. Eng. Res. Technol.* **2015**, *4*, 470–474. [CrossRef]
28. Gheisarnejad, M.; Khooban, M.H. Secondary load frequency control for multi-microgrids: HiL real-time simulation. *Soft Comput.* **2019**, *23*, 5785–5798. [CrossRef]
29. Çelik, E. Design of new fractional order PI–fractional order PD cascade controller through dragonfly search algorithm for advanced load frequency control of power systems. *Soft Comput.* **2020**, *25*, 1–25. [CrossRef]
30. Ionescu, L.-M.; Bizon, N.; Mazare, A.-G.; Belu, N. Reducing the Cost of Electricity by Optimizing Real-Time Consumer Planning Using a New Genetic Algorithm-Based Strategy. *Mathematics* **2020**, *8*, 1144. [CrossRef]
31. Haes Alhelou, H.; Hamedani Golshan, M.E.; Hajiakbari Fini, M. Wind driven optimization algorithm application to load frequency control in interconnected power systems considering GRC and GDB nonlinearities. *Electr. Power Compon. Syst.* **2018**, *46*, 1223–1238. [CrossRef]
32. Attia, A.E.-F.; Mohammed, A.E.-H. Efficient frequency controllers for autonomous two-area hybrid microgrid system using social-spider optimizer. *IET Gener. Transm. Distrib.* **2017**, *11*, 637–648.
33. Nikmanesh, E.; Hariri, O.; Shams, H.; Fasihozaman, M. Pareto design of Load Frequency Control for interconnected power systems based on multi-objective uniform diversity genetic algorithm (MUGA). *Int. J. Electr. Power Energy Syst.* **2016**, *80*, 333–346. [CrossRef]
34. Sahu, P.C.; Mishra, S.; Prusty, R.C.; Panda, S. Improved-salp swarm optimized type-II fuzzy controller in load frequency control of multi area islanded AC microgrid. *Sustain. Energy Grids Netw.* **2018**, *16*, 380–392. [CrossRef]
35. Yang, X.S. *Nature-Inspired Metaheuristic Algorithms*, 2nd ed.; Luniver Press: Beckington, UK, 2008.
36. Rashedi, E.; Nezamabadi-Pour, H.; Saryazdi, S. GSA: A gravitational search algorithm. *Inf. Sci.* **2009**, *179*, 2232–2248. [CrossRef]
37. Kennedy, J.; Eberhart, R. Particle swarm optimization. In Proceedings of the 1995 IEEE International Conference on Neural Networks, Perth, WA, Australia, 27 November–1 December 1995; IEEE: New York, NY, USA; Volume 4, pp. 1942–1948. [CrossRef]
38. Zhu, H.; Wang, Y.; Ma, Z.; Li, X. A Comparative Study of Swarm Intelligence Algorithms for UCAV Path-Planning Problems. *Mathematics* **2021**, *9*, 171. [CrossRef]

Article

Hybrid LQR-PI Control for Microgrids under Unbalanced Linear and Nonlinear Loads

Gerardo Humberto Valencia-Rivera, Luis Ramon Merchan-Villalba, Guillermo Tapia-Tinoco, Jose Merced Lozano-Garcia, Mario Alberto Ibarra-Manzano and Juan Gabriel Avina-Cervantes *,†

Telematics (CA), Engineering Division (DICIS), Campus Irapuato-Salamanca, University of Guanajuato, Carretera Salamanca-Valle de Santiago km 3.5 + 1.8 km, Comunidad de Palo Blanco, Salamanca 36885, Mexico; gh.valenciarivera@ugto.mx (G.H.V.-R.); lr.merchanvillalba@ugto.mx (L.R.M.-V.); g.tapiatinoco@ugto.mx (G.T.-T.); jm.lozano@ugto.mx (J.M.L.-G.); ibarram@ugto.mx (M.A.I.-M.)
* Correspondence: avina@ugto.mx; Tel.: +52-4646-4799-40 (ext. 2400)
† Author thanks the Universidad de Guanajuato by the financial support of the APC.

Received: 11 June 2020; Accepted: 30 June 2020; Published: 4 July 2020

Abstract: A hybrid Linear Quadratic Regulator (LQR) and Proportional-Integral (PI) control for a MicroGrid (MG) under unbalanced linear and nonlinear loads was presented and evaluated in this paper. The designed control strategy incorporates the microgrid behavior, low-cost LQR, and error reduction in the stationary state by the PI control, to reduce the overall energetic cost of the classical PI control applied to MGs. A Genetic Algorithm (GA) calculates the parameters of LQR with high-accuracy fitness function to obtain the optimal controller parameters as settling time and overshoot. The gain values of the classical PI controller were determined through the improved LQR values and geometrical root locus. When MG operates in the grid-tied mode under unbalanced conditions, the controller performance of the Current Source Inverter (CSI) of the MG is considerably affected. Consequently, the CSI operates in a negative-sequence mode to compensate for unbalanced current at the Point of Common Coupling (PCC) between the MG and the utility grid. The study cases involved the reduction of the negative-sequence percentage in the current at the PCC, mitigation of harmonics in the current signal injected by the MG, and close related power quality issues. All these cases have been analyzed by implementing an MG connected at the PCC of a low-voltage distribution network. A numerical model of the MG in Matlab/Simulink was implemented to verify the performance of the designed LQR-PI control to mitigate or overcome the power quality concerns. The extensive simulations have permitted verifying the robustness and effectiveness of the proposed strategy.

Keywords: microgrid; LQR-PI control; grid-tied mode; current imbalance; power quality; genetic algorithms

1. Introduction

Nowadays, fossil fuels are the primary source of energy worldwide, but the extensive use of this natural resource has caused an increase in the average temperature of the earth. Environmental organizations have the aim of gradually decoupling the use of fossil fuels from 70% to 20% in 2050 [1]. Advances in the technology directed on the energy production area, environmental sustainability, and the appearance of small generation systems have opened new opportunities to research in Distributed Energy Resources (DERs). The DERs have raised an alternative solution to efficiently face the actual energy demand, centered on the reliability and energy quality [2]. Many consumers, such as buildings, factories, and residential neighborhoods, are planning to place a Microgrid (MG) considering the cost reductions in the technology associated with the DERs and storage systems, contributing simultaneously to reach a better energy quality [3]. MG is defined as an interconnected load group and DERs with boundaries clearly defined that act as a controllable entity concerning the grid [4]. MG is composed of essential components as the loads, DERs, Static Disconnect Switch (SDS), protections, digital communications, control, and automation systems [5].

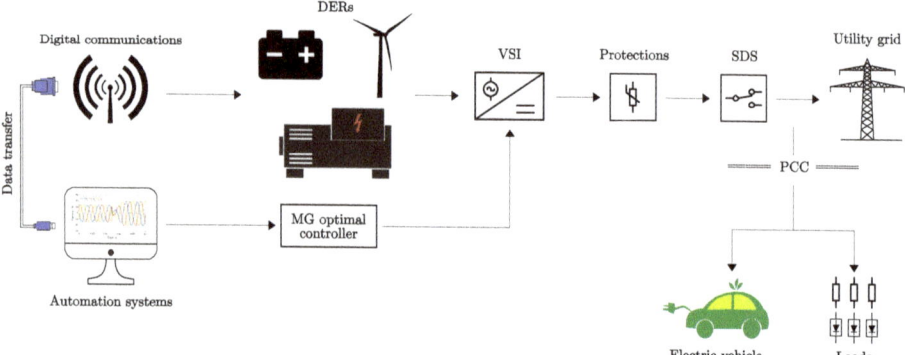

Figure 1. Basic operative elements composing a Microgrid (MG). A network of different power sources is distributed, administrated, and controlled at the same station to provide stable energy to any connected loads.

Figure 1 shows the essential components that integrate an MG. The MGs are defined as the small local distribution systems which promote the use of DERs. DERs are small energy units of generation and storage. These can come from renewable energy sources such as wind turbines, non-renewable energy sources such as diesel generators, or energy storage systems such as batteries. It has linear loads, non-linear loads, or dynamic loads such as electric vehicles. The MG operates in grid-tied mode or stand-alone mode. To achieve this, it has protections and control systems that allow connection and disconnection to the utility grid safely and without affecting its stability. Energy management is established through measurement and control systems. These systems allow the MG to operate with a decentralized or centralized control scheme, which has a direct impact on the MG's power quality and reliability. While MG operates in grid-tied mode, unbalanced loads or grid failures are originated by unbalanced currents and voltage at the Point of Common Coupling (PCC) affecting the energy quality index [6,7]. In our study, the SDS located at the PCC is assumed as closed. Hence, the MG is operating in grid-tied mode.

Likewise, MG voltage depends on the connections associated with the electric grid. Thus, unbalanced loads connected at the PCC provoke that the Current Source Inverter (CSI) may supply three-phase unbalanced currents with components of a negative-sequence directly towards the MG. This effect degrades the CSI performance and the energy quality index due to fluctuations in the inverter's current and power signals. In this way, negative-sequence components are generated in the measured current at the PCC, affecting the current balance percentage, ruled in the IEEE 1159-1995 standard [8]. Moreover, nonlinear loads introduce a band of harmonics into the MG, producing distorted waveforms at the current signal in the utility grid and CSI output. MG must attenuate such harmonics to avoid include perturbations on the system dynamics, as well as in the sensitive loads connected simultaneously at the PCC.

On the other hand, control approaches on MGs are focused on separating current and voltage signals from the negative and positive sequences in the CSI to address these electrical grid issues [7,9,10]. Obtained such a decomposition, a control law is designed for each of these components to increase the CSI control performance [11]. Several studies have proposed to improve the energy quality index by reducing the negative-sequence component of the control signals [12]. Dasgupta et al. worked on a current controller based on the Lyapunov function to control the active and reactive power flow to a three-phase MG system [13]. The Fault Current Limiter (FLC) is another methodology that suppresses fault currents from the utility grid, assuring the good operation of the MG. This tool seems to solve the over-current relay coordination issue in the MG [14]. Surrender et al. proposed a collaborative optimization framework, including the Differential Evolution (DE) and Harmony Search (HS) methods, to obtain the efficient energy resources distribution in the MG. The MG was composed of several renewable energy sources where the optimization processes involved energetic, economic, and environmental factors [15]. Lotfollahzade et al. used an LQR-PID controller optimized by PSO (Particle Swarm Optimization) to compute the proportional, integral, and derivative parameters to obtain an optimal load sharing of an electrical grid [16]. Savage et al. proposed to design hierarchical control techniques to enhance the energy quality in the connected bus at sensitive loads [17]. A primary control took DERs administration, and secondary control drove the voltage levels at the load bus by sending the control signals to the primary block for compensating the unbalances. Shi et al. tested a control strategy under three-phase unbalanced [10]. They proposed a unified three-phase voltage correction through negative-sequence compensation. In practice, dynamic systems need to use robust, versatile, and tunable control strategies, in that sense, the hybrid control is becoming a reliable alternative in automatic systems, especially in MG. Lindiya et al. adopt a conventional multi-variable PID and LQR algorithm in DC-to-DC converters for reducing cross-regulation [18]. For realistic simulation of MG performance, Momoh and Reddy present a platform to simulate basic MG using Hardware-In-Loop (HIL) under different environments [19]. This interesting tool allows testing divers controllers and measures the performance of the MG on delivering the power supply requirements under different scenarios.

In other related areas, Sen et al. tuned a hybrid LQR-PID controller to regulate and monitor the locomotion of a quadruped robot using the Grey-Wolf Optimizer (GWO) [20]. Besides, Nagarkar et al. proposed a PID and LQR control to optimize a nonlinear quarter car suspension system [21]. Ibrahim et al. integrate the dynamic behavior of an LQR-based PID controller applied to a helicopter control with three degrees of freedom [22]. In this article, an efficient control technique based on PI-LQR driven by a Genetic Algorithm (GA) and a high-accuracy fitness function is proposed to regulate the energy provided by an MR. A genetic algorithm warranties that LQR is behaving according to design requirements as settling time and overshoot of the transfer function modeling the MG. To reach the appropriate current and power values to be supplied by the MG, a PI control strategy is included in the optimized transfer function. In such a sense, the required characteristics are a low-cost operation of the system during LQR operation, and the error reduction in the steady-state while using the PI control.

Consequently, an LQR-PI control technique is proposed to conduct the control action of the CSI, so the operation can operate in the negative-sequence mode. Besides, this action helps to mitigate the negative-sequence components of the current signal injected by the MG, phenomena caused by unbalanced linear and nonlinear loads. On the other hand, the proposed control methodology reduces the harmonic distortion generated by nonlinear loads, which is computed by the Total Harmonic Distortion (THD). Moreover, such schemes guarantee the current equilibrium at the PCC, and an acceptable energy quality index according to the norm ruling the MGs. The significant contributions of this paper are to propose the GA with an effective and accurate fitness function that helps to calculate the controller design parameters and to hybridize the properties of PI and LQR controllers applied to the MG to provide the demands of energy. This methodology looks for improving quality energy issues considering the energetic cost of the system during its operation and compares the classical methods used to design PI controllers applied to MG.

2. Microgrid Structure Analysis

The MG configuration used in this work is shown in Figure 2.

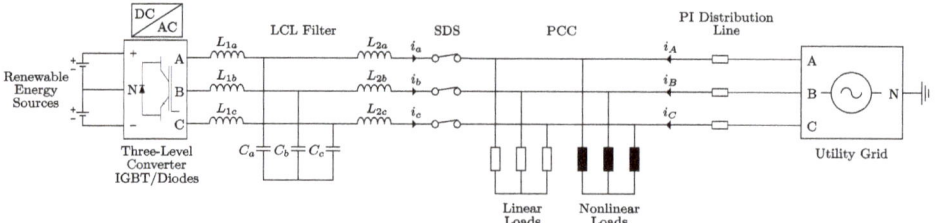

Figure 2. The Microgrid structure used in this work. CSI and LCL filter compound the operational controlled current source.

The utility grid is represented by a three-phase voltage source and its coupling impedance. The linear and nonlinear loads are connected at the common coupling point. In particular, linear loads have a three-phase configuration and operate with a unity power factor. Contrarily, nonlinear loads are modeled as controlled, single-phase current sources. This scheme allows controlling the harmonic content in each of the phases of the MG. Besides, the MG is connected to the PCC employing an SDS switch that is kept closed to operate in a grid-tied mode. Electrically, the MG is represented by an equivalent circuit consisting of a Renewable Energy Source (RES), a converter and a low-pass filter. An ideal voltage source serves to represent the RES because of its energy is directly taken from clean photovoltaic cells. Moreover, there are no associated mechanical components of inertia as in microturbines, wind turbines, among others [23]. The converter is in charge of performing the power transfer between the DC bus and the AC network. An LCL low-pass filter is required to attenuate the currents' high-frequency components provided by the MG (i_a, i_b, i_c).

LCL filters have great advantages considering aspects as a reduced cost and size because the estimated values of the inductors are smaller than L and LC filters topology [24]. Besides, this efficient filter shows better performance on filtering high-frequency harmonics generated by the switching of PWM converters, including grid-connected converters controlled by the current sources [25]. In our study, the SDS that links the MG with the utility grid is considered closed because of MG is operated in grid-tied mode. This analysis considers that the MG provides active power to supply energy into the loads connected to the PCC. In case of failure or instability in the utility grid, the MG works as a support system injecting or consuming reactive power to balance the voltage at PCC. For such a purpose, the methodology seeks to control the currents by a set of inductors located just aside the PCC, considering that these elements deliver the power toward the utility grid directly.

In practice, the Park transformation is used for obtaining a simplified state-space model of the MG to dispose of a decoupled system representing the MG behavior [26]. First, the passive elements (inductors and capacitors) are considered to have the same value for each of the phases. Therefore, the three-phase representation of passive elements is given by

$$L_{\alpha k} = l_k I_3, \quad C_{\alpha k} = c_k I_3, \qquad (1)$$

where I_3 is the identity matrix of order 3, l_k and c_k are the passive components' scalar values in each phase. Second, a framework mapping from the three-phase abc to the dq domain is applied to the passive elements. Such a mapping is made by the well-known current-tension relations for inductors and capacitors as follows,

$$V_{abc}^{L_{\alpha k}} = L_{\alpha k} \frac{d}{dt} I_{abc}^{L_{\alpha k}}, \qquad (2)$$

$$I_{abc}^{C_{\alpha k}} = C_{\alpha k} \frac{d}{dt} V_{abc}^{C_{\alpha k}}. \qquad (3)$$

Third, the Park transform $T_p(\theta)$ is then applied to Equations (2) and (3) to obtain an orthogonal rotating reference frame (dq).

$$V_{dq}^{L_{ak}} = T_p(\theta) L_{ak} \frac{d}{dt}(T_p(\theta)^{-1} I_{dq}^{L_{ak}}), \qquad (4)$$

$$I_{dq}^{C_{ak}} = T_p(\theta) C_{ak} \frac{d}{dt}(T_p(\theta)^{-1} V_{dq}^{C_{ak}}), \qquad (5)$$

where θ represents the axes turning-speed for a determined phase, and in the mapping framework, θ is defined by ωt. Thus, $T_p(\theta)$ operator is well-known as the Park transformation. Finally, Equations (4) and (5) are solved using the chain rule with $\theta = \omega t$ to achieve the dq model for each passive element,

$$v_d^{l_k} = -\omega\, l_k v_q^{l_k} + l_k \frac{d}{dt} I_d^{l_k}, \qquad (6)$$

$$v_q^{l_k} = \omega\, l_k v_d^{l_k} + l_k \frac{d}{dt} I_q^{l_k}, \qquad (7)$$

$$i_d^{c_k} = -\omega\, c_k v_q^{c_k} + c_k \frac{d}{dt} V_d^{c_k}, \qquad (8)$$

$$i_q^{c_k} = \omega\, c_k v_d^{c_k} + c_k \frac{d}{dt} V_q^{c_k}. \qquad (9)$$

These foundations are mathematically represented by the state-space model given in Equations (10) and (11)

$$\frac{d}{dt}\begin{bmatrix} I_d^{L_1} \\ I_d^{L_2} \\ v_d^C \end{bmatrix} = \overbrace{\begin{bmatrix} 0 & 0 & \frac{-1}{L_1} \\ 0 & 0 & \frac{1}{L_2} \\ \frac{1}{C} & \frac{-1}{C} & 0 \end{bmatrix}}^{A} \begin{bmatrix} I_d^{L_1} \\ I_d^{L_2} \\ v_d^C \end{bmatrix} + \overbrace{\begin{bmatrix} \frac{1}{L_1} \\ 0 \\ 0 \end{bmatrix}}^{B} u_d + g(x), \quad y_1 = \overbrace{\begin{bmatrix} 0 & 1 & 0 \end{bmatrix}}^{C} \begin{bmatrix} I_d^{L_1} \\ I_d^{L_2} \\ v_d^C \end{bmatrix}, \qquad (10)$$

$$\frac{d}{dt}\begin{bmatrix} I_q^{L_1} \\ I_q^{L_2} \\ v_q^C \end{bmatrix} = \overbrace{\begin{bmatrix} 0 & 0 & \frac{-1}{L_1} \\ 0 & 0 & \frac{1}{L_2} \\ \frac{1}{C} & \frac{-1}{C} & 0 \end{bmatrix}}^{A} \begin{bmatrix} I_q^{L_1} \\ I_q^{L_2} \\ v_q^C \end{bmatrix} + \overbrace{\begin{bmatrix} \frac{1}{L_1} \\ 0 \\ 0 \end{bmatrix}}^{B} u_q + h(x), \quad y_2 = \overbrace{\begin{bmatrix} 0 & 1 & 0 \end{bmatrix}}^{C} \begin{bmatrix} I_q^{L_1} \\ I_q^{L_2} \\ v_q^C \end{bmatrix}, \qquad (11)$$

where u_d and u_q represent the dq components of the input voltage source. The functions $g(x)$ and $h(x)$ are used to include the system response face to perturbations, which are modeled by Equations (12) and (13),

$$g(x) = \begin{bmatrix} \omega & 0 & 0 \\ 0 & \omega & 0 \\ 0 & 0 & \omega \end{bmatrix} \begin{bmatrix} I_q^{L_1} \\ I_q^{L_2} \\ v_q^C \end{bmatrix} - \begin{bmatrix} 0 \\ \frac{1}{L_2} \\ 0 \end{bmatrix} Vs_d, \qquad (12)$$

$$h(x) = -\begin{bmatrix} \omega & 0 & 0 \\ 0 & \omega & 0 \\ 0 & 0 & \omega \end{bmatrix} \begin{bmatrix} I_d^{L_1} \\ I_d^{L_2} \\ v_d^C \end{bmatrix} - \begin{bmatrix} 0 \\ \frac{1}{L_2} \\ 0 \end{bmatrix} Vs_q, \qquad (13)$$

where the active variables Vs_d and Vs_q represent the dq components of the electrical grid signals. Such perturbations affect the system behavior meaningfully and whose effects must be reduced or eliminated. Indeed, our model is based on the state-space using the Park transformation to simplify the MG analysis so that it is possible to obtain two decoupled systems in the dq framework [27]. The d component controls the active power flow, whereas the q component regulates the reactive power flow, respectively.

All MG interactions are formally described in the state-space system represented by Equations (10)–(13). In such a representation, $\frac{d}{dt} I_d^{L_1}$, $\frac{d}{dt} I_q^{L_1}$, $\frac{d}{dt} I_d^{L_2}$, $\frac{d}{dt} I_q^{L_2}$, $\frac{d}{dt} v_d^C$, and $\frac{d}{dt} v_q^C$ are the voltages in the inductors L_{1k} and L_{2k}, as well as the current in the capacitor C_k into the dq framework, respectively. Similarly, $I_d^{L_1}$, $I_q^{L_1}$, $I_d^{L_2}$, $I_q^{L_2}$ are the inductor currents and v_d^C, v_q^C are the capacitor voltages in the dq reference frame. After estimating the system response, the results allow determining that the modeled MG is critically damped because the transfer function that describes the dynamic system has two complex-conjugated poles on the imaginary axis. In consequence, the method of states-feedback is applied to dispose of a more suitable allocation of the system poles, which will improve the response to the step input. New control law representation and the derived dynamics system are described by Equations (14) and (15)

$$u(t) = r(t) - K \cdot x(t) \tag{14}$$
$$\dot{x}(t) = (A - BK) \cdot x(t) + B \cdot r(t) \tag{15}$$

where $u(t)$ is the system input, $r(t)$ is system reference, and K represents the state feedback gains. The state vector can be calculated by the dominant pole placement technique, which consists of matching the system characteristic polynomial with a theoretical polynomial containing all needed control parameters (i.e., overshoot, rise time, settling time, among others). However, this technique presents a problem tied to the arbitrary pole assignment, which directly affects the control effort with inconvenient or impractical values in the gain matrix K. Therefore, the LQR algorithm was implemented and evaluated to address previous shortcomings in the analysis of the MG. The K matrix is then optimally computed to find the best poles placement of the system [28]. The full state feedback control allows a suitable selection of the components describing the dynamics of the systems [29], in this case focused on controlling the MG where the design is driven by GA. The proposed diagram of full state feedback is shown in Figure 3.

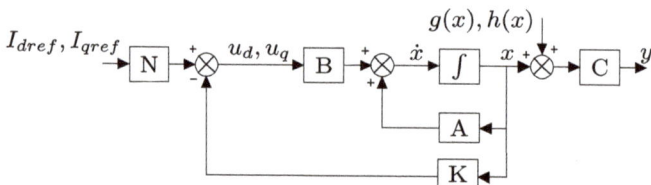

Figure 3. Full state feedback LQR-controller for the decoupled MG system.

In the diagram, matrices $\{A, B, C\}$ are the original system matrices in the state-space representation given by Equations (10) and (11), K is the state feedback matrix, and the precompensation gain N is a scaling factor that multiplies the system reference to achieve the output desired value $y = [y_1, y_2]^T$. Indeed, the N scaling factor is calculated by

$$N = -\frac{1}{C[SI - (A - BK)]^{-1}B}. \tag{16}$$

Additionally, the cost function required by the LQR algorithm to obtain the optimal control parameters is defined as follows,

$$J = \int_0^\infty \left(X^T Q X + u^T R u \right) dt, \tag{17}$$

where $Q \geq 0$, $R > 0$ are positive semi-definite matrices. Q is the state matrix penalization, and R expresses the actuator effort. The cost function J is subject to the next system constraint,

$$\dot{x}(t) = Ax(t) + Bu(t), \tag{18}$$

where $x(t)$ and the $u(t)$ are vectors $\in \mathbb{R}^n$. Thus, the original system input and state vector are theoretically calculated by

$$u(t) = -Kx(t), \qquad (19)$$
$$K = R^{-1}B^T S. \qquad (20)$$

The LQR optimization problem requires firstly to solve the algebraic Riccati equation,

$$A^T S + SA - SBR^{-1}B^T S + Q = 0. \qquad (21)$$

In some cases, the matrices Q and R could be assigned arbitrarily; unfortunately, the control design over the required system response could be compromised or impractical. Hopefully, to address those disadvantages, the Riccati equation allows assigning state penalization values that belong to matrix Q and R to modify the speed response of the LQR controller. Therefore, it is possible to compute the values of the matrix K and the system input $u(t)$ by solving S of Equation (21). However, when faster and more efficient system dynamics are required, the random parameters assignment in the matrix Q and R can lead to a second problem. In that case, the estimated values of matrices Q and R grow out of the allowed range to fulfill the control design requirements (e.g., settling time and overshoot) of the transfer function describing the MG. This issue is efficiently addressed by using an optimization method to compute the matrices Q and R. Hence, a proper cost function, involving the overall design requirements of the transfer function jointly with the parameters of the LQR algorithm, is then minimized. For overcoming the issues of arbitrary pole placement, a genetic optimization algorithm was also proposed in this study, since this optimization technique can operate in parallel to find multiobjective solutions [30].

3. Genetic Algorithms

The genetic algorithms (GA) are metaheuristic algorithms belonging to the family of evolutionary algorithms (EA). All population-based algorithms work with a set of candidate solutions called phenotypes or population and a set of chromosomes representing the model's variables. Each candidate solution in the population is coded in a chain of chromosomes or genotypes. Since these algorithms are inspired in the natural selection, the chromosomes evolve throughout each iteration (i.e., generation) to produce new individuals (i.e., solutions). In each generation, the best chromosomes or individuals are selected by evaluating a suitable fitness function. The next generations are generated by applying a fundamental set of genetic operators until achieving the optimal result. The traditional chain of genetic operators comprises the initialization, crossover, mutation, and selection, as is shown in Figure 4.

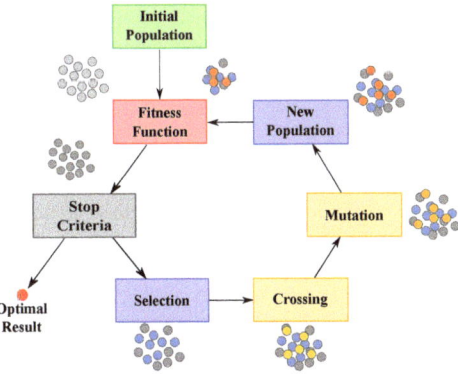

Figure 4. Flowchart of basic GA architecture used for tuning the LQR control.

A new generation is processed when the requirements of a stop condition are not fulfilled [31]. A GA implementation could present many variations, which depends on the particular way of how each genetic operator is applied. The methodology driven by the GA to solve the energy cost issues based on the LQR controller is described by Algorithm 1.

Algorithm 1: Genetic Algorithm.
---
1. Generate a random initial population.
2. Evaluation of each individual in the fitness function.
3. Verify the Stop criteria to detect the optimal solution.
4. Selection or Elitism of the best individuals that will be crossed into the crossover function.
5. Crossing the selected individuals (interbreeding) to generate new progenies or solutions.
6. Mutation of random chromosomes in each individual to diversify the searching space.
7. Applied the genetic operators, the best individuals must repopulate the next generation.
8. The optimal solution is found fulfilling the stopping criteria, otherwise, continue Step 4

The GA starts with a random assignation of the chromosomes in each individual representing the variables of the proposed models. In this manner, each individual is a viable solution that must be further evaluated, and the overall initial individuals will form the initial population. In our implementation, the chromosomes represent the values of each element into the matrices Q and R as described next.

3.1. Chromosome Configuration

The proposed control modeled in real continuous variables led to select a floating-point representation of the chromosomes. In this case, bit string encoding mechanisms can be replaced for floating-point operations in the mutation and crossover functions. The chromosomes represent each variable to identify in the proposed control. In the proposed implementation, four chromosomes were used, three elements to describe the Q matrix, and one element for the variable R that internally are linked to the gains and poles in the system. It is noteworthy that such elements describe the control behavior while interacts with the MG. This controlling behavior is measured at the output by well-known design parameters (e.g., overshoot, settling time, undershoot, stable error, rising time, or natural frequency). In the numerical simulations, settling time and overshoot factor were used in the fitness function. Hence, the chromosomes were coded in floating point, as well as the uniform distributions $U \in [0,1]$ were used to control the action of the morphological operators acting over chromosomes randomly selected. The initial population was fixed to N = 100.

3.2. Mutation and Crossover

Before mutation, the best suitable solutions (elite population) are selected and preserved. The elite population comprises the 5% of the total population, saving the fittest genetic material. The mutation operator using a floating-point representation is defined by

$$x_m = (1-\alpha)*x_j + \alpha*((X_{max} - X_{min})*r_u + X_{min}), \qquad (22)$$

where $\alpha = 0.5$, x_j is the chromosome randomly selected from the entire database X, which is mutated by a bounded interval $[X_{min}, X_{max}]$, and r_u is a random number with uniform distribution $U \in [0,1]$. Hence, the algorithm was tuned to generate 20% of new chromosomes x_m. The mutation should produce new individuals with a probability of about 20% in our implementation. The crossover function used an unbalanced arithmetic operation defined mathematically as

$$x_c = (1 - r_u)*x_i + r_u*x_j, \qquad (23)$$

where r_u is a random number uniformly distributed $U \in [0,1]$, x_i is randomly selected between the best 25% of individuals and x_j is randomly selected from last 75% in the prevailing population. Considering the elitism preserved 5% of the population, and the mutation provided 20%, the crossing probability could be about 75%.

The crossover and mutation probabilities are eventually modified depending on the individuals' fitness to achieve a good trade-off between exploration and exploitation. In fact, the mutation and crossover probabilities should slightly increase when the population is trapped into local optima, and such probabilities decrease when the population is too dispersed. In the order hand, there is no general consensus on how to measure and balance the exploration and exploitation efficiently in genetic algorithms [32]. In our implementation, this trade-off involves direct parameters of the algorithm, the number of generation and the population size, combined with the population diversity controlled by the crossover, mutation, and selection functions. Mutation and crossing explore a new solution with a conjoint probability of $P = 0.2 \times 0.75 = 0.15$, and the exploitation uses basically the elitism function $P = 0.05$, which express certain skew to the exploration in our implementation.

Finally, the GA selects the optimal solution for each chromosome (i.e., optimal matrices Q and R) by using the appropriate cost function and the LQR algorithm, to achieve the desired behavior of the system.

3.3. Fitness Function

The fitness function is related to the error between the actual solution and the optimal solution, represented by the controlled MG behavior (i.e., overshoot and settling time) associated internally to the best chromosomes in the Q y R matrices. In control terms, the fitness function is related to signal error between the output and the set-point responses. However, computationally some parametric error could be used as Mean Absolute Error (MAE), Mean Squared Error (MSE), Mean Absolute Scaled Error (MASE), among others. The commonly used MSE merit function is highly affected by outliers, which produces undesirable results in some applications, such is the case of the proposed design. Therefore, the MAE function was used to cope with this disadvantage obtaining satisfactory results in the implemented GA algorithm. The MAE based on the L_1-norm is mathematically defined as

$$\text{MAE} = \frac{1}{N} \sum_{k=1}^{N} |y_k - \hat{y}_k|, \tag{24}$$

where y_k is the actual value and $\hat{y}_k$ is the estimated behavior vector response of the proposed controlled model. This modest fitness function is really dependent on highly nonlinear parameters of the control interconnecting the grid, loads, and the power sources, which can be dynamically adjusted with the proposed GA optimization algorithm. Including the output control parameters of interest, the fitness function becomes,

$$J = |W^\mathsf{T} * (X_{ref} - X_k)|, \tag{25}$$

where the vector $X_{ref} \in [T_{st} \ M_{os}]^\mathsf{T}$ contains the desired parameters as the settling time (T_{st}) and the overshoot M_{os}, and W is a regularization (or scale) matrix to adjust the contribution and units of each kind of parameters. Besides, X_k is the system's response parameters under the control action, while the chromosomes representing the control operational matrices Q and R. The data used by the fitness function is computing by solving the dynamical model used to represent the control and the MG for each individual in the population. The weight matrix is defined as

$$W = \begin{pmatrix} w_1/T_{st}^o & 0 \\ 0 & w_2/M_{os}^o \end{pmatrix} \tag{26}$$

where $w_1 + w_2 = 1.0$ are the constants allowing to prioritize some of these output features, T_{st}^o, and M_{os}^o are the maximum or actual values of the tested features to avoid dimensionality issues. The chosen fitness function is highly stable for the proposed model. It is worthy to notice that optimal Q and R matrices are highly dependent on the chosen fitness function, which could also be interpreted as an error function.

4. LQR-PI Control Strategy for MG in Grid-Tied Mode

Once computed the matrix K driven by GA, the PI-LQR controller is designed to regulate the power flow from MG toward the utility grid. A robust and performing controller is obtained by combining the optimal properties of the LQR algorithm and a classical PI controller. Such a strategy allows achieving a bounded control action. Figure 5 illustrates the control action for the MG output voltages either to the hybrid PI-LQR controller, PI controller driven by GA, or PI controller tuned by the poles placement method. Control action values are into established parameters by MG voltage.

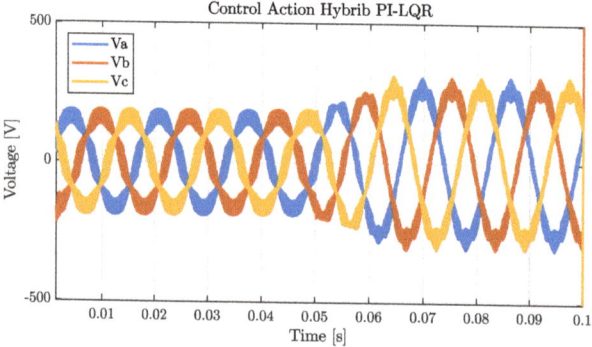

Figure 5. Illustrative voltage control action of the hybrid PI-LQR controller.

It is noticeable that the control action helps the system to recover and operate under established values for MG voltage (see Table 2). This signal controls the three-phases of the CSI switching-pattern to deliver the required power at the electrical grid. By the way, the controller's integral action also reduces the sensitivity face to perturbations, canceling the steady-state error for unit step inputs [33]. The PI control action is mathematically described by

$$G_c(s) = K_p + K_i \frac{1}{s} = K_p \frac{s + K_i/K_p}{s}, \qquad (27)$$

where Kp and Ki are the proportional and integral gains, respectively. All these gains were calculated by using the rlocus method, considering the suitable phase and magnitude conditions into the controller design. Figure 6 shows the optimal control scheme used to regulate the power flow at the PCC.

The instantaneous PQ power theory is used to adjust the input references of the control system in the model shown in Figure 6. Such theory is based on a set of instantaneous powers defined in a temporal framework that imposes no constraints on the voltage or current waveforms. Thus, the same approach can be applied to three-phase systems with or without neutral wire [26]. In the proposed solution, the Park transform should be applied to convert the state-system from a three-phase framework (sinusoidal variables) to the $dq0$ orthogonal reference system (constant values). Park transform allows synthesizing and decoupling the variables and associated states forming the MG. In this sense, the input references' values are obtained using the PQ theory and Park transformation by

$$P = \frac{3}{2} \times (V_d \cdot I_d + V_q \cdot I_q), \qquad (28)$$

$$Q = \frac{3}{2} \times (V_q \cdot I_d - V_d \cdot I_q). \qquad (29)$$

Thus, the control signals are transformed into the three-phase frameworks for regulating the PWM signals into the three-level converters. Since PWM signals control the switching sequence of CSI, the model produces the desired power that the MG must inject toward the utility grid.

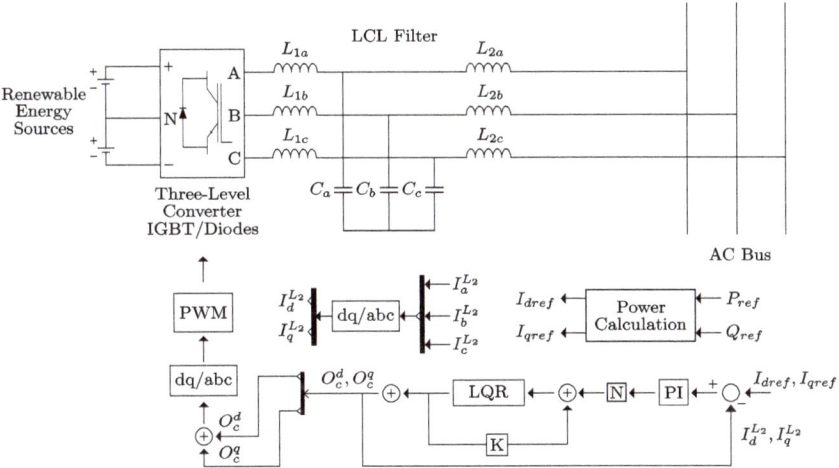

Figure 6. Optimal control scheme of a three-level converter in compensation mode.

5. Energy Quality Index Applied to the Current Signal at the PCC

Low voltage systems include linear and nonlinear single-phase, two-phase, and three-phase loads. These loads cause the system to become unbalanced, and consequently, the current and voltage waveforms of three-phase sources are not identical in magnitude and phase. Unfortunately, some electrical machines depend on a balanced power supply to avoid affectation on their functionality and performance [34]. When the MG operates in grid-tied mode, the failure generally comes from the utility grid and unbalanced loads resulting in three-phase unbalanced voltage at the PCC. However, the unbalanced factor could be quantified in a three-phase unbalanced system by using the symmetrical sequences approach. Such a method decomposes the unbalanced system into three sequence components, called the positive, negative, and homopolar sequences. This unbalanced percentage is calculated by the ratio of the negative and positive-sequences, which is formally known as voltage unbalance factor (*VUF*) and defined by Equation (30),

$$\text{VUF} = \frac{|V_{sec(-)}|}{|V_{sec(+)}|} \times 100 \ [\%], \tag{30}$$

where $V_{sec(-)}$ and $V_{sec(+)}$ are the negative and positive sequence components, respectively; both voltages are measured at the PCC. Similarly, the definition of *VUF* could be adopted to measure the current unbalance factor *CUF*, which is given by

$$\text{CUF} = \frac{|I_{sec(-)}|}{|I_{sec(+)}|} \times 100 \ [\%], \tag{31}$$

where $I_{sec(-)}$ and $I_{sec(+)}$ are the negative and positive sequence components, respectively; both currents are measured at the PCC. The currents I_{sec} are measured at the inductor bank L_{2k} of the MG through the sequence analyzer, where $L_{2k} = \{L_{2a}, L_{2b}, L_{2c}\}$. Our approach uses *CUF* because the designed control strategy is based on a current control loop. Thus, the CUF index is computed with the current signals provided by the MG at the PCC. Otherwise, linear, and nonlinear unbalanced loads may produce excessive levels of unbalanced current and voltages that tend to appear in the MG. These phenomena' main effects are lower performance, energy losses, and instability of the MG [35]. In the presence of negative-sequence components, the power electronics converters and induction motors cannot work or perform poorly [36]. An unbalanced voltage can produce an unbalanced current from 6 to 10 times the magnitude of the unbalanced voltages. Unbalanced currents can provoke excessive heat in the motor windings, leading to permanent damage [37]. The IEEE 1159-2009 standard establishes that the current unbalance recommended for monitoring power electronics, in steady-state, should be between 1% and 30% [8].

Another phenomenon affecting MG behavior is the increased use of nonlinear power electronic devices and sensitive loads [38]. These devices induce harmonics that may degrade the components either in the utility grid or MG. High-frequency harmonics can be filtered by passive or active filters, but the low-frequency harmonics are difficult to filter without reducing the system's operating frequency. There exist methods to mitigate the low-frequency harmonics, but these techniques are expensive and difficult to implement [39]. The current or voltage distortion is measured through Total Harmonic Distortion (THD), such index is applied to the voltage or current as follows,

$$\text{THD}_V = \frac{1}{V_1}\sqrt{\sum_{k=2}^{N} V_k^2} \times 100 \ [\%], \tag{32}$$

$$\text{THD}_I = \frac{1}{I_1}\sqrt{\sum_{k=2}^{N} I_k^2} \times 100 \ [\%]. \tag{33}$$

THD is defined as the ratio of all root-sum-square of all harmonics (excluding the fundamental) divided by the fundamental [40]. The typical limit used in low-tension is about 5% of THD. In this study, the THD index is applied over the current signal delivered by the MG to evaluate the control performance face to harmonics generated by unbalanced nonlinear loads connected at the PCC.

6. MG Control System Design

Three optimal controllers were designed and analyzed to obtain the best response considering criteria as low-cost energy, mitigation of negative-sequence, and harmonics reduction. Low-cost energy criterion is directly associated either with the input source voltage or with MG voltage. In other words, such value expresses the energy required by the system to execute the control action over the injected current by MG, considering the respective estimations of Q and R matrices, the state feedback matrix K, and the control PI parameters K_p and K_i. For each one of the designed controllers, three study cases were set up to evaluate the performance of the control schemes, which fulfilled the design criteria given initially. Proposed three study cases are based on making MG operate under the next conditions:

1. MG faces balanced nonlinear loads, the harmonics and negative sequence attenuation are analyzed.
2. MG works under unbalanced nonlinear loads, the harmonics and negative sequence mitigations are studied.
3. MG handles the unbalanced linear and nonlinear loads simultaneously; the attenuation of harmonics and negative sequence are both quantified.

The LQR algorithm is used to estimate the gains of feedback states driven by the GA method. Likewise, a PI controller was tuned to reach a robust control technique designed in all study cases. The design parameters of the LQR controller include a settling time, T_s = 0.525 ms, and an overshoot of 5%. For simulation purposes, the MATLAB/Simulink environment was used to evaluate the proposed MG control system.

6.1. Hybrid PI-LQR Control Driven by GA

GA implementation starts tuning the initial values of all chromosomes. Each chromosome represents a potential solution to the weights of matrices Q and R, which eventually adjust the controller's desired behavior. In the proposed model, the fitness function (see Section 3.3) performs the quality measure or associated error to each set of chromosomes representing each potential solution. The LCL filter parameters are given in Table 2. Such parameters are constant for all study cases and represented in the state space system given in Equations (10) and (11). The control parameters obtained from LQR driven by the GA, are estimated by,

$$Q = \begin{bmatrix} 1.7042 & 0 & 0 \\ 0 & 8644.6 & 0 \\ 0 & 0 & 7.4115 \end{bmatrix}, \quad K = \begin{bmatrix} 48.2624 \\ 360.3162 \\ 34.4453 \end{bmatrix}^T.$$

Besides determining Q and K matrices, the pre-compensation gain N = 408.5786 and parameter R = 0.0518 were also estimated. Those values were simulated in the proposed model to achieve an output settling time T_s = 0.52454 ms and an overshoot of 4.4643%. The parameters utilized by GA for tuning the established criteria of the LQR algorithm are summarized in Table 1.

Table 1. GA-LQR Simulation Parameters.

GA Parameters	Value/Method
Population Size	100
Max. Generations	200
Stop Criteria	0.01
Elitism	5% of Population Size
Mutation Method	Aleatory Alteration
Crossover Method	Based on a Point

After tuning the parameters of the LQR method, the K_p and K_i PI controller parameters are determined through the rlocus method. A contribution of this study is to combine two control design methodologies to propose a hybrid PI-LQR controller, whose performance was tested on the MG model using Simulink. The utility grid is represented in the simulation model by its Thevenin equivalent (i.e., a three-phase electric source and the coupling impedance). The MG is integrated by a passive filter with topology L-C-L (Inductor-Capacitor-Inductor), a three-phase inverter based on IGBT technology controlled by the current loops (I_d and I_q), and the DC bus, which is powered by a voltage source. For all study cases, the current references for dq components were set at I_d = 20A and I_q = 0. MG simulation parameters for the hybrid PI-LQR controller are shown in Table 2.

Table 2. Parameters of the MG model.

Parameter	Value	Units
MG Voltage (V_{DC} Figure 2)	311	[V]
Filter Inductance	2	[mH]
Filter Capacitance	60	[μF]
Switching Frequency	10	[kHz]
Proportional Constant (K_p)	0.27084	
Integral Constant (K_i)	4289.948	
Balanced 3-phase linear loads	20/20/20	[Ω]
Unbalanced 3-phase linear loads	20/5/1	[Ω]
Fundamental Frequency	60	[Hz]

Figure 7 shows the numerical results corresponding to the MG operation under the action of either (1) the hybrid PI-LQR controller, (2) the PI controller driven by GA, or (3) the PI controller designed by the poles placement technique, under the same operating conditions (linear balanced loads).

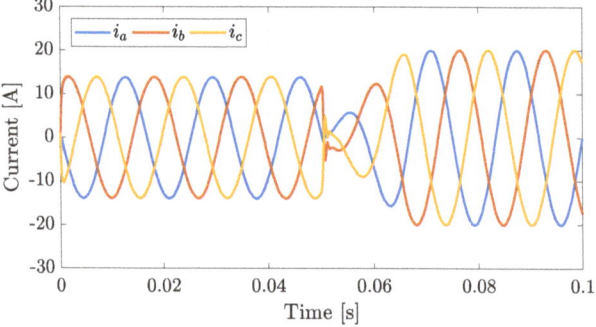

Figure 7. Simulation of three controllers designed to face a balanced linear loads interaction. The analysis is focused on measuring the reactions of the MG currents $\{i_a, i_b, i_c\}$.

The graphical responses of such controllers are equivalent. It is noteworthy that the controller is intentionally inactive in the initial interval of time, $0 < t < 0.05$ s, to have a reference for checking the posterior control action, as shown in Figure 7. The current flows from the utility grid towards MG during this time interval, so the MG is consuming power. The controller starts to operate since $t = 0.05$ s, which produces a transitory response ending at $t = 0.066$ s. In this period, the MG current reaches the reference current ($I_d = 20$ A and $I_q = 0$ A).

Moreover, the MG tracks the set-point until the desired current amplitude is injected into the system ($I_a = 20$A, $I_b = 20$ A, and $I_c = 20$ A). The system was analyzed under ideal balanced conditions, including linear loads from simulated starts. This test is a start-up test of the hybrid PI-LQR controller working under normal operating conditions to demonstrate the method's correct operation. The same simulation is used for all study cases regarding the times of activation and deactivation of the hybrid PI-LQR controller. Figure 8 shows the performance of the action of either hybrid PI-LQR controller, PI controller driven by GA, or PI controller by poles placement under the presence of unbalanced linear loads.

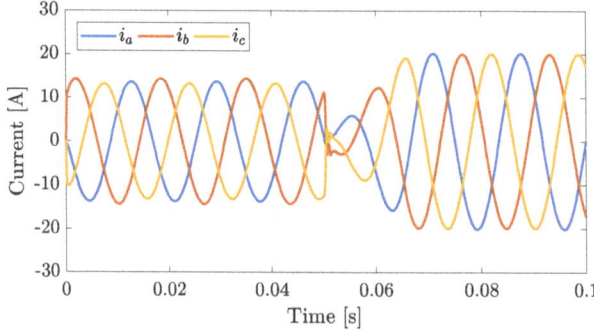

Figure 8. Simulation of three optimal controllers designed to face a unbalanced linear loads interaction. The analysis is focused on measuring the reactions of the MG currents $\{i_a, i_b, i_c\}$.

Considering the same operating conditions and equivalent control responses because of the differences of such controllers are studied, respect the energetic cost, negative sequence mitigation, and harmonics attenuation. Remarkably, a CUF index of 9% was reached when the controllers are disabled, but a CUF index of 1% is obtained when the controllers are under operation for unbalanced linear loads. The MG behavior is evaluated by connecting balanced nonlinear loads at the PCC to test control robustness. The nonlinear loads implemented in the simulations consist of three single-phase uncontrolled rectifiers. For such a case, the magnitude of the harmonics is calculated by $h_c = \frac{I_a}{h_o}$, where I_a is the current of the fundamental component for each phase, and h_o is the harmonic order. Figure 9 shows the current correction of the distorted current waveform, which was affected by operating under balanced nonlinear conditions. The amplitudes of the fundamental nonlinear load currents are $I_{a,1}^L = 20$ A, $I_{b,1}^L = 20$ A, and $I_{c,1}^L = 20$ A for the balanced case.

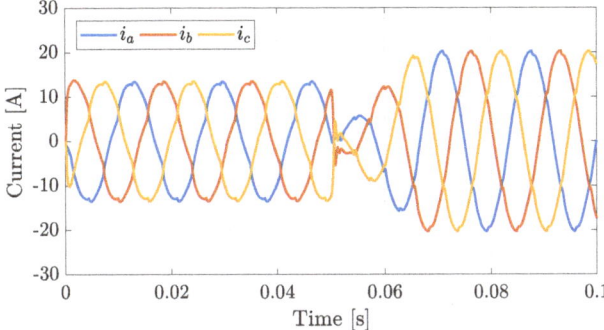

Figure 9. Current correction by the hybrid PI-LQR control action under balanced nonlinear load conditions.

Additionally, an FFT analysis was applied to obtain the spectral component of the current signal for studying the THD_I index behavior for each phase. From this analysis, the $THD(i_a)$ = 5.08%, $THD(i_b)$ = 5.08%, and $THD(i_c)$ = 5.08% indexes are obtained when the controller is disabled. Contrarily, the $THD(i_a)$ = 2.51%, $THD(i_b)$ = 2.51%, and $THD(i_c)$ = 2.51% indexes are reached when the hybrid PI-LQR control is activated. Besides, the harmonics of the 5th and 7th order are substantially attenuated for balanced nonlinear loads; such results are included in the second and third columns in Table 3.

Table 3. Harmonics content attenuated by hybrid PI-LQR controller.

Harmonics (I_a^L, I_b^L, I_c^L)	Balanced Nonlinear Loads		Unbalanced Nonlinear Loads		Unbalanced Linear and Nonlinear Loads	
	Original Currents	Controlled Currents	Original Currents	Controlled Currents	Original Currents	Controlled Currents
Third	0.00, 0.00, 0.00	0.00, 0.00, 0.00	0.58, 1.29, 1.86	0.21, 0.43, 0.52	0.40, 1.00, 1.32	0.20, 0.35, 0.39
Fifth	0.52, 0.52, 0.52	0.31, 0.31, 0.31	0.10, 0.80, 0.74	0.15, 0.44, 0.36	0.21, 0.67, 0.47	0.15, 0.35, 0.21
Seventh	0.38, 0.38, 0.38	0.26, 0.25, 0.25	0.20, 0.43, 0.53	0.16, 0.37, 0.41	0.05, 0.32, 0.37	0.01, 0.24, 0.24
Ninth	0.00, 0.00, 0.00	0.00, 0.00, 0.00	0.26, 0.33, 0.12	0.28, 0.30, 0.14	0.22, 0.27, 0.13	0.22, 0.24, 0.13
Eleventh	0.15, 0.15, 0.15	0.18, 0.18, 0.18	0.12, 0.15, 0.13	0.16, 0.14, 0.11	0.10, 0.17, 0.17	0.11, 0.18, 0.16

In the same context, a study case concerning an unbalanced nonlinear load was analyzed to test the MG performance. Here, the amplitudes of the fundamental unbalanced nonlinear load currents are given by $I_{a,1}^L = 6$ A, $I_{b,1}^L = 24$ A, and $I_{c,1}^L = 120$ A.

In the simulation, the CUF produced by the unbalanced nonlinear loads preserves the CUF ratio considered in the unbalanced linear load's study case. Figure 10 shows the distorted waveforms belonging to the current signals that lost the original sinusoidal shape under the effect of nonlinear loads.

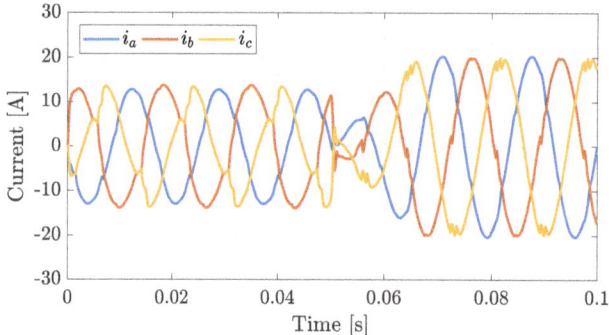

Figure 10. Current correction by the hybrid PI-LQR control action under unbalanced nonlinear load conditions.

However, activated the hybrid PI-LQR controller, the current signal is gradually improved and balanced by the action of the MG control. The CUF index produced by the unbalanced nonlinear loads was 11%, but by using the hybrid control, this index was improved to 1%. Similarly, a harmonics study was implemented to analyze the effects of unbalanced nonlinear loads over the current signal injected by MG. Table 3 (four and five columns) shows the harmonics reduction behavior under unbalanced nonlinear loads, which leads to reduce the THD(i_a), THD(i_b), THD(i_c) indexes from 5.47%, 11.83%, and 18.39% to 2.49%, 4.63%, and 4.61%, respectively.

Finally, the case of unbalanced linear and nonlinear loads connected simultaneously at the PCC is studied. Figure 11 presents the result of the hybrid PI-LQR controller operation under the actions of both loads.

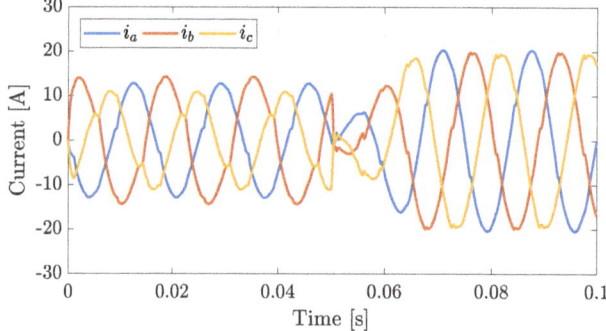

Figure 11. Current correction by the hybrid PI-LQR control action under unbalanced linear and nonlinear load conditions.

In this case, a high CUF = 18% index is obtained while the control is deactivated. Once the controller is activated, the CUF is reduced to 2%. As in the unbalanced nonlinear load case, a significant reduction of harmonics was obtained, and the results are included in Table 3 (six and seven columns). THD indexes (i_a, i_b and i_c) were reduced from 4.39%, 9.15% and 14.31% to 2.11%, 3.39% and 2.97% for this study case.

6.2. PI Controller Driven by GA and Rlocus Design

The GA was implemented for tuning the matrix K in the rlocus method (poles placement method). In this case, only one chromosome (α) was used together with the design parameters as the undamped natural frequency (ω_n), settling time (T_s), damping factor (ζ), and overshoot (O_v) to calculate the components of K. Therefore, the matrix K could be estimated by

$$K = \begin{bmatrix} L_1(\alpha+2)\omega_n\zeta \\ \alpha C_1 L_1 L_2 \omega_n^3 \zeta - K_1 \\ C_1 L_1 (2\alpha\zeta^2+1)\omega_n^2 - \frac{L_1}{L_2} - 1 \end{bmatrix}, \quad (34)$$

where α represents how far the third pole is located according to the dominant poles configuration [41]. The fitness function to determine the matrix K, considering the design parameters, is defined as

$$F = \left| \frac{1}{MO_v} (MO_v - O_v) \right|, \quad (35)$$

where MO_v is the maximum allowed overshot, and O_v is the actual overshot. The control parameters $K = [156.6598 \ 1689 \ 127.9947]^T$ and the precompensation gain $N = 1845.7$ were estimated using this genetic approach. The compensated system achieved a settling time of $T_s = 0.5593$ ms and an overshoot of 4.9109%. Table 4 summarizes the parameters used in the GA for tuning the poles placement method.

Table 4. GA and poles placement method.

GA Parameters	Value or Method
Population Size	100
Max. Generations	100
Stop Criteria	0.01
Elitism	5% of Population
Mutation	Aleatory Alteration
Crossover	Based on a Point
Undamped natural freq. (ω_n)	11040 [rad/s]
Damping Factor (ζ)	0.6901

Finished the tuning process, the PI controller is designed through the poles placement method, and the simulation tests are then analyzed. The simulation parameters are given in Table 5.

Table 5. Parameters of MG model for PI controller.

Parameters	Value	Units
MG Voltage (V_{DC} Figure 2)	1000	[V]
Filter Inductance	2	[mH]
Filter Capacitance	60	[µF]
Switching Frequency	10	[kHz]
Proportional Constant (K_p)	0.82045	
Integral Constant (K_i)	6668.4268	
Balanced 3-phase linear loads	20/20/20	[Ω]
Unbalanced 3-phase linear loads	20/5/1	[Ω]
Fundamental Frequency	60	[Hz]

The simulation results were carried out on a set of interconnected nonlinear loads, which allowed testing the PI controller response under balanced and unbalanced nonlinear conditions. Figure 12 presents the PI controller performance driven by GA.

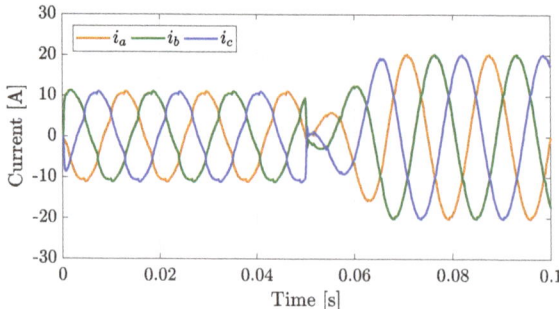

Figure 12. Compensation of PI controller driven by GA under balanced nonlinear conditions.

The control was able to compensate for the current distorted waveform, produced by balanced nonlinear loads. This methodology reduced the THD(i_a), THD(i_b), THD(i_c) indexes from 5.26%, 5.26% and 5.26% to 1.35%, 1.35% and 1.35%. Likewise, the harmonic content reduction is shown in Table 6 (two and three columns) for this study case. Figure 13 shows a study case considering unbalanced nonlinear loads to verify the control quality face to harmonics and unbalanced actions of the MG control.

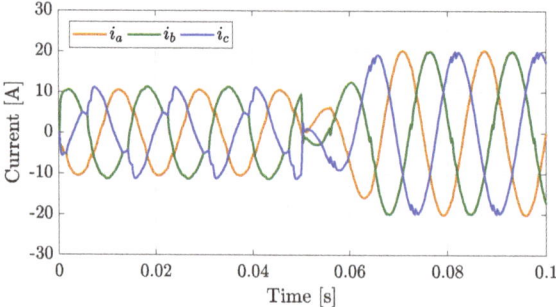

Figure 13. Compensation of PI controller driven by GA under balanced nonlinear conditions.

The numerical results generated a CUF=11% for unbalanced nonlinear loads, but the index is reduced to 0.6% once the controller is activated. Similarly, the THD(i_a), THD(i_b), and THD(i_c) indexes of the current signals injected by MG are reduced from 5.61%, 12.09%, and 18.69% to 1.30%, 2.50%, and 2.55%, respectively. In Table 6 (four and five columns), the harmonic content reduction corresponding to this study case is shown. In Figure 14, unbalanced linear and nonlinear loads were connected and simulated, which gave an improved performance for an unbalanced current compensation from 18% to 2%.

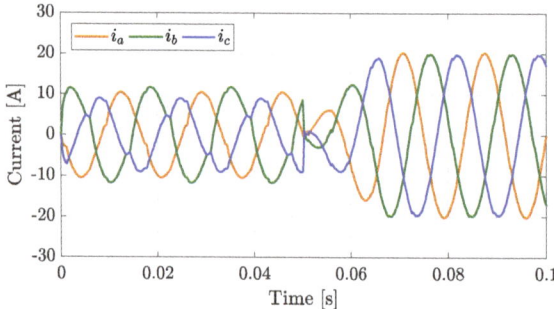

Figure 14. Compensation of PI controller driven by GA under unbalanced linear and nonlinear conditions.

The harmonics were significantly reduced, which is depicted in Table 6 (six and seven columns). Likewise, a THD reduction was carried out, attenuating the THD(i_a), THD(i_b) and THD(i_c) indexes from 4.56%, 9.34%, and 14.53% to 1.14%, 1.73%, and 1.53%, respectively.

Table 6. Harmonics content attenuated by PI controller driven by GA.

Harmonics (I_a^L, I_b^L, I_c^L)	Balanced Nonlinear Loads		Unbalanced Nonlinear Loads		Unbalanced Linear and Nonlinear Loads	
	Original Currents	Controlled Currents	Original Currents	Controlled Currents	Original Currents	Controlled Currents
Third	0.00, 0.00, 0.00	0.00, 0.00, 0.00	0.48, 1.06, 1.53	0.11, 0.22, 0.27	0.33, 0.83, 1.09	0.10, 0.18, 0.20
Fifth	0.44, 0.44, 0.44	0.15, 0.15, 0.15	0.09, 0.67, 0.62	0.07, 0.21, 0.18	0.18, 0.57, 0.39	0.08, 0.17, 0.11
Seventh	0.32, 0.32, 0.32	0.12, 0.12, 0.12	0.17, 0.37, 0.46	0.08, 0.18, 0.20	0.04, 0.27, 0.31	0.01, 0.12, 0.12
Ninth	0.00, 0.00, 0.00	0.00, 0.00, 0.00	0.22, 0.28, 0.22	0.13, 0.15, 0.08	0.19, 0.23, 0.12	0.11, 0.12, 0.07
Eleventh	0.13, 0.13, 0.13	0.09, 0.09, 0.09	0.11, 0.13, 0.11	0.08, 0.07, 0.04	0.08, 0.15, 0.15	0.05, 0.09, 0.08

6.3. PI Control Design by the Poles Placement Method

In this design, the LQR algorithm was implemented to estimate the K matrix of feedback states. Besides, the penalization matrix Q and effort control R were assigned according to the requirements initially established using the poles placement method. The LQR control matrices and parameters assigned by the designer were determined as R=0.002 and pre-compensation gain N=707.9018,

$$Q = \begin{bmatrix} 2.25 & 0 & 0 \\ 0 & 1000 & 0 \\ 0 & 0 & 0.04 \end{bmatrix}, \text{ and } K = \begin{bmatrix} 67.2952 \\ 640.6066 \\ 51.0547 \end{bmatrix}^T. \tag{36}$$

Numerical results gave a settling time T_s = 0.525 ms and an overshoot = 6.39%. Next, the PI controller's K_p and K_i constants were calculated, combining the properties of LQR and PI controllers. The simulation parameters for this case study are shown in Table 7.

Table 7. Parameters of MG model for PI-LQR controller.

Parameters	Value	Units
MG Voltage (V_{DC} Figure 2)	500	[V]
Filter Inductance	2	[mH]
Filter Capacitance	60	[μF]
Switching Frequency	10	[kHz]
Proportional Constant (K_p)	0.15093	
Integral Constant (K_i)	4003.3102	
Balanced 3-phase linear loads	20/20/20	[Ω]
Unbalanced 3-phase linear loads	20/5/1	[Ω]
Fundamental Frequency	60	[Hz]

Similarly to the last two controllers, the MG behavior under balanced nonlinear loads, unbalanced nonlinear loads, and unbalanced linear and nonlinear loads connected at PCC was analyzed.

Figure 15 shows the controller performance under balanced nonlinear conditions, reducing the THD(i_a), THD(i_b) and THD(i_c) indexes from 5.27%, 5.27% and 5.27% to 2.30%, 2.30% and 2.30% respectively. Harmonic reduction through PI controller by dominant poles is shown in Table 8 (two and three columns).

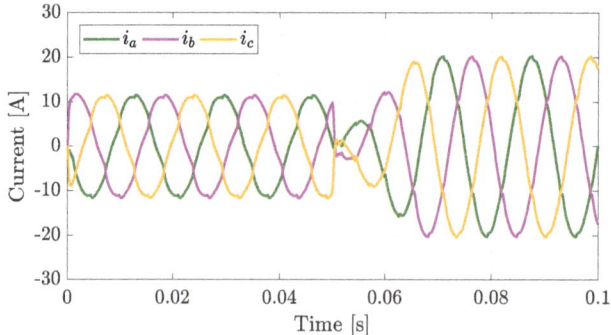

Figure 15. Compensation of PI controller tuned by poles placement method under balanced nonlinear conditions.

In the presence of unbalanced nonlinear loads, a CUF = 11% index is initially obtained, but those indexes are significantly reduced to 1% under the MG control. Such an effect is shown in Figure 16.

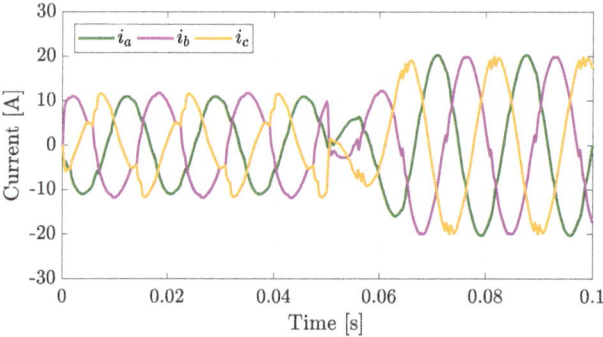

Figure 16. Compensation of PI controller tuned by poles placement method under unbalanced linear and nonlinear conditions.

In the same sense, a considerable harmonics mitigation was induced, which is given in Table 8 (four and five columns). Additionally, THD indexes of three phases were attenuated from 5.60%, 12.05%, and 18.65% to 2.27%, 4.22%, and 4.22% for this study case. Figure 17 shows an unbalanced linear and nonlinear study case to verify the controller performance.

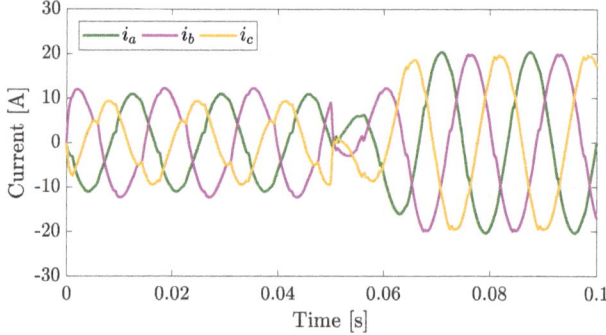

Figure 17. Compensation of PI controller tuned by poles placement method under unbalanced linear and nonlinear conditions.

Table 8. Harmonics content attenuated by PI controller tuned by poles placement method.

Harmonics (I_a^L, I_b^L, I_c^L)	Balanced Nonlinear Loads		Unbalanced Nonlinear Loads		Unbalanced Linear And Nonlinear Loads	
	Original Currents	Controlled Currents	Original Currents	Controlled Currents	Original Currents	Controlled Currents
Third	0.00, 0.00, 0.00	0.00, 0.00, 0.00	0.50, 1.11, 1.60	0.20, 0.39, 0.47	0.35, 0.86, 1.14	0.18, 0.32, 0.36
Fifth	0.46, 0.46, 0.46	0.28, 0.28, 0.28	0.09, 0.70, 0.64	0.13, 0.39, 0.32	0.19, 0.59, 0.40	0.15, 0.32, 0.20
Seventh	0.34, 0.34, 0.34	0.23, 0.23, 0.23	0.18, 0.38, 0.48	0.14, 0.33, 0.37	0.04, 0.28, 0.32	0.01, 0.22, 0.22
Ninth	0.00, 0.00, 0.00	0.00, 0.00, 0.00	0.23, 0.29, 0.11	0.25, 0.28, 0.13	0.19, 0.24, 0.12	0.20, 0.22, 0.12
Eleventh	0.13, 0.14, 0.13	0.16, 0.16, 0.16	0.11, 0.13, 0.11	0.15, 0.13, 0.10	0.09, 0.16, 0.15	0.10, 0.17, 0.14

Under these conditions, the numerical results determine that MG reduced the current unbalanced index *CUF* from 18% to 2%. The respective harmonic reduction is shown in Table 8 (six and seven columns). Besides, THD indexes (i_a, i_b, and i_c) were reduced from 4.53%, 9.30%, and 14.30% to 1.94%, 3.08%, and 2.68%, respectively.

It is noteworthy that the harmonic content shown in Table 9 is extracted from the harmonic content from Tables 3, 6 and 8 when the controller in operation (see columns 3, 5, and 7). The energetic cost is associated with the MG Voltage (V_{DC}) parameter highlighted in Line 2 from Tables 2, 5 and 7. Numerical results in Table 9 allows determining that PI controller driven by GA obtained the best THD_I factor for the three study cases. Nevertheless, to reach this low distortion was necessary to use the higher DC voltage (i.e., 1000 V), which could be highly restrictive and expensive in practice. On the other hand, the hybrid LQR-PI reached the lower energetic cost consuming only 311 V_{DC}. Additionally, the measured harmonic distortion under this control action fulfills the allowed THD limits of 5% and is pretty close to the best results given by the PI controller driver by GA. Finally, the positive impact of saving energy by using the hybrid LQR-PI is fundamental in the selection criteria in an efficient MG system.

Table 9. Comparative analysis of THD and energetic cost obtained from the evaluated controllers.

Control	Balanced Nonlinear Loads (THD)			Unbalanced Nonlinear Loads (THD)			Unbalanced Linear and Nonlinear Loads (THD)			MG Voltage (V_{DC})
	i_a	i_b	i_c	i_a	i_b	i_c	i_a	i_b	i_c	
Hybrid LQR-PI	2.51	2.51	2.51	2.49	4.63	4.61	2.11	3.39	2.97	311
PI Driven by GA	1.35	1.35	1.35	1.3	2.5	2.55	1.14	1.73	1.53	1000
PI+Poles Placement	2.3	2.3	2.3	2.27	4.22	4.22	1.94	3.08	2.68	500

Finally, Table 10 shows the comparison between research works found in literature and the proposed approach using related hybrid PI-LQR controllers.

This study proposed an efficient optimal control technique, combining the performance of LQR and PI controllers tuned by Genetic Algorithms using a reliable fitness function. The particular discriminant definition of the fitness function associated with the MG control scheme and appropriate implementation of the GA algorithm were fundamental to reach the required accuracy for estimating the controller design parameters. The proposal contributions were focused on using the hybrid PI-LQR controller driven by GA to regulate the energy supplied by the MG, showing the positive effects over typical compensating scenarios involving quality energy issues. Moreover, other controllers were designed to compare and evaluate which control scheme had the lowest energetic cost in its operation. The results can corroborate the effectiveness, robustness, and proper fitting of the hybrid PI-LQR controller according to the criteria as the simultaneous reduction on the negative sequence, harmonics, and energetic cost.

Table 10. Functional comparison of hybrid PI-LQR controllers.

Author	Fitness Function (F.F.)	Optimized Variables	Optimizer
Lindiya et al. [18]	F.F. based on T_s, O_v, rising time (T_r), peak time (T_p), and undershoot (U_v)	Cross regulation	GA
Sen et al. [20]	F.F. based on mean squared error using the robot coordinates	Coordinates of the reference foot trajectory (X_{ref} and Y_{ref})	GA, PSO, and GWO
Nagarkar et al. [21]	F.F. based on minimize control force, RMS tyre deflection and RMS suspension travel	RMS acceleration (A_w), and fourth power vibration (VDV)	GA
Ibrahim et al. [22]	F.F. based on T_s, O_v, T_r and error steady state (ess)	Elevation, pitch, and travel axis	GA
Proposed method	F.F. based on T_s, O_v	MG voltage or Input Source Voltage	GA

7. Conclusions

In this paper, three study cases were analyzed by considering important electrical features as low energetic cost, reduction of unbalanced currents, and harmonics attenuation. The three proposed tuning strategies allowed determining the correct controller parameters, as were requested by the parameter design. The second study case (PI controller driven by GA and rlocus design) obtained the best results considering mitigation of unbalanced current and harmonics reduction but demanding a high energetic cost that would require considerable photovoltaics configurations. In comparison, the parameters showed in Tables 2, 5 and 7, as well as the obtained results related to mitigation of harmonics and current unbalances, the proposed PI-LQR controller driven by GA, allowed fulfilling the international energy quality index normative and the design specifications. Besides, the proposed approach improved the MG power quality and accomplished a considerable magnitude reduction in the LQR parameters, K matrix of feedback states, and MG voltage. Simulated results showed the effectiveness and robustness of the PI-LQR hybrid controller tuned by the GA, achieving an equilibrium on the initial electrical features established in the study cases.

Author Contributions: Conceptualization, G.H.V.-R. and L.R.M.-V.; Methodology, J.G.A.-C. and L.R.M.-V.; Software, G.T.-T., L.R.M.-V. and J.G.A.-C.; Validation, J.M.L.-G., M.A.I.-M. and G.T.-T.; Formal analysis, J.G.A.-C., M.A.I.-M. and L.R.M.-V.; Investigation, G.H.V.-R., L.R.M.-V. and J.G.A.-C.; Data curation, L.R.M.-V., M.A.I.-M. and G.T.-T.; Visualization, G.H.V.-R., M.A.I.-M. and J.M.L.-G.; Writing—original draft preparation, L.R.M.-V., and G.H.V.-R.; Writing—review and editing, J.G.A.-C. and J.M.L.-G. and G.T.-T.; Funding acquisition, J.G.A.-C. and J.M.L.-G. All authors have read and agreed to the published version of the manuscript.

Funding: The APC was funded by the Universidad de Guanajuato.

Acknowledgments: This project was fully supported by the Electronics and Electrical Departments of the Universidad de Guanajuato under the Program POA 2020, grant NUA 824646, and the Mexican Council of Science and Technology CONACyT, grant number 863547/491067.

Conflicts of Interest: The funders had no role in the design of the study; in the collection, analyses, or interpretation of data; in the writing of the manuscript, or in the decision to publish the results.

References

1. Hirsch, A.; Parag, Y.; Guerrero, J. Microgrids: A review of technologies, key drivers, and outstanding issues. *Renew. Sustain. Energy Rev.* **2018**, *90*, 402–411, doi:10.1016/j.rser.2018.03.040. [CrossRef]
2. Banerji, A.; Sen, D.; Bera, A.K.; Ray, D.; Paul, D.; Bhakat, A.; Biswas, S.K. Microgrid: A review. In Proceedings of the 2013 IEEE Global Humanitarian Technology Conference: South Asia Satellite (GHTC-SAS), Trivandrum, India, 23–24 August 2013; pp. 27–35, doi:10.1109/GHTC-SAS.2013.6629883. [CrossRef]
3. Blaabjerg, F.; Teodorescu, R.; Liserre, M.; Timbus, A.V. Overview of Control and Grid Synchronization for Distributed Power Generation Systems. *IEEE Trans. Ind. Electron.* **2006**, *53*, 1398–1409, doi:10.1109/TIE.2006.881997. [CrossRef]
4. DOE. *2012 DOE Microgrid Workshop*; Summary Report; U.S. Department of Energy: Chicago, IL, USA, 2012.
5. Parhizi, S.; Lotfi, H.; Khodaei, A.; Bahramirad, S. State of the art in research on microgrids: A review. *IEEE Access* **2015**, *3*, 890–925, doi:10.1109/ACCESS.2015.2443119. [CrossRef]
6. Li, Y.W.; Vilathgamuwa, D.M.; Loh, P.C. A grid-interfacing power quality compensator for three-phase three-wire microgrid applications. In Proceedings of the 2004 IEEE 35th Annual Power Electronics Specialists Conference (IEEE Cat. No.04CH37551), Aachen, Germany, 20–25 June 2004; Volume 3, pp. 2011–2017, doi:10.1109/PESC.2004.1355426. [CrossRef]
7. Wang, F.; Duarte, J.L.; Hendrix, M.A.M. Grid-Interfacing Converter Systems With Enhanced Voltage Quality for Microgrid Application—Concept and Implementation. *IEEE Trans. Power Electron.* **2011**, *26*, 3501–3513, doi:10.1109/TPEL.2011.2147334. [CrossRef]
8. IEEE Std 1159-2009 (Revision of IEEE Std 1159-1995). *IEEE Recommended Practice for Monitoring Electric Power Quality*; Standard; IEEE Power & Energy Society: New York, NY, USA, 2009; pp. 1–94, doi:10.1109/IEEESTD.2009.5154067. [CrossRef]
9. Hamzeh, M.; Karimi, H.; Mokhtari, H. Harmonic and Negative-Sequence Current Control in an Islanded Multi-Bus MV Microgrid. *IEEE Trans. Smart Grid* **2014**, *5*, 167–176. [CrossRef]
10. Shi, H.; Zhuo, F.; Yi, H.; Geng, Z. Control strategy for microgrid under three-phase unbalance condition. *J. Mod. Power Syst. Clean Energy* **2016**, *4*, 94–102, doi:10.1007/s40565-015-0182-3. [CrossRef]
11. Azevedo, G.; Rocabert Delgado, J.; Cavalcanti, M.; Neves, F.; Rodriguez, P. A Negative-sequence Current Injection Method To Mitigate Voltage Imbalances In Microgrids. *Eletrônica de Potência* **2011**, *16*, 296–303, doi:10.18618/REP.20114.296303. [CrossRef]
12. Foad, N.; Mohsen, H.; Matthias, F. Unbalanced Current Sharing Control in Islanded Low. *Energies* **2018**, *11*, 2776, doi:10.3390/en11102776. [CrossRef]
13. Dasgupta, S.; Mohan, S.N.; Sahoo, S.K.; Panda, S.K. Lyapunov Function-Based Current Controller to Control Active and Reactive Power Flow From a Renewable Energy Source to a Generalized Three-Phase Microgrid System. *IEEE Trans. Ind. Electron.* **2013**, *60*, 799–813. [CrossRef]
14. Sinha, T.; Ray, P.; Salkuti, S. Protection Coordination in Microgrid Using Fault Current Limiters. *J. Green Eng.* **2018**, *8*, 125–150. doi:10.13052/jge1904-4720.822. [CrossRef]
15. Surender Reddy, S.; Park, J.; Jung, C. Optimal Operation of Microgrid Using Hybrid Differential Evolution and Harmony Search Algorithm. *Front. Energy* **2016**, *10*, 355–362, doi:10.1007/s11708-016-0414-x. [CrossRef]
16. Lotfollahzade, M.; Akbarimajd, A.; Javidan, J. Design LQR and PID Controller for Optimal Load Sharing of an Electrical Microgrid. *Int. Res. J. Appl. Basic Sci.* **2013**, *4*, 704–712.
17. Savaghebi, M.; Jalilian, A.; Vasquez, J.C.; Guerrero, J.M. Secondary control for voltage quality enhancement in microgrids. *IEEE Trans. Smart Grid* **2012**, *3*, 1893–1902, doi:10.1109/TSG.2012.2205281. [CrossRef]
18. Lindiya, S.; Subashini, N.; Vijayarekha, K. Cross Regulation Reduced Optimal Multivariable Controller Design for Single Inductor DC-DC Converters. *Energies* **2019**, *12*, 477, doi:10.3390/en12030477. [CrossRef]
19. Momoh, J.; Reddy, S. Value of Hardware-In-Loop for Experimenting Microgrid Performance System Studies. In Proceedings of the IEEE PES PowerAfrica Conference, Livingstone, Zambia, 28 June–3 July 2016; pp. 199–203, doi:10.1109/PowerAfrica.2016.7556600. [CrossRef]

20. Şen, M.A.; Kalyoncu, M. Grey Wolf Optimizer Based Tuning of a Hybrid LQR-PID Controller for Foot Trajectory Control of a Quadruped Robot. *Gazi Univ. J. Sci.* **2019**, *32*, 674–684.
21. Nagarkar, M.; Bhalerao, Y.; Patil, G.V.; Patil, R.Z. Multi-Objective Optimization of Nonlinear Quarter Car Suspension System—PID and LQR Control. *Procedia Manuf.* **2018**, *20*, 420–427. doi:10.1016/j.promfg.2018.02.061. [CrossRef]
22. Mohammed, I.; Abdullah, A.I. Elevation, pitch and travel axis stabilization of 3DOF helicopter with hybrid control system by GA-LQR based PID controller. *Int. J. Electr. Comput. Eng. (IJECE)* **2020**, *10*, 1868, doi:10.11591/ijece.v10i2.pp1868-1884. [CrossRef]
23. Verdugo, C.; Tarraso, A.; Candela, J.I.; Rocabert, J.; Rodriguez, P. Synchronous Frequency Support of Photovoltaic Power Plants with Inertia Emulation. In Proceedings of the 2019 IEEE Energy Conversion Congress and Exposition (ECCE), Baltimore, MD, USA, 29 September–3 October 2019; pp. 4305–4310, doi:10.1109/ECCE.2019.8913200. [CrossRef]
24. Liu, F.; Zha, X.; Zhou, Y.; Duan, S. Design and research on parameter of LCL filter in three-phase grid-connected inverter. In Proceedings of the 2009 IEEE 6th International Power Electronics and Motion Control Conference, Wuhan, China, 17–20 May 2009; pp. 2174–2177.
25. Shen, G.; Xu, D.; Cao, L.; Zhu, X. An Improved Control Strategy for Grid-Connected Voltage Source Inverters With an LCL Filter. *IEEE Trans. Power Electron.* **2008**, *23*, 1899–1906. [CrossRef]
26. Gamit, B.R.; Vyas, S.R. Harmonic Elimination in Three Phase System By Means of a Shunt Active Filter. *Int. Res. J. Eng. Technol. (IRJET)* **2018**, *5*, 313–322.
27. Escudero, R.; Noel, J.; Elizondo, J.; Kirtley, J. Microgrid fault detection based on wavelet transformation and Park's vector approach. *Electric Power Syst. Res.* **2017**, *152*, 401–410, doi:10.1016/j.epsr.2017.07.028. [CrossRef]
28. Lewis, F.L.; Draguna Vrabie, V.L.S. *Optimal Control*, 3rd ed.; John Wiley & Sons, Inc.: Hoboken, NJ, USA, 2012; p. 553.
29. Fadali, M.S.; Visioli, A. State Feedback Control. In *Digital Control Engineering*, 2nd ed.; Fadali, M.S., Visioli, A., Eds.; Academic Press: Boston, MA, USA, 2013; Chapter 9, pp. 351–397, doi:10.1016/B978-0-12-394391-0.00009-5. [CrossRef]
30. White, M.S.; Flockton, S.J. *Genetic Algorithms for Digital Signal Processing*; Springer: New York, NY, USA, 1994; Volume 865, pp. 291–303, doi:10.1007/3-540-58483-8_22. [CrossRef]
31. Robandi, I.; Nishimori, K.; Nishimura, R.; Ishihara, N. Optimal feedback control design using genetic algorithm in multimachine power system. *Int. J. Electr. Power Energy Syst.* **2001**, *23*, 263–271, doi:10.1016/S0142-0615(00)00062-4. [CrossRef]
32. Črepinšek, M.; Liu, S.H.; Mernik, M. Exploration and Exploitation in Evolutionary Algorithms: A Survey. *ACM Comput. Surv.* **2013**, *45*, doi:10.1145/2480741.2480752. [CrossRef]
33. Ogata, K. *Modern Control Engineering*, 5th ed.; Prentice Hall: Upper Saddle River, NJ, USA, 2010; p. 905.
34. Chen, T.-H.; Yang, C.-H.; Hsieh, T.-Y. Case Studies of the Impact of Voltage Imbalance on Power Distribution Systems and Equipment. In Proceedings of the 8th WSEAS International Conference on Applied Computer and Applied Computational Science, 2009; Volume 8, pp. 461–465. Available online: https://www.scribd.com/document/49242179/Case-Studies-of-the-Impact-of-Voltage-Imbalance-on-Power (accessed on 28 May 2020).
35. Mousavi, S.Y.M.; Jalilian, A.; Savaghebi, M.; Guerrero, J.M. Flexible compensation of voltage and current unbalance and harmonics in microgrids. *Energies* **2017**, *10*, 1568, doi:10.3390/en10101568. [CrossRef]
36. Savaghebi, M.; Jalilian, A.; Vasquez, J.C.; Guerrero, J.M. Autonomous Voltage Unbalance Compensation in an Islanded Droop-Controlled Microgrid. *IEEE Trans. Ind. Electron.* **2013**, *60*, 1390–1402, doi:10.1109/TIE.2012.2185914. [CrossRef]
37. ANSI/NEMA MG 1-2016 *Motors and Generators*; Standard; American National Standard Institute: Rosslyn, VI, USA, 2016.
38. Share Pasand, M.M. Harmonic Aggregation Techniques. *J. Electr. Electron. Eng.* **2015**, *3*, 117–120, doi:10.11648/j.jeee.20150305.13. [CrossRef]
39. Mazin, H.E.; Xu, W. Harmonic cancellation characteristics of specially connected transformers. *Electr. Power Syst. Res.* **2009**, *79*, 1689–1697, doi:10.1016/j.epsr.2009.07.006. [CrossRef]

40. IEC 61000-4-7:2002 *Testing and Measurement Techniques—General Guide on Harmonics and Interharmonics Measurements and Instrumentation, for Power Supply Systems and Equipment Connected Thereto*; Standard; International Electrotechnical Commission: Geneva, Switzerland, 2002.
41. Nise, N.S. *Control Systems Engineering*, 6th ed.; John Wiley & Sons, Inc.: Hoboken, NJ, USA, 2011; p. 1001.

© 2020 by the authors. Licensee MDPI, Basel, Switzerland. This article is an open access article distributed under the terms and conditions of the Creative Commons Attribution (CC BY) license (http://creativecommons.org/licenses/by/4.0/).

Article

Differential Flatness-Based Cascade Energy/Current Control of Battery/Supercapacitor Hybrid Source for Modern e–Vehicle Applications

Burin Yodwong [1,2], Phatiphat Thounthong [1,3,*], Damien Guilbert [4] and Nicu Bizon [5,6,*]

1. Renewable Energy Research Centre (RERC), King Mongkut's University of Technology North Bangkok, 1518, Pracharat 1 Road, Bangsue, Bangkok 10800, Thailand; burin.y@tfii.kmutnb.ac.th
2. Thai-French Innovation Institute (TFII), King Mongkut's University of Technology North Bangkok, 1518, Pracharat 1 Road, Bangsue, Bangkok 10800, Thailand
3. Department of Teacher Training in Electrical Engineering (TE), Faculty of Technical Education, King Mongkut's University of Technology North Bangkok, Bangkok 10800, Thailand
4. Groupe de Recherche en Energie Electrique de Nancy (GREEN), Université de Lorraine, F-54000 Nancy, France; damien.guilbert@univ-lorraine.fr
5. Faculty of Electronics, Communication and Computers, University of Pitesti, 110040 Pitesti, Romania
6. ICSI Energy, National Research and Development Institute for Cryogenic and Isotopic Technologies, 240050 Ramnicu Valcea, Romania
* Correspondence: phatiphat.t@fte.kmutnb.ac.th (P.T.); nicu.bizon@upit.ro (N.B.)

Received: 31 March 2020; Accepted: 21 April 2020; Published: 2 May 2020

Abstract: This article proposes a new control law for an embedded DC distributed network supplied by a supercapacitor module (as a supplementary source) and a battery module (as the main generator) for transportation applications. A novel control algorithm based on the nonlinear differential flatness approach is studied and implemented in the laboratory. Using the differential flatness theory, straightforward solutions to nonlinear system stability problems and energy management have been developed. To evaluate the performance of the studied control technique, a hardware power electronics system is designed and implemented with a fully digital calculation (real-time system) realized with a MicroLabBox dSPACE platform (dual-core processor and FPGA). Obtained test bench results with a small scale prototype platform (a supercapacitor module of 160 V, 6 F and a battery module of 120 V, 40 Ah) corroborate the excellent control structure during drive cycles: steady-state and dynamics.

Keywords: battery; capacitor; differential flatness; double-layer capacitor; electric vehicle; energy management; interleaved converter; nonlinear control; second order equation; supercapacitor

---

1. Introduction

The crisis of continuously growing fossil fuel costs has provoked transportation industries to advance more efficient automobiles technology. Another solution is to transform technology into other sources including biodiesel or energy from ethanol, etc. Electric vehicles, or hybrid plug-in vehicles, or hybrid vehicles (which are mainly supplied by battery) are a promising solution. Therefore, the electric vehicles (e-vehicle, EVs) industries have designed and developed the technology to progress the extension of EVs [1–3]. A lot of research works have been conducted on future vehicle technology [1–3]. In EVs, the powertrains are composed of batteries, power converters, and AC electrical motors, such as permanent magnet synchronous motors due to their high energy efficiency, suitable torque-to-weight ratio, and long life span [4,5].

The request for power from battery during dynamic operations can decrease drastically its lifetime [6]. To cope with this challenging issue, the hybridization of batteries with supercapacitors (SC or "ultracapacitor" or "double-layer capacitor") is an attractive solution. Indeed, SCs feature a high power density compared to batteries, which enables them to respond quickly to dynamic operations [7]. Besides, SCs are compact and have high energy efficiency, particularly fit for automotive applications. However, the combination of SC and battery requires a good control algorithm between these two sources, which is helpful in order to decrease the battery size and to enhance its life span [6,7].

The linear control is generally employed for energy management of the hybrid system. Generally, proportional–integral (PI) compensation is used for energy stability [8–12]. In [13], Marzouguia et al. have proposed a control technique of the hybrid network for a hydrogen electric vehicle (fuel cell car) based on three estimation approaches: first, "a fuzzy logic estimation"; second, "a differential flatness control approach" (model-based technique); third "rule-based algorithm", making complex the energy management strategy. Indeed, the fuzzy logic controller is employed to manage the energy flows between the main source (i.e., fuel cell) and storage devices (i.e., battery, SC); whereas the flatness controller is used to regulating the DC bus voltage, allowing ensuring the stability of the microgrid. Finally, a rule-based algorithm enables controlling the state-of-charge of SC to keep a good operation of charge/discharge cycles. As a result, the three controllers must interact with each other and their implementations are more complex.

On the other side, batteries and SCs are interfaced with the DC bus through classic buck–boost converters, making them unavailable in case of power switch failures. These converters are controlled based on PI-current control laws. The parameters for PI controllers have been designed and tuned in agreement with the linear optimum technique, requiring classic linear approximations contingent on the defined equilibrium point. Hence, the performances can be guaranteed only for specific operating cases.

Since the hybrid system includes bidirectional DC–DC converters linked to storage devices, the power converter model is nonlinear behavior. Then, it is important to use a nonlinear model-based control approach to the balance of the nonlinearity of the power electronics network [13]. In [14], Song et al. have developed an energy management algorithm for an electric vehicle supplied by batteries and SCs. Two algorithms are used: one based on Lyapunov-function regulation to stabilize the DC bus, and another based on a sliding mode approach to regulate both classic 2-quadrant converters connected to power sources, making them less reliable in case of electrical failures. It has to be noted that availability and reliability are currently major concerns so that EVs must access the mass automotive market. On one hand, the use of sliding mode controllers allows ensuring excellent performances to control both the charge/discharge of batteries and SC. On the other hand, the development of the Lyapunov function to make it stable for any operating scenario is a challenging issue.

In [15], Zhang et al. have applied a real-time unified speed regulation and control technique of a hybrid car supplied by batteries and SCs. The developed strategy is based on the Lyapunov nonlinear control technique. However, only simulation results have been reported to validate the developed control strategy. Furthermore, compared to the work reported in [14], the Lyapunov-based controller has several objectives such as the speed control of the AC motor and the energy management of batteries and SC (i.e., reducing battery stress, and extending battery lifetime). Since the controller must meet both objectives, its stability must be analyzed thoroughly to ensure good dynamic performances.

Next, Fliess et al. [16] were the first to develop differential flatness estimation (nonlinear approach). This approach has enabled the system to be an alternative representative, of which motion planning and regulator tuning is clear-cut. This theory has lately been utilized in a variety of networks in different scientific domains [17–23]. Compared to the nonlinear algorithm (i.e., sliding mode, Lyapunov, fuzzy logic) reported in [13–15], nonlinear algorithms based on differential flatness require the use of trajectory planning to implement the control laws. This trajectory planning aims at controlling different variables (e.g., currents of converters, stored energy in the DC bus and SC) to manage the energy in an EV while optimizing the performance of the system for any operating point. The use of this algorithm allows ensuring the robustness of energy management to meet the requirements of EVs (e.g., dynamic performances, the extension of the lifetime of storage devices).

For clarity, Figure 1 presents experimental results from the laboratory comparing the nonlinear differential flatness estimation and the classic PI control law during the great changed current set-point [17] of 3-phase inverter control. From these test-bench results, the differential flatness-based estimation approach presents the excellent response of the current control to its set-point i_{qREF} from step 1 A to 6 A. For this reason, it can be concluded that differential flatness control offers better dynamics than the traditional PI regulator.

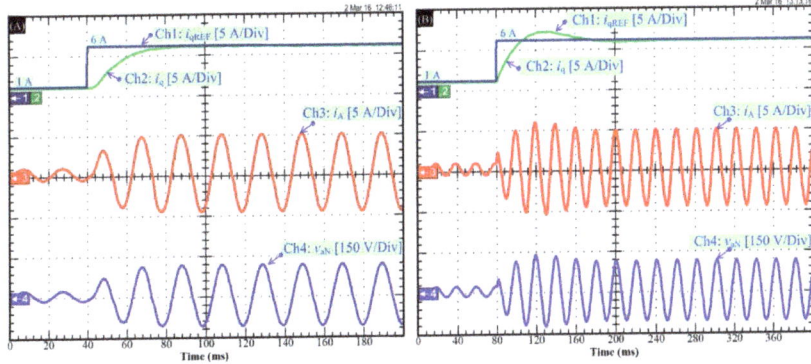

Figure 1. Experimental results: evaluation of current control of three-phase inverter drive during great changed current reference: (**A**) the differential flatness control; (**B**) a traditional PI regulation (vector control) [17].

So far, algorithms based on differential flatness have been successfully applied to power converters (e.g., 3-phase inverter and rectifier, interleaved boost converter, modular multilevel converter) [17,18,20,21], permanent magnet synchronous motor and AC servomotor [19,22,23]. Based on these previous works, the purpose of this article is to extend the use of differential flatness algorithm in an embedded DC microgrid (i.e., EV powertrain) to manage optimally its operation during static and dynamic operations. It has to be noted that the implementation of this algorithm is challenging since several variables have to be controlled to meet some expectations from the dynamic performances and stability point of view.

In this work, following the introduction part in section 1, Section 2 is detailed on the presentation of the hybrid power source–Battery/SC devices: power converter circuits and system equations. Afterward, in Section 3, energy management strategy (inner current control loops and outer energy control loops) and control laws are provided. Finally, in Section 4, an experimental test bench results are given to corroborate the proposed control law.

2. Hybrid Power Source

2.1. Power Converter Structure

SC and Battery power modules are frequently combined with buck–boost DC–DC converters (or 2-quadrant converters) to allow the charge and discharge of the storage devices. However, these converter cells are restricted when increasing power scale or when a high voltage gain is requested. Besides, the availability and reliability of electric vehicles is an important issue, which cannot be met by using a classic buck–boost converter. Therefore, the parallel power converters (parallel multi-phase converters as shown in Figure 2) with the interleaved technique are particularly suitable to meet the abovementioned issues [24–26]. The load at the DC bus is a 3-phase inverter driving a three AC motor [induction motor or permanent magnet synchronous motor (PMSM)], as a vehicle traction drive.

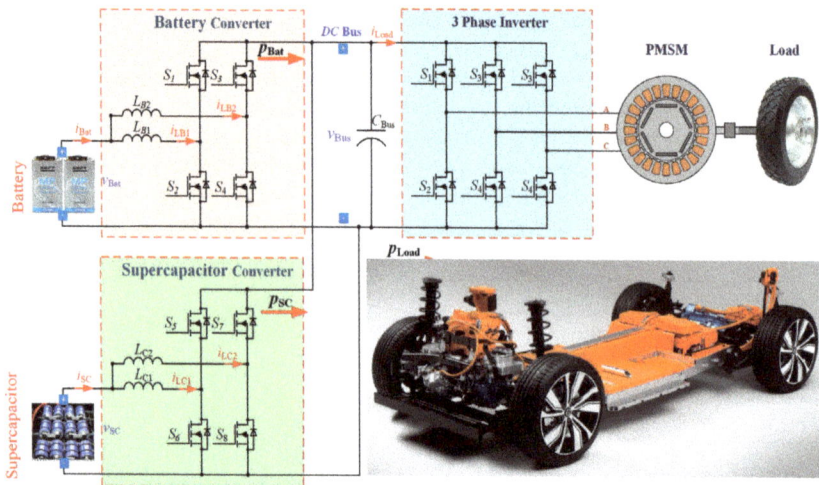

Figure 2. Proposed DC–DC converter circuits for e-Vehicle applications based on SC/battery hybrid power source.

2.2. Power Plant Modeling

The interleaved switching algorithm consists of the phase shift control signal of multiple converter modules (N) connecting in parallel [24–26]. In this article, two-phase interleaved buck–boost converters (N = 2) have been chosen to carry out this work. Indeed, by choosing only two phases, the shift control signal is equal to 180° and be easily achieved by using a prototyping dSPACE board (employed for experiment purposes). The differential equations of the two-phase buck–boost converters for SC and battery modules when the converter operates in continuous conduction mode may be expressed as [27]:

$$\frac{di_{LB1}}{dt} = \frac{1}{L_{LB1}}[v_{Bat} - (1-d_1) \times v_{Bus} - r_{LB1} \times i_{LB1}] \qquad (1)$$

$$\frac{di_{LB2}}{dt} = \frac{1}{L_{LB2}}[v_{Bat} - (1-d_2) \times v_{Bus} - r_{LB2} \times i_{LB2}] \qquad (2)$$

$$\frac{di_{LC1}}{dt} = \frac{1}{L_{LC1}}[v_{SC} - (1-d_3) \times v_{Bus} - r_{LC1} \times i_{LC1}] \qquad (3)$$

$$\frac{di_{LC2}}{dt} = \frac{1}{L_{LC2}}[v_{SC} - (1-d_4) \times v_{Bus} - r_{LC2} \times i_{LC2}] \qquad (4)$$

$$\frac{dv_{Bus}}{dt} = \frac{1}{C_{Bus}}[(1-d_1) \times i_{LB1} + (1-d_2) \times i_{LB2} + (1-d_3) \times i_{LC1} + (1-d_4) \times i_{LC2} - i_{Load}] \qquad (5)$$

where the subscripts B1, B2, C1, C2 are parameters of each cell connected to the battery (B) or SC (C); i_{Load} is the load current; v_{SC} is the SC voltage; v_{Bat} is the battery voltage; i_L is the inductor current; C_{Bus} is the total capacitance at the DC grid; L is the inductance, r_L is the parasitic resistor of the inductor; and d is the controlled duty cycle of the pulse width modulation (PWM) for power circuit. This model is simplified to carry out this work since it does not take into account some type of losses generally met in DC–DC converters (dynamics losses, switching dead-time, etc . . .) [28].

The SC and battery currents are assumed to follow their desired set-points completely. In consequence,

$$i_{Bat} = i_{BatREF} = \frac{P_{Bat}}{v_{Bat}} = \frac{P_{BatREF}}{v_{Bat}} \qquad (6)$$

$$i_{SC} = i_{SCREF} = \frac{P_{SC}}{v_{SC}} = \frac{P_{SCREF}}{v_{SC}} \qquad (7)$$

The DC grid electrostatic energy E_{Bus} and the SC electrostatic energy E_{SC} is given by [13]:

$$E_{Bus} = \frac{1}{2} \times C_{Bus} \times v^2_{Bus} \qquad (8)$$

$$E_{SC} = \frac{1}{2} \times C_{SC} \times v^2_{SC} \qquad (9)$$

The total stored energy E_T in the SC C_{SC} and in the DC bus capacitor C_{Bus} can be expressed by the following expression:

$$E_T = \frac{1}{2} \times C_{Bus} \times v^2_{Bus} + \frac{1}{2} \times C_{SC} \times v^2_{SC} \qquad (10)$$

Based on Figure 2, the differential equation of power balance is given as follows [6]:

$$\dot{E}_{Bus} = P_{Bato} + P_{SCo} - P_{Load} \qquad (11)$$

where

$$P_{Bato} = P_{Bat} - r_{Bat}\left(\frac{P_{Bat}}{v_{Bat}}\right)^2 \qquad (12)$$

$$P_{SCo} = P_{SC} - r_{SC}\left(\frac{P_{SC}}{v_{SC}}\right)^2 \qquad (13)$$

$$P_{Load} = v_{Bus} \times i_{Load} = \sqrt{\frac{2E_{Bus}}{C_{Bus}}} \times i_{Load} \qquad (14)$$

$$P_{SC} = v_{SC} \times i_{SC} = \sqrt{\frac{2E_{SC}}{C_{SC}}} \times i_{SC} \qquad (15)$$

3. Control Structure and Control Laws

3.1. Inner Current Regulations

To evaluate if the studied network is flat [19,20], one defines the flat vector output variables: y_1, y_2, y_3, y_4; state vector variables: x_1, x_2, x_3, x_4; and control vector variables: u_1, u_2, u_3, u_4 as:

$$y_1 = i_{LB1}; y_2 = i_{LB2}; y_3 = i_{LC1}; y_4 = i_{LC2} \qquad (16)$$

$$u_1 = d_1; u_2 = d_2; u_3 = d_3; u_4 = d_4 \qquad (17)$$

$$x_1 = i_{LB1}; x_2 = i_{LB2}; x_3 = i_{LC1}; x_4 = i_{LC1} \qquad (18)$$

Hence, the state vector variables: x_1, x_2, x_3, x_4 may be expressed as:

$$x_1 = \varphi_1(y_1); x_2 = \varphi_2(y_2); x_3 = \varphi_3(y_3); x_4 = \varphi_4(y_4) \qquad (19)$$

From (1) to (4) and (16) to (18), the control vector variables of u are assessed from the flat output variables y and its time derivative [19]:

$$u_1 = d_1 = 1 + \frac{1}{v_{Bus}}\left(L\dot{y}_1 - v_{Bat} + r_{LB1} \times y_1\right) = \psi_1(y_1, \dot{y}_1) \qquad (20)$$

$$u_2 = d_2 = 1 + \frac{1}{v_{Bus}}\left(L\dot{y}_2 - v_{Bat} + r_{LB1} \times y_2\right) = \psi_2(y_2, \dot{y}_2) \qquad (21)$$

$$u_3 = d_3 = 1 + \frac{1}{v_{Bus}}(L\dot{y}_3 - v_{SC} + r_{LC1} \times y_3) = \psi_3(y_3, \dot{y}_3) \tag{22}$$

$$u_4 = d_4 = 1 + \frac{1}{v_{Bus}}(L\dot{y}_4 - v_{SC} + r_{LC1} \times y_4) = \psi_4(y_4, \dot{y}_4) \tag{23}$$

The desired references of inductor current of each phase i_{LB1}, i_{LB2}, i_{LC1}, i_{LC2} are defined by y_{1REF} (=i_{LB1REF}), y_{2REF} (=i_{LB2REF}), y_{3REF} (=i_{LC1REF}), y_{4REF} (=i_{LC2REF}). Control laws (feedback regulation) reaching an exponential following of the references are written as [19,29]:

$$(\dot{y}_1 - \dot{y}_{1REF}) + K_{i11}(y_1 - y_{1REF}) + K_{i12}\int_0^t (y_1 - y_{1REF})\,d\tau = 0 \tag{24}$$

$$(\dot{y}_2 - \dot{y}_{2REF}) + K_{i21}(y_2 - y_{2REF}) + K_{i22}\int_0^t (y_2 - y_{2REF})\,d\tau = 0 \tag{25}$$

$$(\dot{y}_3 - \dot{y}_{3REF}) + K_{i31}(y_3 - y_{3REF}) + K_{i32}\int_0^t (y_3 - y_{3REF})\,d\tau = 0 \tag{26}$$

$$(\dot{y}_4 - \dot{y}_{4REF}) + K_{i41}(y_4 - y_{4REF}) + K_{i42}\int_0^t (y_4 - y_{4REF})\,d\tau = 0 \tag{27}$$

where K_{i11}, K_{i12}, K_{i21}, K_{i22}, K_{i31}, K_{i32}, K_{i41}, and K_{i42}, are the regulation parameters. A set dynamic polynomial can set the following as [30]:

$$p_1(s) = s^2 + 2\zeta_1\omega_{n1}s + \omega_{n1}^2 \tag{28}$$

$$p_2(s) = s^2 + 2\zeta_2\omega_{n2}s + \omega_{n2}^2 \tag{29}$$

$$p_3(s) = s^2 + 2\zeta_3\omega_{n3}s + \omega_{n3}^2 \tag{30}$$

$$p_4(s) = s^2 + 2\zeta_4\omega_{n4}s + \omega_{n4}^2 \tag{31}$$

$$K_{i11} = 2\zeta_1\omega_{n1}; K_{i12} = \omega_{n1}^2 \tag{32}$$

$$K_{i21} = 2\zeta_2\omega_{n2}; K_{i22} = \omega_{n2}^2 \tag{33}$$

$$K_{i31} = 2\zeta_3\omega_{n3}; K_{i32} = \omega_{n3}^2 \tag{34}$$

$$K_{i41} = 2\zeta_4\omega_{n4}; K_{i42} = \omega_{n4}^2 \tag{35}$$

where ζ_1, ζ_2, ζ_3, ζ_4, ω_{n1}, ω_{n2}, ω_{n3}, and ω_{n4} are the chosen damping ratio and defined natural frequency. Therefore, new variables are determined $\lambda_1 = \dot{y}_1$; $\lambda_2 = \dot{y}_2$; $\lambda_3 = \dot{y}_3$; $\lambda_4 = \dot{y}_4$.

The flatness-based control requires trajectory planning to implement the control law. Hence, a second-order filter has been chosen to set the battery and SC currents, dynamics commands i_{BatCOM}, i_{SCCOM} as the following equations [19]:

$$\frac{i_{BatREF}(s)}{i_{BatCCOM}(s)} = \frac{1}{\left(\frac{s}{\omega_{nt1}}\right)^2 + \frac{2\zeta_{t1}}{\omega_{nt1}}s + 1} \tag{36}$$

$$\frac{i_{SCREF}(s)}{i_{SCCCOM}(s)} = \frac{1}{\left(\frac{s}{\omega_{nt2}}\right)^2 + \frac{2\zeta_{t2}}{\omega_{nt2}}s + 1} \tag{37}$$

where ω_{nt1}, ω_{nt2}, ζ_{t1}, and ζ_{t2}, are again the desired natural frequency and dominant damping ratio, refer to Figure 3.

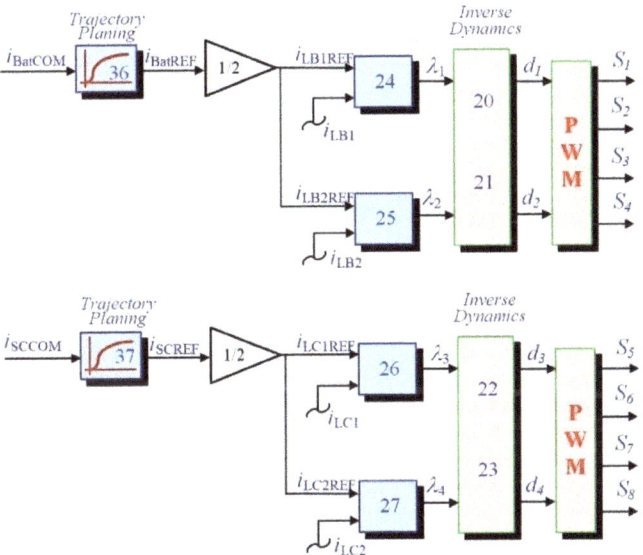

Figure 3. Proposed inner current regulation loops based on the differential flatness approach.

3.2. Outer Energy Controls

The energy control law for the studied system shown in Figure 2 consists of using two energy control laws which include DC link energy E_{Bus} and SC energy E_{SC} by two voltage variables to be regulated. Thus, based on the literature review [6,31,32], the first priority variable to be regulated is E_{Bus} and E_{SC} is a secondary variable. Given that the fastest dynamic power source of the studied system is the SC, it has been decided to use this device to supply energy to the DC link. On the other side, since the slowest dynamic power source is the battery, the latter has been chosen to provide the energy to both the SC C_{SC} and the DC bus capacitor C_{Bus} to store energy.

The flat output y_5, y_6, state variable x_5, x_6 and control variable u_5, u_6 can be expressed as follows [6]:

$$y_5 = E_{Bus}; y_6 = E_T \tag{38}$$

$$u_5 = P_{SCREF}; u_6 = P_{SCREF} \tag{39}$$

$$x_5 = v_{Bus}; x_6 = i_{SC} \tag{40}$$

$$x_5 = \sqrt{\frac{2y_5}{C_{Bus}}} = \varphi_5(y_5) \tag{41}$$

$$x_6 = \sqrt{\frac{(2y_6 - y_5)}{C_{SC}}} = \varphi_2(y_5, y_6) \tag{42}$$

By using (8)–(15), the control input vector u may be computed from the flat output variable y and its time derivatives [6]:

$$u_5 = 2P_{SCMax}\left[1 - \sqrt{1 - \frac{\dot{y}_5 + \sqrt{\frac{2y_5}{C_{Bus}}} \times i_{Load} - P_{Bato}}{P_{SCMax}}}\right] = \psi_5(y_5, \dot{y}_5) = P_{SCREF} \tag{43}$$

$$u_6 = 2p_{TMax}\left[1 - \sqrt{1 - \frac{\dot{y}_6 + \sqrt{\frac{2y_6}{C_{Bus}}} \times i_{Load}}{p_{TMax}}}\right] = \psi_6(y_6, \dot{y}_6) = p_{TREF} \qquad (44)$$

where

$$p_{SCMax} = \frac{v^2_{SC}}{4r_{SC}}, p_{TMax} = \frac{v^2_T}{4r_T} \qquad (45)$$

In this case, p_{TMax} and p_{SCMax} correspond to the set limited power of the SC and battery devices (maximum power), respectively.

In the first energy control law, the set-point for the DC link energy is defined by y_{5REF}. The closed-loop control law is written by the following expression:

$$\dot{y}_5 - \dot{y}_{5REF} + K_{v1}(y_5 - y_{5REF}) + K_{v2}\int_0^t (y_5 - y_{5REF})dt = 0 \qquad (46)$$

where K_{v1} and K_{v2} are the controller parameters. The suitable way to tune these parameters is achieved by corresponding the desired dynamic polynomial p(s), with set root positions. One can set the following equations:

$$p_5(s) = s^2 + 2\zeta_5\omega_{n5}s + \omega_{n5}^2 \qquad (47)$$

$$K_{v1} = 2\zeta_5\omega_{n5}; K_{v2} = \omega_{n5}^2 \qquad (48)$$

where ω_{n5} and ζ_5 are the chosen natural frequency and dominant damping ratio.

The control law of the DC link energy regulation detailed previously is displayed in Figure 4. The proposed control law generates an SC power desired variable p_{SCREF}. Next, this signal is divided by the SC voltage v_{SC} and restricted to keep the SC voltage within the gap [maximum V_{SCMax}, minimum V_{SCMin}] by limiting the SC module discharging current or charging current, as shown in the block "SuperC Current Limitation Function" [32]. Then, this becomes SC current command i_{SCCOM}.

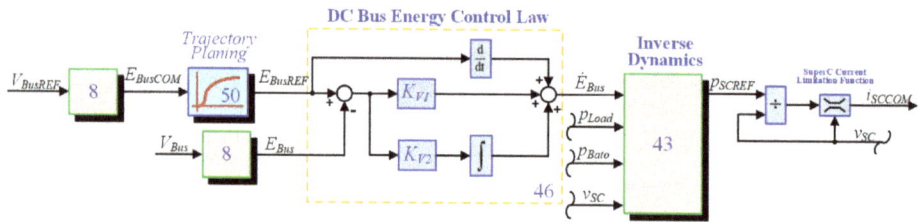

Figure 4. Proposed DC bus energy stabilization.

Second, for the total energy control law (charging SC), the set-point is represented by y_{6REF}. Indeed, the SC energy has been set as slower dynamics than the DC link energy and the SC device features a high energy storage capacity. Again, the feedback control law is expressed as follows:

$$\dot{y}_6 - \dot{y}_{6REF} + K_{v3}(y_6 - y_{6REF}) = 0 \qquad (49)$$

Refer to Figure 5, the proposed control law based on the differential flatness approach estimates the battery power set-point p_{BatREF}. Then, it is divided by the battery sensor voltage v_{Bat} and generates the battery current command i_{BatCOM}, limited within i_{BatMax} and i_{BatMin} (=0 A).

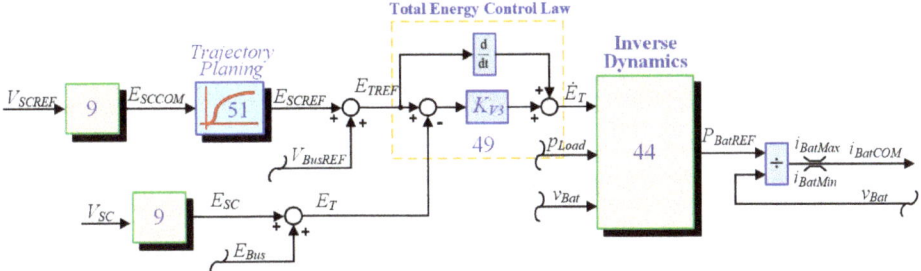

Figure 5. Proposed total energy stabilization.

Finally, the soft-start system in the smooth transition point of view, the energy command has to be generated for the converter and restricted set-point profiles for smooth transforms between operating points. The trajectory motion planning for the reference signals E_{BusREF} and E_{TREF} are written as [19]:

$$\frac{y_{BusREF}(s)}{y_{BusCOM}(s)} = \frac{1}{\left(\frac{s}{\omega_{nt3}}\right)^2 + \frac{2\zeta_{t3}}{\omega_{nt3}}s + 1} \quad (50)$$

$$\frac{y_{SCREF}(s)}{y_{SCCOM}(s)} = \frac{1}{\left(\frac{s}{\omega_{nt4}}\right)^2 + \frac{2\zeta_{t4}}{\omega_{nt4}}s + 1} \quad (51)$$

where ω_{n3}, ω_{n4}, ζ_{t3}, and ζ_{t4} are the chosen natural frequency and dominant damping ratio.

4. Performance Validation

4.1. Test Bench Setup and Flatness Control Parameters

To validate the effectiveness of the studied control algorithm for system management in an embedded DC microgrid, an experimental platform has been realized in the Renewable Energy Research Centre (RERC) at King Mongkut's University of Technology North Bangkok, as presented in Figure 6. The DC–DC converter circuit parameters are provided in Table 1. The SC module is 160 V, 6 F, (BMOD0006 E160 B02—Maxwell Technologies Company) and the battery module is 40 Ah, 120 V (Panasonic Technology). The studied DC link voltage is 310 V, meeting the high DC grid voltage requested for automotive applications. The inner current regulation parameters are given in Table 2. Parameters related to the outer energy control parameters are shown in Tables 3 and 4, respectively. Additionally, the battery current slope control can be seen in Table 4. This number has been approved by experimental results to have the highest slope of the battery device. Besides, the proposed control algorithm (based on Figures 3–5), which generates desired duty cycle signals d for both interleaved buck–boost converters, and regulates the total stored energy (including the DC bus and SC), has been realized in MATLAB®—Simulink environment. Then, it has been implemented into the real-time board DS1202 dSPACE–MicroLabBox (2 GHz dual-core real-time microprocessor and user-programmable Field-Programmable Gate Array FPGA) with the sampling frequency (timer interrupt) of 25,000 Hz. This value is related to the high switching frequency of both interleaved buck–boost DC–DC converters.

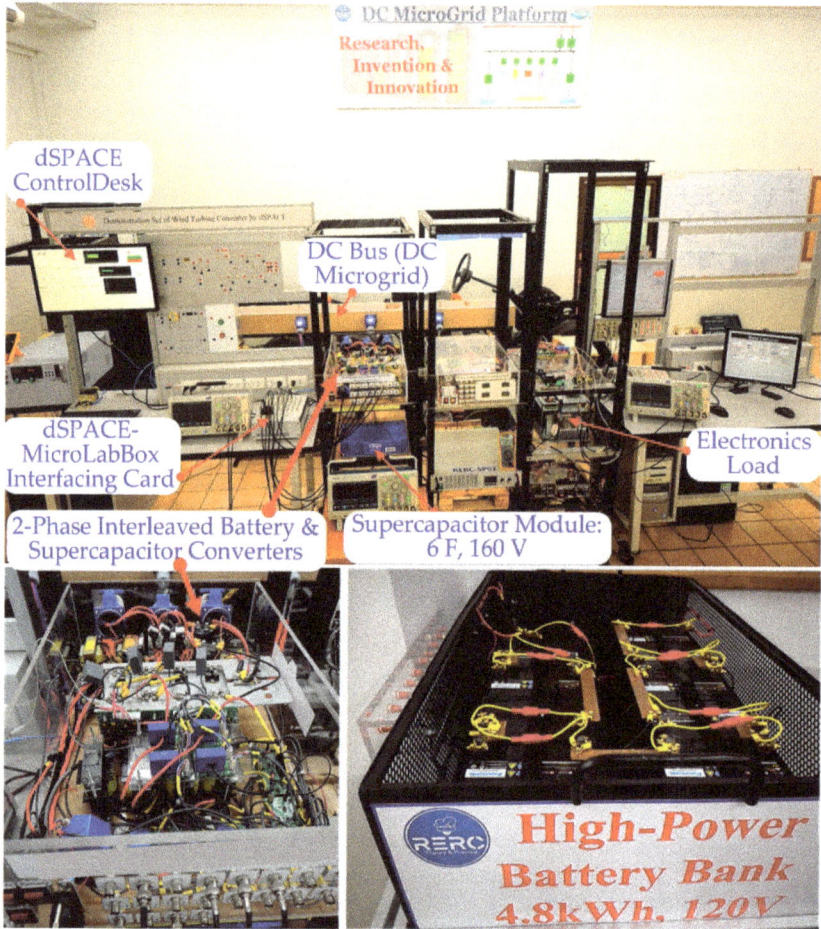

Figure 6. Hybrid test bench platform at the laboratory RERC–KMUTNB.

Table 1. Converter parameters.

Parameters	Value
Rated DC grid voltage, v_{Bus}	310 V
Nominal battery voltage, v_{Bat}	120 V
Nominal SC voltage, v_{SC}	140 V
Inductor $L_{B1} = L_{B2} = L_{C1} = L_{C2}$	200 µH
Equivalent serial resistances $R_{LB1} = R_{LB2} = R_{LC1} = R_{LC2}$	0.06 Ω
Total DC Bus Capacitors	2000 µF, 900 V
Power MOSFETs Switching Frequency, f_S	25 kHz

Table 2. Current control parameters.

Parameters	Value
$\zeta_1 = \zeta_2 = \zeta_3 = \zeta_4$	0.7
$\omega_{n1} = \omega_{n2} = \omega_{n3} = \omega_{n4}$	8000
$K_{i11} = K_{i21} = K_{i31} = K_{i41}$	11,200
$K_{i12} = K_{i22} = K_{i32} = K_{i42}$	64,000,000

Table 3. DC link energy regulation parameters.

Parameters	Value	Parameters	Value
v_{BusREF}	310 V	p_{SCMax}	+3600 W
ζ_5	0.7	p_{SCMin}	−3600 W
ω_{n5}	80 rad·s^{-1}	v_{SCMax}	160 V
K_{v1}	112	V_{SCMin}	70 V
K_{v2}	6400	$i_{SCRated}$	30 A

Table 4. Total energy regulation parameters.

Parameters	Value	Parameters	Value
v_{SCREF}	140 V	P_{BatMax}	+2100 W
C_{SC}	6 F	P_{BatMin}	0 W
ζ_6	1	I_{BatMax}	+18 A
ω_{n6}	0.8 rad·s^{-1}	I_{BatMin}	0 A
K_{v3}	0.1		

Firstly, the oscilloscope waveforms in Figures 7 and 8 show the steady-state switching behaviors of the studied interleaved 2-quadrant DC–DC converters for the battery and SC module at different current references. In Figure 7, the following signals are available:

- Ch1: the battery current set-point i_{BatREF} at +20 A (battery discharging mode);
- Ch2: the measured battery current i_{Bat};
- Ch3: the 1st inductor current i_{LB1};
- Ch4: the 2nd inductor current iLB2.

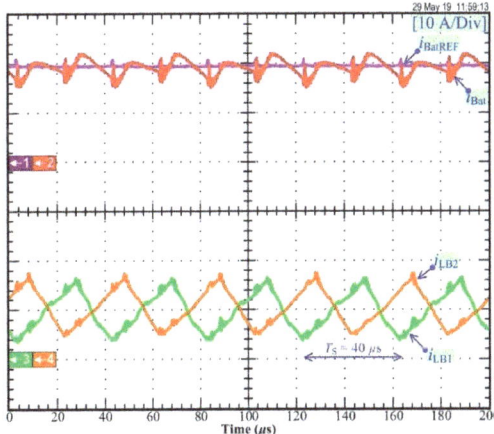

Figure 7. Steady-state waveforms of the battery converter at a discharge of 20 A.

The obtained results show i_{LB1} and i_{LB2} where their average values are equal to $i_{BatREF}/2$ (i.e., 10 A). It can be observed that the current of the i_{Bat} battery is the sum of i_{LB1} and i_{LB2}. It is equal to 20 A according to i_{BatREF}, but there is a small current ripple due to the use of an interleaved buck–boost converter.

Then, Figure 8 presents the experimental results in charge mode of the SC at 15 A or i_{SCREF} = −15 A. In Figure 8, the following measurements are available:

- Ch1: the SC current set-point i_{SCREF};

- Ch2: the measured SC current i_{SC};
- Ch3: the 1st inductor current i_{LC1};
- Ch4: the 2nd inductor current i_{LC2}.

The current i_{SC} is equal to i_{SCREF} and is close to a pure DC current. The i_{LC1} and i_{LC2} have a very small ripple, with an average value of $i_{SCREF}/2$ equal to −7.5 A.

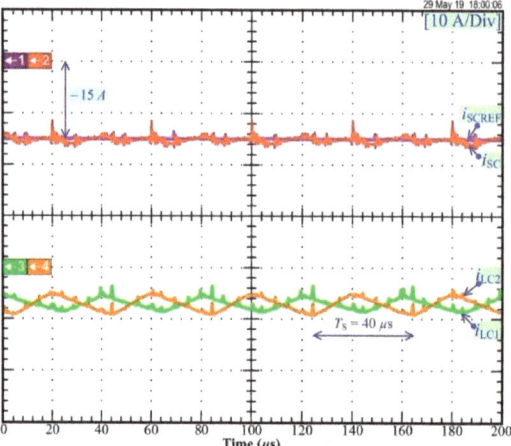

Figure 8. Steady-state waveforms of the SC converter at −15 A.

In Figure 9, it is shown the dynamics response when the battery reference i_{BatREF} increases instantaneously from initial +5 A to final +15 A. The following signals are available:

- Ch1: the battery reference i_{BatREF};
- Ch2: the measured input battery current i_{Bat};
- Ch3: the 1st inductor current i_{LB1};
- Ch4: the 2nd inductor current i_{LB2}.

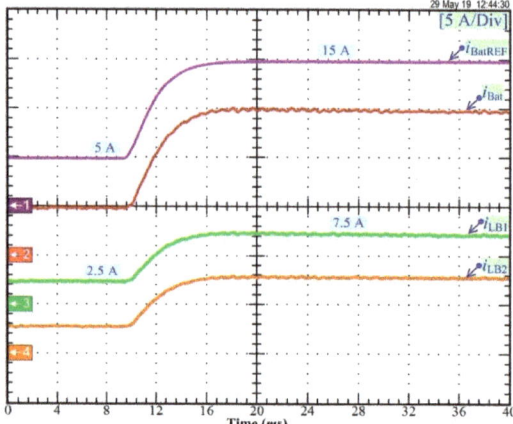

Figure 9. Experimental results: battery current response during i_{BatREF} changing from 5 A to 15 A.

First of all, it can be seen that the current i_{Bat} follows perfectly the reference i_{BatREF}. As a result of the operating conditions change, the response of the current i_{Bat} is damped due to the use of a 2nd order filter equation. Like in the previous results, the battery current i_{Bat} ($=i_{BatREF}$) is the summation of i_{LB1} and i_{LB2}.

Finally, Figure 10 shows the dynamics response by modifying the equilibrium points of the SC current from 5 A to −5 A. The following signals are available:

- Ch1: the SC current reference i_{SCREF};
- Ch2: the measured SC current i_{SC};
- Ch3: the 1st inductor current i_{LC1};
- Ch4: the 2nd inductor current i_{LC2}.

It can be noted that the reference i_{SCREF} has a steep slope (2nd order filter characteristics) and the current i_{SC} follows i_{SCREF} completely, with a low settling time of 10 ms.

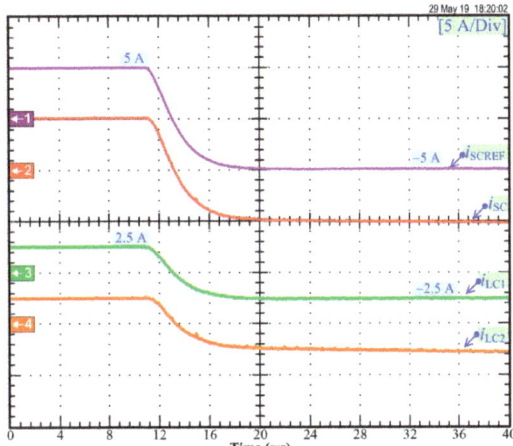

Figure 10. Experimental results: SC current response during i_{SCREF} changing from 5 A (discharging) to −5 A (charging).

4.2. Hybrid Power Plant Load Cycles

To assess the performance of the differential flatness-based controller in regulating the DC bus energy by using SC, dynamic tests were carried out by modifying the load power from 0 W to 3 kW. The obtained results are shown in Figure 11, providing the DC bus voltage v_{Bus}, the SC voltage v_{SC}, the load power (disturbance) p_{Load}, and the SC power p_{SC} in transient and steady-state operation. Given that the first operating condition does not consider any load (i.e., from 0 to 120 ms), the storage device is full of charge ($v_{SCREF} = v_{SC} = 140$ V), and the DC link voltage is controlled at 310 V ($v_{BusREF} = v_{Bus} = 310$ V). Hence, the SC and battery powers are zero. Since the battery power is set to $p_{BatREF} = 0$ (see Figure 5), the role of SC module to maintain the DC bus voltage stability can be examined.

After that, at t = 120 ms, the load power (disturbance) instantly changes from 0 W to 3 kW (positive transition ↑). It can be observed that the SC module provides the steady-state and dynamics load power demand. The DC bus voltage is slightly influenced by the large load disturbance by utilizing the nonlinear differential flatness-based estimation for the proposed system.

Then, Figure 12 presents experimental results during a load drive cycle. Here, the electronic load has been changed to emulate the electric vehicle characteristics: overload, positive power (acceleration mode) or negative power (regenerative braking), and positive and negative transients. This Figure depicts the DC bus voltage v_{Bus}, the battery voltage v_{Bat}, the load power p_{Load}, the battery power p_{Bat}, the SC power p_{SC}, the battery current i_{Bat}, the SC current i_{SC}, and the SC voltage v_{SC} (represents the

SC state-of-charge). Like in Figure 11, it can be noted that the DC bus voltage is not impacted by the large perturbation.

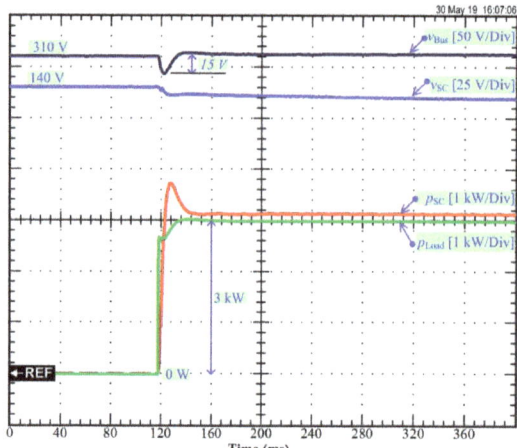

Figure 11. Experimental results: DC bus voltage stabilization of the studied hybrid power plant during load step from 0 to 3 kW.

At the beginning, the load power = 600 W and the SC module is likewise full-of-charge ($v_{SC} = v_{SCREF} = 140$ V); consequently, the battery power is equal to 600 W for the load; while the SC is the second source, of which its power is equal to zero.

At t_1, the load power changes from 600 W to the 3600 W (overload, high energy requested by the load). The following explanation can be made:

1. The SC provides power most of the dynamic large load of 3600 W.
2. Concurrently, the battery power goes up to a limited level (maximum value setting) of 2100 W at t_2.
3. The SC device provides most of the power dynamics that are requested during the load step and continue in discharge mode.

After that at t_3, the load power demand decreases drastically from 3600 to 600 W; consequently, the SC module changes its operating mode from discharging to charging. It can be noted that:

1. The battery remains constant supplying its maximum power (limited power) of around 2100 W. It means the battery provides powers to load and charge the SC module.
2. At t_4 ($v_{SC} = 130$ V), the SC module is almost charged at 140 V, and afterward, the SC power decreases. Accordingly, the battery power is reduced gradually.
3. At t_5, the SC is full of charge at 140 V; after that, the SC current is zero. Synchronously, the battery main source provides only energy to the load 600 W.

Afterward, at t_6, the load power changes from 600 W to −600 W to emulate vehicle braking. The SC is extremely charged and recovers the energy at the DC bus; concurrently, the battery power declines (with a limited slope) to zero.

At t_7, the SC absorbs the negative power provided only by the load. Thus, the SC is in overcharged state, i.e., $v_{SC} > v_{SCREF} = 140$ V.

Subsequently, at t_8, the load power changes immediately from −600 W to +600 W, where $v_{SC} = 155$ V (overcharged); therefore, the SC modules changes its operating mode from charging to discharging and the battery remains in idle state (the current limitation at 0 A). Therefore, the requested load power is provided by only the SC source.

At t_9, when v_{SC} reaches v_{SCREF} equal to 140V, the SC power drops to 0 W; then, the battery current increases to provide the power for the load requested.

At t_{10}, the battery power remains at a constant level of 600 W.

Finally, at t_{11}, the load power changes from 600 W to zero (stop mode). The SC recovers the energy and the battery power decreases to zero. It can be concluded that the hybrid network enables keeping energy balance by using the proposed energy control law.

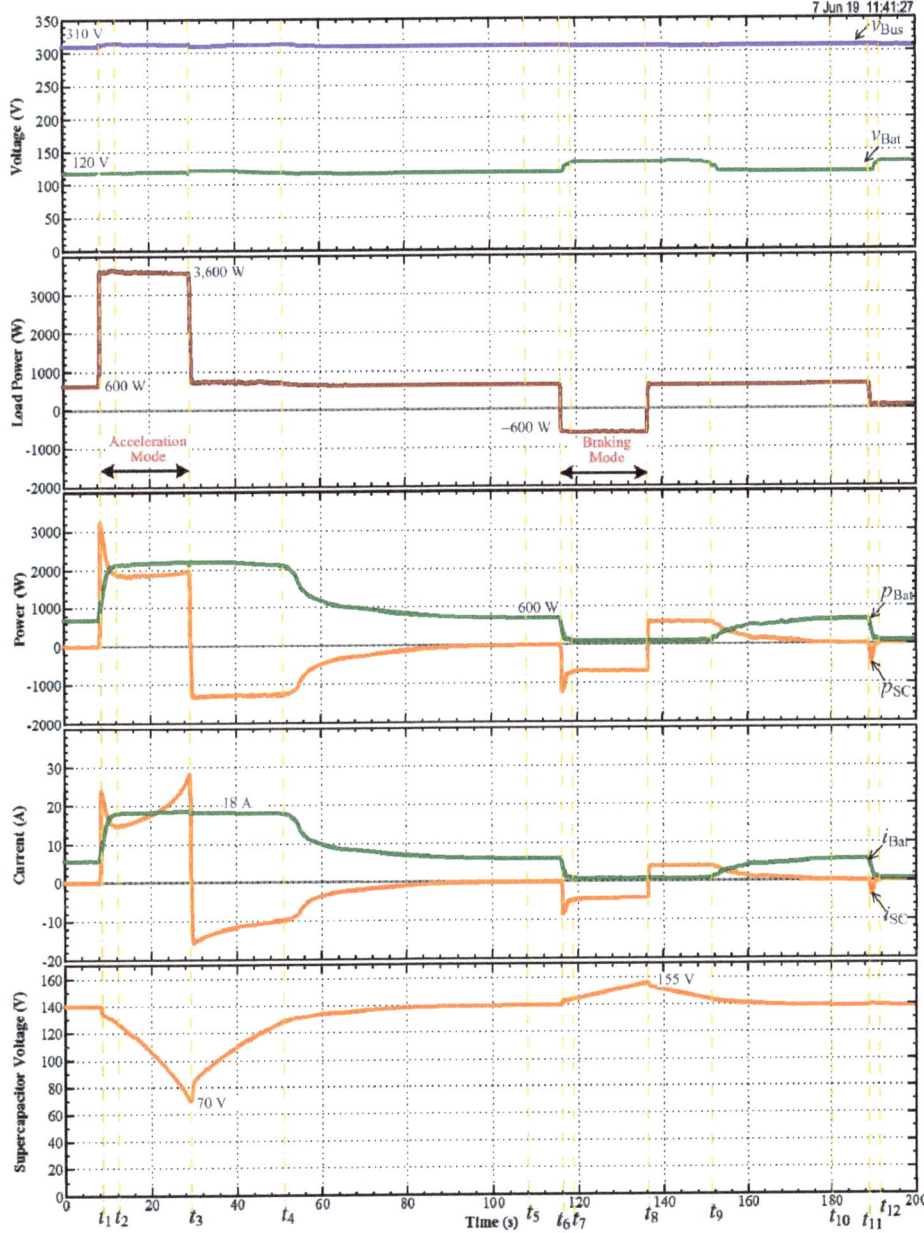

Figure 12. Hybrid source response during load drive cycles.

4.3. Comparison of the Performances Compared to the Previous Works

As highlighted in the introduction, the previous works reported in [13–15] have been focused on the energy management of electric vehicles based on various nonlinear algorithms (e.g., sliding mode, Lyapunov, fuzzy logic). However, only in [13,14], the developed control algorithms have been validated experimentally on load power profile. Both load power profiles present the same dynamic operations (i.e., three acceleration modes and one braking mode); whereas the chosen load power profile in this work includes one acceleration and braking operation as shown in Figure 12. The dynamic performances obtained both for batteries and SC are close to those obtained in [14]. In comparison, reported results in [13] show that the responses of sources (i.e., batteries, SC) to different dynamics are slower. Indeed, since PI current controllers are used to control both buck–boost converters connected to the sources, the required time to track the different references are longer. In conclusion, the algorithms based on Lyapunov-function and sliding mode controllers or differential flatness offer excellent dynamic performance while preserving the good operation of the storage devices against fast dynamics and keeping the stability of the DC bus. Besides, the use of interleaved buck–boost converters allows reducing the current ripple (as shown in Figures 9 and 10) and ensuring the availability of both converters in case of power switch failures.

5. Conclusions

The key objective of this work is to study new energy management of supercapacitor/battery hybrid sources for modern electric transportation applications. The combination of battery and ultracapacitor is suitable for the hybridization network since it offers high power and high energy densities. The control law allows avoiding the fast dynamic current transient of battery and decreasing the battery stresses. For this reason, the proposed hybrid system allows optimizing its life span. However, this issue does not come within the scope of this article to reveal the battery's lifetime.

The developed control strategy has been tested with an experimental prototype platform implemented in the laboratory, including a battery bank (120 V, 140 Ah—Panasonic) and a supercapacitor module (6 F, 160 V—Maxwell Technologies). The obtained experimental results have enabled validating the outstanding performances of the developed control strategy during the steady-state and dynamic state.

The differential flatness control theory is principally a model-based concept. It is mandatory to identify model parameters (such as r_{LB1}, r_{LB2}, etc.) to determine the flatness property, accurately [17]. To enhance the proposed control strategy, some parameter observers (or online state observers) will be studied in future works.

Author Contributions: Conceptualization, Methodology, and Writing—Original Draft Preparation: B.Y., D.G.; formal analysis and investigation: D.G.; Validation and Supervision: P.T., D.G.; Writing-Review and Editing: P.T., D.G., N.B. All authors have read and agreed to the published version of the manuscript.

Funding: This work was supported by the international research cooperation program of the "UL-KMUTNB International Research Partnerships: Electrical Engineering Thai-French Research Center (EE-TFRC)" between Groupe de Recherche en Energie Electrique de Nancy (GREEN), Université de Lorraine (UL) and Renewable Energy Research Centre (RERC), King Mongkut's University of Technology North Bangkok (KMUTNB) under Grant No. KMUTNB-61-GOV-01-02.

Acknowledgments: The authors would like to thank Pongsiri Mungporn for operating the test bench system during experimentations.

Conflicts of Interest: The authors declare no conflict of interest.

References

1. Tu, H.; Feng, H.; Srdic, S.; Lukic, S. Extreme fast charging of electric vehicles: A technology overview. *IEEE Trans. Transp. Electrif.* **2019**, *5*, 861–878. [CrossRef]
2. Skouras, T.A.; Gkonis, P.K.; Ilias, C.N.; Trakadas, P.T.; Tsampasis, E.G.; Zahariadis, T.V. Electrical Vehicles: Current State of the Art, Future Challenges, and Perspectives. *Clean Technol.* **2020**, *2*, 1–16. [CrossRef]

3. Sun, X.; Li, Z.; Wang, X.; Li, C. Technology development of electric vehicles: A review. *Energies* **2020**, *13*, 90. [CrossRef]
4. Hoai, H.-K.; Chen, S.-C.; Than, H. Realization of the sensorless permanent magnet synchronous motor drive control system with an intelligent controller. *Electronics* **2020**, *9*, 365. [CrossRef]
5. De Santis, M.; Agnelli, S.; Patanè, F.; Giannini, O.; Bella, G. Experimental study for the assessment of the measurement uncertainty associated with electric powertrain efficiency using the back-to-back direct method. *Energies* **2018**, *11*, 3536. [CrossRef]
6. Sikkabut, S.; Mungporn, P.; Ekkaravarodome, C.; Bizon, N.; Tricoli, P.; Nahid-Mobarakeh, B.; Pierfederici, S.; Davat, B.; Thounthong, P. Control of high-energy high-power densities storage devices by li-ion battery and supercapacitor for fuel cell/photovoltaic hybrid power plant for autonomous system applications. *IEEE Trans. Ind. Appl.* **2016**, *52*, 4395–4407. [CrossRef]
7. Mutarraf, M.U.; Terriche, Y.; Niazi, K.A.K.; Vasquez, J.C.; Guerrero, J.M. Energy storage systems for shipboard microgrids—A review. *Energies* **2018**, *11*, 3492. [CrossRef]
8. Manandhar, U.; Wang, B.; Zhang, X.; Beng, G.H.; Liu, Y.; Ukil, A. Joint control of three-level DC–DC converter interfaced hybrid energy storage system in DC microgrids. *IEEE Trans. Energy Convers.* **2019**, *34*, 2248–2257. [CrossRef]
9. Mukherjee, N.; Strickland, D. Control of cascaded DC–DC converter-based hybrid battery energy storage systems—part I: Stability issue. *IEEE Trans. Ind. Electron.* **2016**, *63*, 2340–2349. [CrossRef]
10. Roche, M.; Shabbir, W.; Evangelou, S.A. Voltage control for enhanced power electronic efficiency in series hybrid electric vehicles. *IEEE Trans. Veh. Technol.* **2017**, *66*, 3645–3658. [CrossRef]
11. Vargas, U.; Lazaroiu, G.C.; Tironi, E.; Ramirez, A. Harmonic modeling and simulation of a stand-alone photovoltaic-battery-supercapacitor hybrid system. *Int. J. Electr. Power Energy Syst.* **2019**, *105*, 70–78. [CrossRef]
12. Hu, J.; Shan, Y.; Xu, Y.; Guerrero, J.M. A coordinated control of hybrid ac/dc microgrids with PV-wind-battery under variable generation and load conditions. *Int. J. Electr. Power Energy Syst.* **2019**, *104*, 583–592. [CrossRef]
13. Marzougui, H.; Kadri, A.; Martin, J.; Amari, M.; Pierfederici, S.; Bacha, F. Implementation of energy management strategy of hybrid power source for electrical vehicle. *Energy Convers. Manag.* **2019**, *195*, 830–843. [CrossRef]
14. Song, Z.; Hou, J.; Hofmann, H.; Li, J.; Ouyang, M. Sliding-mode and Lyapunov function-based control for battery/supercapacitor hybrid energy storage system used in electric vehicles. *Energy* **2017**, *122*, 601–612. [CrossRef]
15. Zhang, L.; Ye, X.; Xia, X.; Barzegar, F. A real-time energy management and speed controller for an electric vehicle powered by a hybrid energy storage system. *IEEE Trans. Ind. Inform.* **2020**. [CrossRef]
16. Fliess, M.; Lévine, J.; Martin, P.; Rouchon, P. Flatness and defect of nonlinear systems: Introductory theory and examples. *Int. J. Control* **1995**, *61*, 1327–1361. [CrossRef]
17. Mungporn, P.; Thounthong, P.; Sikkabut, S.; Yodwong, B.; Chunkag, V.; Kumam, P.; Bizon, N.; Nahid-Mobarakeh, B.; Pierfederici, S. Dynamics improvement of 3-phase inverter with output LC-filter by using differential flatness based control for grid connected applications. In Proceedings of the IEEE 19th International Conference on Electrical Machines and Systems (ICEMS), Chiba, Japan, 13–16 November 2016; pp. 1–6.
18. Poonnoy, N.; Mungporn, P.; Thounthong, P.; Sikkabut, S.; Yodwong, B.; Boonseng, A.; Ekkaravarodome, C.; Kumam, P.; Bizon, N.; Nahid-Mobarakeh, B.; et al. Differential flatness based control of 3-phase AC/DC converter. In Proceedings of the IEEE 2017 European Conference on Electrical Engineering and Computer Science (EECS), Bern, Switzerland, 17–19 November 2017; pp. 136–141.
19. Thounthong, P.; Sikkabut, S.; Poonnoy, N.; Mungporn, P.; Yodwong, B.; Kumam, P.; Bizon, N.; Nahid-Mobarakeh, B.; Pierfederici, S. Nonlinear differential flatness-based speed/torque control with state-observers of permanent magnet synchronous motor drives. *IEEE Trans. Ind. Appl.* **2018**, *54*, 2874–2884. [CrossRef]
20. Mehrasa, M.; Pouresmaeil, E.; Taheri, S.; Vechiu, I.; Catalão, J.P.S. Novel control strategy for modular multilevel converters based on differential flatness theory. *IEEE J. Emerg. Sel. Top. Power Electron.* **2018**, *6*, 888–897. [CrossRef]
21. Huangfu, Y.; Li, Q.; Xu, L.; Ma, R.; Gao, F. Extended state observer based flatness control for fuel cell output series interleaved boost converter. *IEEE Trans. Ind. Appl.* **2019**, *55*, 6427–6437. [CrossRef]

22. Sriprang, S.; Nahid-Mobarakeh, B.; Pierfederici, S.; Takorabet, N.; Bizon, N.; Kumam, P.; Mungporn, P.; Thounthong, P. Robust flatness control with extended Luenberger observer for PMSM drive. In Proceedings of the 2018 IEEE Transportation Electrification Conference and Expo, Asia-Pacific (ITEC Asia-Pacific), Bangkok, Thailand, 6–9 June 2018; pp. 1–8.
23. Sriprang, S.; Nahid-Mobarakeh, B.; Takorabet, N.; Pierfederici, S.; Bizon, N.; Kuman, P.; Thounthong, P. Permanent magnet synchronous motor dynamic modeling with state observer-based parameter estimation for AC servomotor drive application. *Appl. Sci. Eng. Prog.* **2019**, *12*, 286–297. [CrossRef]
24. Ma, R.; Xu, L.; Xie, R.; Zhao, D.; Huangfu, Y.; Gao, F. Advanced robustness control of DC–DC converter for proton exchange membrane fuel cell applications. *IEEE Trans. Ind. Appl.* **2019**, *55*, 6389–6400. [CrossRef]
25. Mungporn, P.; Thounthong, P.; Sikkabut, S.; Yodwong, B.; Ekkaravarodome, C.; Kumam, P.; Junkhiaw, S.T.; Bizon, N.; Nahid-Mobarakeh, B.; Pierfederici, S. Differential flatness-based control of current/voltage stabilization for a single-phase PFC with multiphase interleaved boost converters. In Proceedings of the 2017 IEEE European Conference on Electrical Engineering and Computer Science (EECS 2017), Bern, Switzerland, 17–19 November 2017; pp. 124–130. [CrossRef]
26. Thammasiriroj, W.; Chunkag, V.; Phattanasak, M.; Pierfederici, S.; Davat, B.; Thounthong, P. Nonlinear model based single-loop control of interleaved converters for a hybrid source system. *ECTI Trans. Electr. Eng. Electron. Commun.* **2017**, *15*, 19–31.
27. Bougrine, M.; Benalia, A.; Delaleau, E.; Benbouzid, M. Minimum time current controller design for two-interleaved bidirectional converter: Application to hybrid fuel cell/supercapacitor vehicles. *Int. J. Hydrogen Energy* **2018**, *43*, 11593–11605. [CrossRef]
28. Erickson, R.; Maksimović, D. *Fundamentals of Power Electronics*, 2nd ed.; Kluwer Academic: New York, NY, USA, 2004.
29. Sriprang, S.; Nahid-Mobarakeh, B.; Pierfederici, S.; Takorabet, N.; Bizon, N.; Kumam, P.; Mungporn, P.; Thounthong, P. Robust flatness-based control with state observer-based parameter estimation for PMSM drive. In Proceedings of the 2018 IEEE International Conference on Electrical Systems for Aircraft, Railway, Ship Propulsion and Road Vehicles and International Transportation Electrification Conference (ESARS-ITEC 2018), Nottingham, UK, 7–9 November 2018; pp. 1–6. [CrossRef]
30. Cordero, A.; Maimó, J.G.; Torregrosa, J.R.; Vassileva, M.P. Iterative methods with memory for solving systems of nonlinear equations using a second order approximation. *Mathematics* **2019**, *7*, 1069. [CrossRef]
31. Thammasiriroj, W.; Chunkag, V.; Phattanasak, M.; Pierfederici, S.; Davat, B.; Thounthong, P. Simplified single-loop full-flatness control of a hybrid power plant. In Proceedings of the 2016 IEEE SICE International Symposium on Control Systems (ISCS 2016), Nagoya, Japan, 7–10 March 2016. [CrossRef]
32. Thounthong, P.; Raël, S.; Davat, B. Control strategy of fuel cell/supercapacitors hybrid power sources for electric vehicle. *J. Power Sources* **2006**, *158*, 806–814. [CrossRef]

 © 2020 by the authors. Licensee MDPI, Basel, Switzerland. This article is an open access article distributed under the terms and conditions of the Creative Commons Attribution (CC BY) license (http://creativecommons.org/licenses/by/4.0/).

Article

Electricity Cost Optimization in Energy Storage Systems by Combining a Genetic Algorithm with Dynamic Programming

Seung-Ju Lee and Yourim Yoon *

Department of Computer Engineering, Gachon University, 1342 Seongnamdaero, Sujeong-gu, Seongnam-si, Gyeonggi-do 13120, Korea; poketred13@gmail.com
* Correspondence: yryoon@gachon.ac.kr; Tel.: +82-31-750-5326

Received: 3 August 2020; Accepted: 4 September 2020; Published: 7 September 2020

Abstract: Recently, energy storage systems (ESSs) are becoming more important as renewable and microgrid technologies advance. ESSs can act as a buffer between generation and load and enable commercial and industrial end users to reduce their electricity expenses by controlling the charge/discharge amount. In this paper, to derive efficient charge/discharge schedules of ESSs based on time-of-use pricing with renewable energy, a combination of genetic algorithm and dynamic programming is proposed. The performance of the combined method is improved by adjusting the size of the base units of dynamic programming. We show the effectiveness of the proposed method by simulating experiments with load and generation profiles of various commercial electricity consumers.

Keywords: energy storage systems; renewable energy sources; genetic algorithms; dynamic programming

1. Introduction

An energy storage system (ESS) is a system that can store energy and provide it for consumer use for a certain time period at an acceptable level. In an electrical grid system, the ESS can be used to adjust the electricity usage and charge. The ESS is charged and discharged when the electricity usage is low and high, respectively. In other words, the overall energy efficiency of the system is improved and the energy flow from the electrical grid connected to the system is stabilized. Reliability is the key to the effective use of smart grid systems and new renewable energy sources [1]. Thus, the demand for ESSs is increasing [2–5].

The ESS acts as buffer between energy generation and load. New renewable energy sources often generate electricity even when the electrical energy usage is low. To avoid the waste of energy, the energy can be stored in the ESS and withdrawn from the ESS when needed, thereby increasing the energy efficiency. Energy providers benefit from more predictable power generation requirements. The ESS provides reliable and high-quality electricity to all industrial, commercial, and residential users [6,7].

Dynamic electricity pricing has been used with new technologies such as smart meters. In dynamic electricity pricing, the electricity charges vary depending on the time of day and time-of-use (TOU) tariffs. Energy providers can set high prices during times of high energy use, which encourages the consumers to avoid the overuse of energy, thereby preventing emergencies such as power outages [8,9]. In general, the TOU pricing consists of two or three pricing tiers (e.g., light load, heavy load, and overload). The price depends on the time of day. Many utilities in various countries such as the US energy company, Pacific Gas and Electric (PG&E) [10], the Canadian energy company, Hydro Ottawa (HO) [11], Korea Electric Power Corporation (KEPCO) [12], and Taiwan Power Company (TPC) [13] offer TOU pricing for commercial and industrial customers.

In a pricing system in which the prices vary depending on the time of day, consumers can reduce electricity costs by using energy during times with low electricity prices. The ESS plays a crucial role in the dynamic pricing policy. By storing energy during low load periods and using the stored energy during a high pricing tier period, consumers can avoid high electricity bills. To maximize electricity bill savings based on dynamic pricing, various studies have been carried out regarding the scheduling of the charge/discharge amount of the ESS [14–17] or consumer electricity planning solution [18]. These studies focused on various optimization methods such as dynamic, linear, nonlinear, and mixed integer linear programming as well as stochastic and particle swarm optimization and genetic algorithms.

The most widely used method is dynamic programming (DP), which was first introduced by Maly and Kwan [19] who focused on minimizing electrical energy usage costs without reducing the battery life. Van de Ven et al. [20] focused on minimizing the installation costs of the ESS. They emphasized the user demand and price, such as the Markov decision process, which can be solved by DP. Koutsopoulos et al. [21] proposed an optimal ESS control system from the viewpoint of facility providers and solved the offline problem in a limited time period using DP. Romaus et al. [22] suggested stochastic DP for the energy management of the hybrid ESS for electric vehicles.

In this paper, we propose a method that solves an ESS scheduling problem for electricity cost optimization for enterprise ESSs with dynamic pricing and renewable energy sources. We suggest a DP approach that considers the forecasts of the power generation and load for 24 h. We also aim to improve the performance of the optimization method by combining DP with a genetic algorithm (GA). Although several studies focused on DP, it has some problems when applied to commercial electrical systems. In the case of residential electrical systems, the amount of power used is small such that the memory and time constraints for DP are relatively low, so DP can be a reasonable choice for this case. However, the amount of power used in commercial electrical systems is large. In that case, DP uses large memory sizes and time resources for finding good solutions. We can increase the size of the base unit in DP for reducing memory and time resources. However, in this case, errors will likely increase. To resolve this problem, we employ a genetic algorithm, which is one of the metaheuristic methods that can be used to identify near-optimal values (not the optimal values). By using the solutions of DP with a large base unit as the initial population of a GA, the memory and time constraints of DP can be satisfied.

In addition to charges based on the electrical energy usage, electricity bills may include a demand charge, which is determined by the highest amount of power (kW) during the billing period multiplied by the relevant demand charge rate ($/kW). The demand charge rate is usually fixed when a commercial or industrial customer signs the contract [23]. By this demand charge, utility companies can charge customers consuming large amount of power more fees for their use of extra resources associated with the power maintenance [24]. In a customer's point of view, the larger the highest amount of power used during the billing period, the larger the demand charge. Hence, customers with demand charge should try to reduce the highest amount of power during the billing period to decrease their electricity bills. Since it is difficult to measure the exact amount of power practically, the highest amount of power is usually measured by calculating the electrical energy drawn during a predetermined time interval. We consider scenarios with demand charge and those without demand charge in this study.

The DP without demand charge produces reasonable dynamic pricing results. However, the design of DP is difficult if there are demand charges because the objective functions become much more complex. A metaheuristic method, such as a GA, can perform better than DP in such a case. In this paper, we compare the performances of DP and a GA for cases with demand charge and those without demand charge, and we propose more effective algorithms for each case by combining DP with the GA.

There have also been studies on the optimization of operation of ESS considering renewable energy in microgrid using various strategies including other metaheuristic algorithms. Wang and Huang [2] proposed a two-period stochastic programming program for the joint optimization of investment and operation of a microgrid, taking the impact of energy storage, renewable energy integration, and demand response into consideration. Mozafari and Mohammadi [3] applied bee swarm optimization algorithm to optimize the operation strategies and capacities of ESS considering various factors. In the study of Li et al. [4], a new optimal scheduling of ESS based on chance-constrained programming has been proposed for minimizing the operating costs of an isolated microgrid. Tushar et al. [5] proposed a real-time decentralized demand-side management in the integration of electric vehicles (EVs), ESSs, and renewable energy sources by formulating a game with mixed strategy between customers. In this paper, we explore a large variety of optimization approaches to energy storage systems especially using the combination of dynamic programming (DP) and genetic algorithms (GA). We conduct more extensive simulations than previous work, with various 18 scenarios, with and without demand charge.

We summarize our contributions in this study as follows: (i) we propose a combined method of DP and GA for electric cost optimization with renewable energy and ESS under TOU with/without demand charge; (ii) we improve the performance of the proposed method by adjusting the size of the base units of DP; (iii) we perform comparative experiments on the proposed method for various industrial electricity load and renewable energy generation profile; and (iv) finally we show that our combined method is effective for both cases with and without demand charge in terms of cost saving and time.

The remainder of the paper is organized as follows: the problem, optimization method used for ESS scheduling, and DP operation process are described in Section 2. Our method combining DP with a GA is presented in Section 3. The savings and computing time associated with DP, a pure GA, and the combination of DP and GA depending on various DP base unit sizes are compared in Section 4. We draw conclusions in Section 5.

2. Dynamic Programming for ESS Scheduling

2.1. Problem Formulation

We formally define ESS scheduling problems with demand charge and that without demand charge in this section. The definitions are similar to those presented in previous work [25]. The load l_i refers to the amount of energy used during the ith time interval and g_i refers to the amount of energy generated during the ith time interval. The variable x_i refers to the amount of energy stored in the ESS at the ith time interval. Instead of SOC (state of charge), which is the level of charge of an electric battery relative to its capacity, we used the amount of energy stored in the ESS as a variable to be optimized. There is a relationship such that SOC at the ith time interval is the same as $\frac{x_i}{C} \times 100\%$, where C is the capacity of ESS. So optimizing the value of x_i can be considered the same as optimizing that of SOC after multiplying some coefficients.

Each time interval is defined to be one hour in this study. The amount of energy provided to the ESS at the ith time interval is $x_i - x_{i-1}$; thus, the net energy required from the power grid, E_i can be calculated as follows:

$$E_i = x_i - x_{i-1} + l_i - g_i \tag{1}$$

That is, if the electricity price at the ith time interval is p_i, the electrical energy charge amount of the ith time interval is $E_i \cdot p_i$. If E_i is negative, it means that electricity is sent back to the grid. Although there may be several pricing policies for this feed-in electricity, in this study, we assumed that there is no compensation of the feed-in electricity. That is, the amount of net energy E_i is negative, the cost at that time interval only becomes 0. Thus, the sum of the costs in T time intervals can be represented as $\sum_{i=1}^{T} I_{\mathbb{R}+}(E_i) \cdot \{E_i \cdot p_i\}$, where $I_{\mathbb{R}+}(x)$ is the indicator function that returns 1 if x is a positive real number, otherwise, returns 0. This ensures that the sum is not negative, although the costs in several time intervals can be negative. Therefore, the following equation is the formulation of this ESS scheduling problem.

Minimize:

$$\sum_{i=1}^{T} I_{\mathbb{R}+}(E_i) \cdot \{E_i \cdot p_i\} \tag{2}$$

subject to

$$0 \leq x_i \leq C, \ i = 1, 2, \ldots, T \tag{3}$$

$$-P_d \leq x_i - x_{i-1} \leq P_c, \ i = 1, 2, \ldots, T, \tag{4}$$

where C is the capacity of a battery, P_d is the amount of maximum battery discharge in an hour, and P_c is the amount of maximum battery charge in an hour. This means that x_i cannot exceed the capacity of the battery and $x_i - x_{i-1}$ must range between $-P_d$ and P_c.

The objective function, Equation (2) is the sum of hourly electrical energy costs, and each hourly cost is calculated by multiplying the amount of electrical energy from the power grid during an hour and the electricity price at that time. Only when the amount of electrical energy from the power grid is negative, the cost of that time interval is 0. This property is represented with indicator function I. Equation (2) does not have a linear nor a quadratic property because of the existence of the function I. The function only produces 0 or 1. Moreover, the objective function is not convex. At some points, gradients cannot be calculated. So general linear or quadratic programming cannot be applied to this problem.

Equation (2) is the objective function when we assume that the battery efficiency can be 100%. In fact, recent battery technology has developed a lot, and it is becoming possible to develop a battery with an efficiency of 99% or more with the lithium-ion battery (Li-ion) [26], lithium-sulfur battery (Li-S) [27], and vanadium redox flow battery (VRFB) [28]. However, this high efficiency can be achieved in an ideal environment, so in practice, there would be battery charge and discharge loss. These losses are likely to lead to some different simulation results. So, in our experiments, we used modified objective function considering battery efficiency α. In this case, E'_i, which is the net energy of the ith time interval considering battery efficiency α, is calculated as follows:

$$E'_i = \alpha^{-1}(x_i - x_{i-1}) + l_i - g_i, \tag{5}$$

where $0 < \alpha < 1$ is battery efficiency. That is, to increase the amount of energy stored in the battery from x_{i-1} to x_i, the amount of $\alpha^{-1}(x_i - x_{i-1})$ is required to charge the battery. The objective function of the problem considering battery efficiency can be written using the modified amount of net energy as follows:

$$\sum_{i=1}^{T} I_{\mathbb{R}_+}(E'_i) \cdot \{E'_i \cdot p_i\} \tag{6}$$

With regard to the pricing including the demand charge, the total electrical energy cost is the sum of the energy and demand charges, which is the product of the fixed rate p^* and peak demand and can thus be written as: $\max_{1 \leq i \leq T} E'_i \cdot p^*$ [29]. Peak demand refers to the highest amount of power during the billing period and is represented as kW. However, in practical, the highest amount of power is usually measured by calculating the electrical energy drawn during a predetermined time interval. So, in this study, we define peak demand as the largest hourly electrical energy required from the power grid during the billing period T. The problem related to minimizing the total electrical energy cost can then be formulated as follows:

Minimize:

$$\sum_{i=1}^{T} I_{\mathbb{R}_+}(E'_i) \cdot \{E'_i \cdot p_i\} + \max_{1 \leq i \leq T} E'_i \cdot p^* \tag{7}$$

Equations (2) and (7), which are the objective functions in the case without and with demand charge respectively, do not have a linear or a quadratic property and are not convex. At some points, gradients cannot be calculated. So simple mathematical optimization methods using some gradients cannot be applied to these problems. In this study, we adopted DP and GA to solve the problems because they have some characteristics to fit these problems.

DP is based on splitting the problem into smaller subproblems and there should be an equation that describes the relationship between these subproblems. The problems defined in this study have those properties. The relationship between subproblems are addressed in the next subsection. GA can also be applied to these kinds of problems. It can be easily applied regardless of the type of objective function. GA is a sort of a metaheuristic, which is a higher-level procedure or heuristic designed to find, generate, or select a heuristic that may provide a sufficiently good solution to an optimization problem, especially with incomplete or imperfect information or limited computation capacity [30,31].

2.2. Assumptions and Limitations of the Proposed Problem Formulation

In this subsection, assumptions and limitations of the proposed problem formulation are discussed.

- In relation with the demand charge, the amount of the largest hourly electrical energy required from the power grid during the billing period is used for the value of peak demand instead of the exact amount power. In practical, power is usually measured by calculating the electrical energy drawn during a predetermined time interval. We used an hour as this time interval in this study, however, if we use a shorter time interval, such as five minutes, the result will be more accurate.
- For the electricity that is sent back to the grid, we assumed that there is no compensation of the feed-in electricity. If the other pricing policies for this feed-in electricity are applied, the problem formulation should be modified.
- Another existing study [32] has modeled the problem of scheduling the charge/discharge power of ESS considering power balance constraint. The problem formulation of our study has slightly different view. In our study, we optimize the amount of charge/discharge energy during the unit time interval instead of optimizing the power of the ESS. Therefore, energy balance among generation, load, grid, and ESS is considered instead of power balance. Both models can be applied to ESS scheduling problem considering the other environments.
- In this study, we experimented the proposed method assuming that actual generation and load completely follow the certain predetermined patterns. However, in practical, generation and load may not follow the same pattern every day, so the proposed method should be applied with some predicted generation and load patterns to be used in the field. There have been a number of recent studies on day-ahead prediction of photovoltaic (PV) output [33,34], wind power generation [35], and load [36–39]. It is expected that the proposed method of combining GA and DP will show a good performance when an ideal prediction algorithm with great accuracy is adopted as we simulated in this study. However, generation and load predictions will usually have errors and simulation results may be different from this study. A statistical analysis of the day-ahead (and two-days-ahead) load forecasting errors have been made in [40] and economic impact assessment of load forecast errors have been discussed in [41]. In the ESS scheduling problem addressed in this paper, if the net energy (the difference between renewable energy and load) is underforecasted, excessive electrical energy may be accumulated in the battery uselessly by the predetermined schedule. On the other hand, the net energy is overforecasted, the energy contained in the battery may be used up in advance, so the consumer may have to buy energy from the grid even when the price is high. In both cases, there can be some economic inefficiencies in practical.

2.3. Dynamic Programming

If there is no demand charge and the price is determined only by energy charge with TOU, near-optimal ESS schedules can be obtained using DP. The DP is a technique that improves the algorithm performance by storing previously computed optimal solutions of subproblems in the memory and reducing the computations based on the information stored in the memory if necessary.

We define the two-dimensional table $D[i, w]$ as the minimum electrical energy cost when an amount w is stored in the ESS at the ith time interval. The $D[i, w]$ should select the minimum value of $D[i-1, x] + cost(i, x, w)$ for all possible values of the residual amount of the battery at the $(i-1)$th time interval and x's, where $cost(i, x, w)$ is the elecricity cost when the residual amount of the battery becomes w at the ith time interval from x at the $(i-1)^{th}$ time interval. Possible x values are in the range of $\max(0, w - P_c) \leq x \leq \min(C, w + P_d)$ and $cost(i, x, w)$ is calculated as $(x - w + l_i - g_i) \cdot p_i$ using the objective function. Therefore, the recurrence equation used in DP is as follows:

$$D[i, w] = \min_{w - P_c \leq x \leq w + P_d} (D[i-1, x] + cost(i, x, w)) \tag{8}$$

Therefore, the minimum electrical energy cost is one of $D[T, w]$'s in the last time interval T. Based on backward tracing, the path toward obtaining the optimal value can be found. The pseudo-code for a scheduling algorithm according to T time intervals is given in Figure 1.

---

for $w \leftarrow 0$ **to** P_c
    $D[1, w] \leftarrow w + l_i - g_i$;

for $i \leftarrow 2$ **to** T
    for $w \leftarrow 0$ **to** C
        for $k \leftarrow \max(0, w - P_d)$ **to** $\min(C, w + P_c)$
            $value \leftarrow w - k + l_i - g_i$;   // $x_i - x_{i-1} + l_i - g_i$
            if $value \geq 0$
                $D[i, w] \leftarrow \min(D[i, w], D[i-1, k] + value \cdot p_i)$;

$result \leftarrow D[T, 0]$;
for $w \leftarrow 0$ **to** C
    $result \leftarrow \min(result, D[T, w])$;

return $result$;

Figure 1. Pseudocode of the proposed dynamic programming.

If load and power generation can be predicted accurately, the DP can achieve near-optimal solutions. However, in reality, it is very difficult to accurately predict the load and power generation. In addition, based on the proposed method, the algorithm can be performed in a short time period when the capacity of the battery is low such as in the residential power system, but it cannot be performed in a short time period when the capacity of the battery is high such as in the enterprise power system. One way to solve this issue is to use a large base unit in DP. For example, assuming that the capacity of the battery is 1000 kWh, the algorithm can be run faster if a base unit of 10 kWh is used instead of the default unit 1 kWh. Although the use of a large base unit makes the solution less accurate, it has an advantage in terms of the computing time.

Examples of DP with different base units are shown in Figure 2. In this example, DP in Figure 2a has three states (0, 5, and 10 kWh) and that in Figure 2b has 11 states (0–10 kWh). Both have four time intervals, the charging and discharging power is 5 kW, and the initial battery is empty.

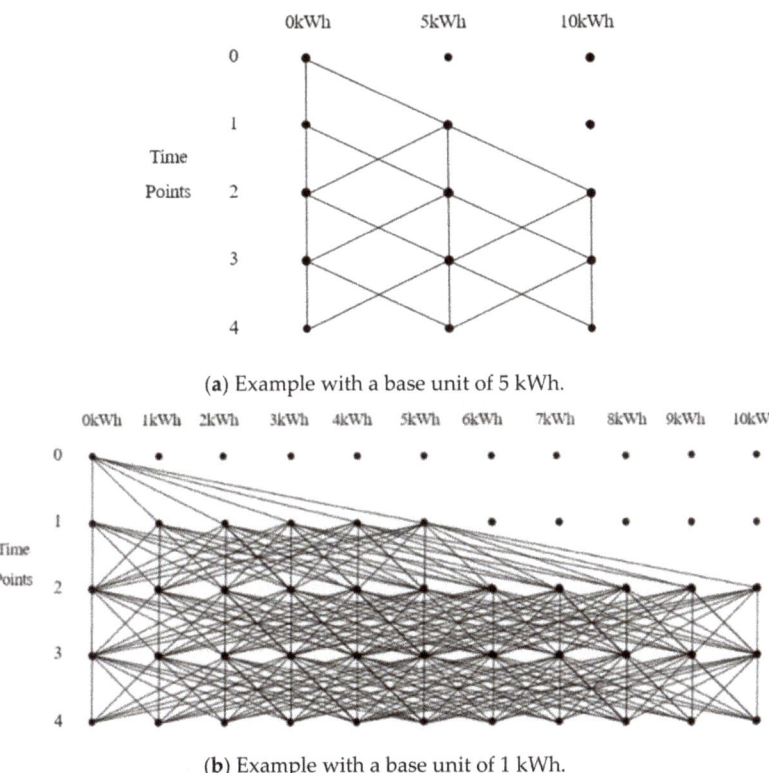

(a) Example with a base unit of 5 kWh.

(b) Example with a base unit of 1 kWh.

Figure 2. Schematic application of dynamic programming according to different base units.

The DP in Figure 2a is based on a base unit of 5 kWh and that in Figure 2b is based on a base unit of 1 kWh. The path from each point at the time interval t to each point at the time interval $t+1$ is calculated for each possible pair, consisting of the states in the time intervals t and $t+1$. This simple example shows that DP with smaller base unit is more complex and has higher computational cost than that with larger base unit.

The size of the base unit is important in the design of DP both in terms of solution quality and temporal performance. For example, if the capacity of ESS is 3.5 kWh and we set 1 kWh as the base unit of DP, the proposed DP algorithm can only deal with 0 kWh, 1 kWh, 2 kWh, and 3 kWh as the value of x_i, and the amount of 0.5 kWh is not considered. In this case, the proposed DP algorithm is not efficient. On the other hand, if the capacity of ESS is 500.5 kWh and the base unit is also 1 kWh, the left amount of 0.5 kWh is not so critical compared with the former case. If we set 0.1 kWh as the base unit, the obtained solution will be more accurate. As shown in these examples, the smaller the base unit compared with the amount of capacity, the more efficiently the DP algorithm performs. If we set the base unit small enough, the proposed DP can produce near-optimal solution. However, DP with small base unit may have high computational cost so the size of the base unit should be carefully determined considering both the quality of solutions and time cost.

For the pseudocode of the proposed DP in Figure 1, it is easy to calculate the time complexity: $O(T \cdot C \cdot (P_c + P_d))$. Because P_c and P_d values are proportional to C, the complexity can be written as $O(T \cdot C^2)$. This time complexity is valid for the DP with a base unit of 1 kWh. However, if a base unit of b kWh is used, the time complexity becomes $O(T \cdot (C/b)^2)$.

The proposed DP algorithm yields a near-optimal solution for the case without a demand charge. Moreover, if the domain is limited to set of integers with a given precision, it produces an optimal solution. For example, if we limit possible values of each x_i to only integers that are multiples of 10, an optimal solution is derived by the proposed DP with base unit 10. However, the objective function becomes more complex when a demand charge is included. Thus, in that case, it is difficult to achieve the desired performance with a similar DP method described above. However, DP has a strong advantage of optimizing energy charge represented as the first term in Equation (7), which is the objective function of the problem with demand charge, because the first term is exactly the same as the objective function of the problem without demand charge. DP can optimize the energy charge, but it cannot optimize demand charge. If we combine DP with other appropriate methods to optimize demand charge, we can get good solutions of the problem. GA is adopted as the method to optimize demand charge in this study and the method will be explained in the next section.

3. Genetic Algorithm Process

The GA [42] is a metaheuristic technique that expresses information about solutions in a genetic form and optimizes a given objective function using an evolution process such as crossover and mutation. In addition, GAs for real-valued representations are called real-coded GAs (RCGAs). Real-valued representations were first used to generate a metaoperator and identify the most appropriate parameters related to chemometric problems. Nowadays, RCGA is mainly being used for numerical optimization problems in continuous domains [43–46].

In this study, the population size of the GA is set to 100. Both parents are randomly selected from the population, and crossover and mutation operations are performed. Subsequently, the worst solution in the population is replaced by the offspring generated by crossover and mutation. This process is repeated up to a maximum of 100,000 generations. The pseudocode of the proposed GA is shown in Figure 3.

- Encoding: in the proposed RCGA, a real-number vector is encoded, with a length of the number T of the maximum time intervals. Unlike in general real encoding, the value of the gene x_i of a solution vector is limited by the value of the gene x_{i-1} of the previous index. Therefore, the range of x_i is as follows:

$$0 \leq x_i \leq C, \tag{9}$$

$$-P_d \leq x_i - x_{i-1} \leq P_c \leftrightarrow x_{i-1} - P_d \leq x_i \leq x_{i-1} + P_c \tag{10}$$

$$\Rightarrow \max(0, x_{i-1} - P_d) \leq x_i \leq \min(C, x_{i-1} + P_c) \tag{11}$$

- Evaluation: if there is no demand charge, the objective function of this problem is the same as Equation (2) in Section 2. If there is a demand charge, the function in Equation (4) is used. The lower the function value is, the higher is the possibility to be selected as parents.
- Initialization: an initial population of 100 individuals is generated and the encoding constraint in Equation (11) should be adhered to. The individuals are randomly generated and the limit is not exceeded.
- Crossover operator: in this study, blend crossover (BLX$_\alpha$), one of the crossover techniques for real-valued chromosomes, is used, where α is a non-negative real number. This crossover operation randomly determines genes within the range $[C_{min} - \alpha I, C_{max} + \alpha I]$, where $C_{max} = \max(x, y)$ and $C_{min} = \min(x, y)$. The parameter α used in this study is 0.5 and $I = C_{max} - C_{min}$. This study includes additional constraints because the encoding conditions should not be violated. Therefore, the range of the ith gene of the offspring of the crossover should be set to $[\max(0, x_{i-1} - P_d), \min(C, x_{i-1} + P_c)]$.
- Mutation: the mutation transforms a part of the offspring generated via crossover such that more diverse solutions are generated during the genetic process. Mutation is not performed always but depending on the probability value. In this study, the probability is set to 0.2. The mutation

process selects a part of the chromosome index and changes the corresponding part, but it assigns values uniformly and randomly within the range of the encoding constraint.

---

  max_generation ← 100,000;
  initialize population;
  res ← min(population);
  for generation ← 1 to max_generation
    randomly select two parents;
    offspring ← BLX$_\alpha$ crossover(p_1, p_2);
    offspring into population;
    p ← a random real number between 0.0 and 1.0;
    if $p \leq 0.2$
      mutate(offspring);
    remove the worst vector from the population;
    res ← min(res, offspring) ;
  return res;

Figure 3. Pseudocode of the proposed genetic algorithm.

We described the process of a pure GA in the above. We also examine the performance of a combination of GA and DP. In the proposed combined method, DP can help the GA to identify a better solution than that obtained with a standalone of pure GA or DP. For the combined method, firstly, the solution derived from DP is calculated. After that, the solution obtained by DP is included in the GA population, when constructing an initial population. The solution by DP and other solutions in the GA population evolves to better solutions by GA process through generations. So we can improve the solution quality by using this combined method regardless of base unit of DP. That is, we can obtain near-optimal solutions with more accurate precision, and the obtained solutions are always better than those by stand-alone DP.

As the ESS capacity increases, the temporal performance of DP degrades. However, the temporal performance can be improved by adjusting the size of the base unit in DP. To obtain economic efficiency and reasonable solution quality, the optimal solution of DP with a large base unit is included in the population of GA.

DP algorithm proposed in Section 2.2 produces reasonable results for the problem without a demand charge. However, when a demand charge is included, it is difficult to achieve the desired performance with the proposed DP, because the DP is designed to optimize the energy charge without considering the demand charge. If we combine the proposed DP with GA addressed in this section, we can get good solutions of the problem even for the case with demand charge. GA is a kind of metaheuristics, so it can find appropriate solutions that are fit for given objective functions.

4. Experimental Results

4.1. Experiment Data

The electricity load data used in this study were obtained from the Office of Energy Efficiency and Renewable Energy (EERE) [47] and include information about electricity load of United States (industrial and residential). This dataset contains hourly load profile data for commercial and residential buildings. Hourly load profiles are available for all TMY3 locations in the United States. We only used commercial building load profiles for Anchorage in our experiments. There are three types of commercial buildings (hospital, restaurant, and office). The photovoltaic (PV) watts calculator [48] was developed by the National Renewable Energy Laboratory (NREL). The PV watts calculator uses past PV data and calculates the energy generated from the grid-connected PV system. We used six types of PV generation data considering the combinations of three weather types (cloudy, rainy, and sunny) and two season types (summer and winter). Table 1 shows information about each test case.

Table 1. Detailed information for each case.

Case	Season	Weather	Building
1	Summer	Cloudy	Hospital
2	Summer	Rainy	Hospital
3	Summer	Sunny	Hospital
4	Winter	Cloudy	Hospital
5	Winter	Rainy	Hospital
6	Winter	Sunny	Hospital
7	Summer	Cloudy	Office
8	Summer	Rainy	Office
9	Summer	Sunny	Office
10	Winter	Cloudy	Office
11	Winter	Rainy	Office
12	Winter	Sunny	Office
13	Summer	Cloudy	Restaurant
14	Summer	Rainy	Restaurant
15	Summer	Sunny	Restaurant
16	Winter	Cloudy	Restaurant
17	Winter	Rainy	Restaurant
18	Winter	Sunny	Restaurant

Typical TOU prices were generated by simulations using three price levels for summer and winter based on the TOU pricing models of several utility companies. The TOU pricing model that was constructed in this study is given in Table 2.

Table 2. Hourly pricing for summer and winter.

Hour (from-to)	Summer (Cents/kWh)	Winter (Cents/kWh)
0–1	5	5
1–2	5	5
2–3	5	5
3–4	5	5
4–5	5	5
5–6	5	5
6–7	5	5
7–8	10	15
8–9	10	15
9–10	10	15
10–11	10	15
11–12	15	10
12–13	15	10
13–14	15	10
14–15	15	10
15–16	15	10
16–17	15	10
17–18	10	15
18–19	10	15
19–20	5	5
20–21	5	5
21–22	5	5
22–23	5	5
23–24	5	5

Demand charge rate, $p^* = 20$.

The ESS capacity used for the experiments differs for each building. The capacity of hospitals and offices is 500 kWh. The capacity of restaurants is 250 kWh. The P_c and P_d values are assumed to be one-fifth of the capacity.

4.2. Performace Comparison for the Case without Demand Charge

Table 3 and Figure 4 show the comparison of the temporal and economic performances of DP1 (DP with a base unit 1 kWh), DP10 (DP with a base unit of 10 kWh), GA, GA+DP1 (the combined method of GA and DP with a base unit 1 kWh), and GA+DP10 (the combined method of GA and DP with a base unit 10 kWh) without demand charge. To compare the performance of the proposed method with other existing methods, we also performed the Harmony Search (HS) algorithm previously proposed in [49]. HS is a kind of metaheuristic algorithm and can be applied to optimization problems instead of GA. We similarly implemented HS, HS+DP1 (combined method of HS and DP with a base unit 1 kWh), and HS+DP10 (combined method of GA and DP with a base unit 10 kWh) using HS instead of GA. For GAs and HSs, average values and standard deviations over 100 runs were given. Economic performances were measured by calculating cost savings for each case. Cost savings are expressed as a percentage of the cost when there is no ESS. That is, the cost saving of Algorithm A is calculated by the formula, $100 \times (Cost_{NO-ESS} - Cost_A)/Cost_{NO-ESS}$, where $Cost_A$ is the electrical energy cost incurred by Algorithm A. Computing time of a single run of each algorithm, which is expressed in seconds, is also provided in Table 3.

For the DP method, we conducted experiments with base units 1 kWh and 10 kWh. In Table 3, as the base unit increases, the economic performance decreases, but the time performance increases. In terms of the temporal performance, DP1 is slower than DP10. The combined method of GA and DP is affected by the time consumed, hence, GA+DP1 is slower than GA+DP10. GA is slower than DP10 but considerably faster than DP1. The economic performance of GA+DP1 is outstanding: the performance of DP1 is better than that of DP10, and the performance of GA+DP1 is better than that of GA+DP10. Consequently, the combination of GA and DP leads to better solutions.

HS performed slightly worse than GA overall, however, as in GA, it was the same that combining with DP produced better results than a standalone method. The performances of HS+DP1 and HS+DP10 are better than that of HS.

We conducted a *t*-test to compare the performances of GA+DP1 and GA+DP10. The results have a significance level of $1.2\,e^{-1}$, that is, the performance of GA+DP10 is significantly similar to that of GA+DP1, but GA+DP10 is faster than GA+DP1, which shows that GA+DP10 is preferable for practical application when compared with GA+DP1.

Table 3. Comparison of cost savings of the proposed methods without demand charge.

Case	DP1	DP10	GA	GA+DP1	GA+DP10	HS	HS+DP1	HS+DP10
1	10.32% (2.924)	10.29% (0.052)	8.75% (0.16%) (0.192)	10.33% (0.00%) (3.201)	10.29% (0.04%) (0.257)	8.37% (0.17%) (0.224)	10.32% (0.00%) (3.212)	10.31% (0.05%) (0.281)
2	**7.85%** (2.854)	7.79% (0.048)	6.61% (0.11%) (0.184)	**7.85%** (0.00%) (3.433)	**7.85%** (0.00%) (0.262)	5.22% (0.13%) (0.237)	**7.85%** (0.00%) (3.238)	7.79% (0.01%) (0.243)
3	7.36% (2.994)	7.36% (0.047)	6.05% (0.10%) (0.164)	7.52% (0.50%) (3.452)	7.36% (0.01%) (0.273)	5.40% (0.10%) (0.242)	7.36% (0.00%) (3.228)	7.48% (0.02%) (0.201)
4	**3.60%** (3.014)	**3.60%** (0.051)	2.96% (0.14%) (0.171)	**3.60%** (0.00%) (3.321)	**3.60%** (0.00%) (0.281)	1.93% (0.12%) (0.204)	**3.60%** (0.00%) (3.216)	**3.60%** (0.01%) (0.239)
5	**3.58%** (2.962)	**3.58%** (0.039)	2.95% (0.13%) (0.199)	**3.58%** (0.01%) (3.361)	**3.58%** (0.00%) (0.299)	1.88% (0.15%) (0.218)	**3.58%** (0.00%) (3.276)	**3.58%** (0.00%) (0.267)
6	**3.96%** (2.940)	**3.96%** (0.043)	3.28% (0.06%) (0.201)	**3.96%** (0.00%) (3.399)	**3.96%** (0.00%) (0.244)	2.09% (0.06%) (0.291)	**3.96%** (0.00%) (3.268)	**3.96%** (0.00%) (0.227)
7	13.35% (3.092)	13.35% (0.045)	11.64% (0.08%) (0.175)	13.50% (0.05%) (3.417)	13.45% (0.05%) (0.263)	10.07% (0.16%) (0.272)	13.50% (0.05%) (3.213)	**13.52%** (0.02%) (0.298)
8	**15.17%** (2.885)	15.10% (0.038)	11.97% (0.15%) (0.161)	**15.17%** (0.00%) (3.363)	15.10% (0.03%) (0.251)	9.30% (0.14%) (0.271)	**15.17%** (0.00%) (3.232)	15.11% (0.01%) (0.277)
9	17.66% (2.911)	17.64% (0.050)	14.46% (0.17%) (0.134)	**17.67%** (0.00%) (3.367)	17.64% (0.02%) (0.259)	12.07% (0.12%) (0.208)	**17.67%** (0.01%) (3.268)	17.64% (0.01%) (0.214)
10	**8.30%** (2.923)	**8.30%** (0.055)	6.85% (0.13%) (0.156)	**8.30%** (0.00%) (3.393)	**8.30%** (0.00%) (0.262)	4.49% (0.17%) (0.275)	**8.30%** (0.00%) (3.252)	**8.30%** (0.01%) (0.227)
11	**8.19%** (2.924)	**8.19%** (0.058)	6.80% (0.16%) (0.161)	**8.19%** (0.00%) (3.264)	**8.19%** (0.00%) (0.241)	4.43% (0.13%) (0.221)	**8.19%** (0.01%) (3.206)	**8.19%** (0.00%) (0.234)
12	11.58% (2.975)	11.58% (0.052)	10.35% (0.10%) (0.183)	11.60% (0.04%) (3.251)	**11.61%** (0.03%) (0.268)	8.38% (0.10%) (0.258)	11.59% (0.02%) (3.214)	11.60% (0.00%) (0.242)
13	21.17% (2.963)	18.48% (0.053)	15.95% (0.11%) (0.199)	**21.19%** (0.03%) (3.411)	19.33% (0.07%) (0.290)	10.95% (0.11%) (0.251)	21.17% (0.00%) (3.289)	18.48% (0.01%) (0.271)
14	19.63% (2.989)	19.19% (0.045)	15.22% (0.08%) (0.181)	**19.67%** (0.05%) (3.252)	19.40% (0.02%) (0.284)	10.11% (0.13%) (0.279)	19.63% (0.04%) (3.266)	19.19% (0.05%) (0.226)
15	23.83% (2.931)	20.17% (0.047)	18.86% (0.16%) (0.179)	**23.85%** (0.04%) (3.245)	20.91% (0.16%) (0.285)	14.29% (0.10%) (0.249)	23.83% (0.02%) (3.240)	20.51% (0.06%) (0.274)
16	**11.80%** (2.938)	**11.80%** (0.049)	9.86% (0.13%) (0.169)	**11.80%** (0.00%) (3.273)	**11.80%** (0.00%) (0.279)	6.41% (0.06%) (0.230)	**11.80%** (0.03%) (3.217)	**11.80%** (0.03%) (0.264)
17	**11.77%** (2.946)	11.77% (0.045)	9.75% (0.12%) (0.185)	**11.77%** (0.00%) (3.293)	**11.77%** (0.00%) (0.268)	6.50% (0.15%) (0.263)	**11.77%** (0.01%) (3.279)	11.77% (0.00%) (0.276)
18	**12.26%** (2.974)	**12.26%** (0.054)	10.17% (0.09%) (0.193)	**12.26%** (0.00%) (3.283)	**12.26%** (0.00%) (0.282)	6.51% (0.13%) (0.278)	**12.26%** (0.00%) (3.208)	**12.26%** (0.00%) (0.264)

For GA, GA+DP1, GA+DP10, HS, HS+DP1, and HS+DP10, average values and standard deviations over 100 runs are shown. The computing time in seconds is given in the second line of each cell. For each case, the best result among compared eight cells is shown in bold type.

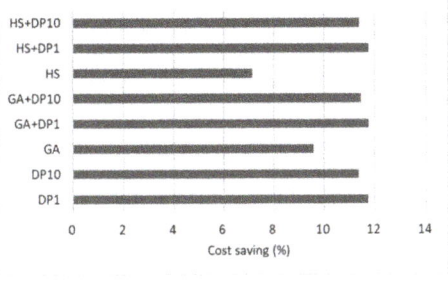 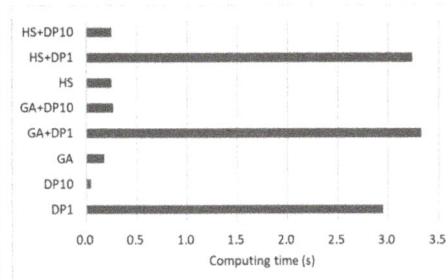

(a) Average cost saving over all cases   (b) Average computing time over all cases

Figure 4. Average cost savings and computing times of the proposed methods without demand charge.

4.3. Comparison for the Case with Demand Charge

Table 4 and Figure 5 show the experimental results for the case including demand charge. In this case, the proposed DP methods (DP1 and DP10) optimize only the charge by the TOU but not the demand charge, which often results in bad performances. However, the combined methods of GA and DP can perform better than the individual DP or GA methods. The combined methods of HS and DP also performed better than DP, however, their results were slightly worse than the combined methods of GA and DP. The combined methods of GA and DP (GA+DP1 and GA+DP10) performed better than the GA for almost all cases except two ones. In Table 4, there does not seem to be much difference between GA+DP1 and GA+DP10. Statistically, the p-value obtained from t-test was $7.3\ e^{-1}$, which showed that the performance of GA+DP1 had no significant differences from that of GA+DP10.

In terms of the temporal performance, DP10 is faster than DP1 because DP10 searches fewer solutions than DP1. Computing times of GA and HS are similar, though GA obtained better results than HS. For this case with demand charge, GA and HS are not only faster than DP1, but also showed better performance in terms of cost savings. The results by GA and HS can be improved by combining DP, as shown by the results of GA+DP1, GA+DP10, HS+DP1, and HS+DP10. However, they are slower than a standalone of GA or HS method. GA+DP10 takes shorter time than GA+DP1. In the above, we have shown that the performance of GA+DP1 has no significant differences from that of GA+DP10 by t-test, so we can conclude that GA+DP10 is the most practical method among compared eight ones.

Although the performances of DP1 and DP10 are poor for this case with demand charge, GA+DP1 and GA+DP10 show relatively good performances. This means that the performance of DP can be improved by combining GA, which helps to optimize the demand charge while DP only optimizes energy charge with TOU but not the demand charge.

Table 4. Comparison of cost savings of the proposed methods with varying base units including a demand charge.

Case	DP1	DP10	GA	GA+DP1	GA+DP10	HS	HS+DP1	HS+DP10
1	5.87% (2.924)	5.76% (0.052)	6.71% (0.25%) (0.192)	6.75% (0.19%) (3.201)	6.74% (0.02%) (0.257)	5.94% (0.15%) (0.224)	6.73% (0.05%) (3.212)	6.61% (0.05%) (0.281)
2	4.38% (2.854)	4.33% (0.048)	4.51% (0.10%) (0.184)	4.62% (0.15%) (3.433)	**4.65%** (0.04%) (0.262)	3.66% (0.12%) (0.237)	4.64% (0.06%) (3.238)	4.61% (0.04%) (0.243)
3	3.60% (2.994)	3.60% (0.047)	4.05% (0.09%) (0.164)	4.08% (0.17%) (3.452v)	4.12% (0.03%) (0.273)	3.58% (0.08%) (0.242)	**4.26%** (0.10%) (3.228)	4.25% (0.07%) (0.201)
4	2.22% (3.014)	2.21% (0.051)	2.42% (0.12%) (0.171)	2.75% (0.20%) (3.321)	2.77% (0.00%) (0.281)	1.84% (0.10%) (0.204)	2.30% (0.06%) (3.216)	2.29% (0.08%) (0.239)
5	2.21% (2.962)	2.19% (0.039)	2.39% (0.14%) (0.199)	2.70% (0.15%) (3.361)	2.72% (0.01%) (0.299)	1.83% (0.12%) (0.218)	2.29% (0.08%) (3.276)	2.28% (0.10%) (0.267)
6	2.42% (2.94)	2.40% (0.043)	2.62% (0.14%) (0.201)	2.95% (0.25%) (3.399)	2.97% (0.00%) (0.244)	1.99% (0.09%) (0.291)	2.52% (0.07%) (3.268)	2.49% (0.15%) (0.227)
7	2.50% (3.092)	2.50% (0.045)	**13.54%** (0.13%) (0.175)	13.20% (0.27%) (3.417)	13.43% (0.00%) (0.263)	9.71% (0.12%) (0.272)	10.12% (0.12%) (3.213)	10.29% (0.02%) (0.298)
8	12.30% (2.885)	12.26% (0.038)	12.32% (0.11%) (0.161)	12.69% (0.24%) (3.363)	12.76% (0.05%) (0.251)	11.14% (0.15%) (0.271)	13.48% (0.13%) (3.232)	13.39% (0.05%) (0.277)
9	6.21% (2.911)	6.15% (0.05)	**14.81%** (0.08%) (0.134)	14.61% (0.20%) (3.367)	14.60% (0.04%) (0.259)	10.88% (0.10%) (0.208)	11.18% (0.04%) (3.268)	11.10% (0.14%) (0.214)
10	5.18% (2.923)	5.18% (0.055)	4.28% (0.09%) (0.156)	5.70% (0.24%) (3.393)	5.72% (0.00%) (0.262)	3.13% (0.13%) (0.275)	5.19% (0.05%) (3.252)	5.18% (0.07%) (0.227)
11	5.12% (2.924)	5.12% (0.058)	4.18% (0.10%) (0.161)	5.58% (0.21%) (3.264)	5.63% (0.00%) (0.241)	3.21% (0.12%) (0.221)	5.12% (0.07%) (3.206)	5.12% (0.13%) (0.234)
12	7.73% (2.975)	7.73% (0.052)	7.92% (0.13%) (0.183)	**8.74%** (0.18%) (3.251)	8.51% (0.05%) (0.268)	6.95% (0.11%) (0.258)	7.98% (0.05%) (3.214)	7.94% (0.09%) (0.242)
13	10.01% (2.963)	7.79% (0.053)	11.09% (0.15%) (0.199)	12.26% (0.19%) (3.411)	12.28% (0.06%) (0.29)	8.14% (0.10%) (0.251)	11.4% (0.10%) (3.289)	9.63% (0.12%) (0.271)
14	9.74% (2.989)	11.57% (0.045)	10.32% (0.14%) (0.181)	11.69% (0.15%) (3.252v)	12.60% (0.04%) (0.284)	7.38% (0.11%) (0.279)	10.19% (0.14%) (3.266)	11.63% (0.10%) (0.226)
15	11.16% (2.931)	8.20% (0.047)	13.06% (0.12%) (0.179)	**13.82%** (0.21%) (3.245)	13.48% (0.12%) (0.285)	11.06% (0.13%) (0.249)	12.28% (0.10%) (3.24)	11.73% (0.04%) (0.274)
16	7.20% (2.938)	7.20% (0.049)	6.17% (0.15%) (0.169)	**8.38%** (0.24%) (3.273)	7.87% (0.00%) (0.217)	4.63% (0.06%) (0.23)	7.20% (0.07%) (3.217)	7.21% (0.06%) (0.264)
17	7.16% (2.946)	7.16% (0.045)	6.11% (0.11%) (0.185)	**8.39%** (0.16%) (3.293)	7.83% (0.00%) (0.268)	4.58% (0.10%) (0.263)	7.16% (0.08%) (3.279)	7.17% (0.08%) (0.276)
18	7.45% (2.974)	7.45% (0.054)	6.33% (0.13%) (0.193)	**8.54%** (0.18%) (3.283)	8.30% (0.00%) (0.282)	4.72% (0.09%) (0.278)	7.48% (0.06%) (3.208)	7.45% (0.10%) (0.264)

For GA, GA+DP1, GA+DP10, HS, HS+DP1 and HS+DP10, average values and standard deviations over 100 runs are shown. The computing time in seconds is given in the second line of each cell. For each case, the best result among compared eight ones is shown in bold type.

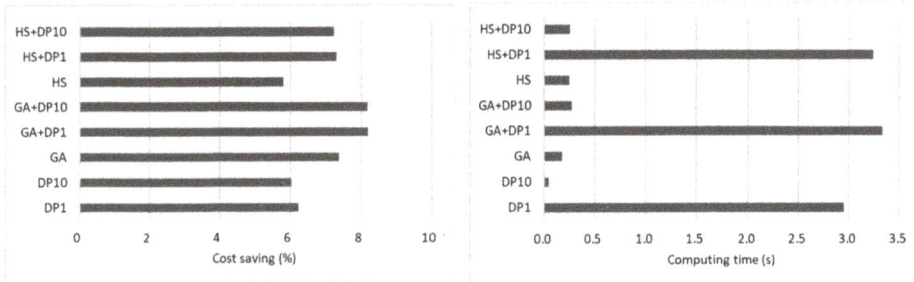

(a) Average cost saving over all cases   (b) Average computing time over all cases

Figure 5. Average cost savings and computing times of the proposed methods with demand charge.

4.4. Experiments with Various Sizes of Base Unit

In the experiments in Sections 4.2 and 4.3, we have only experimented the combined methods of GA and DP with only two kinds of base units, 1 kWh and 10 kWh. In this subsection, we have investigated the performance of the combined methods of GA and DP with various sizes of base unit, 0.5 kWh, 1 kWh, 10 kWh, 20 kWh, and 50 kWh. Figure 6 shows those results.

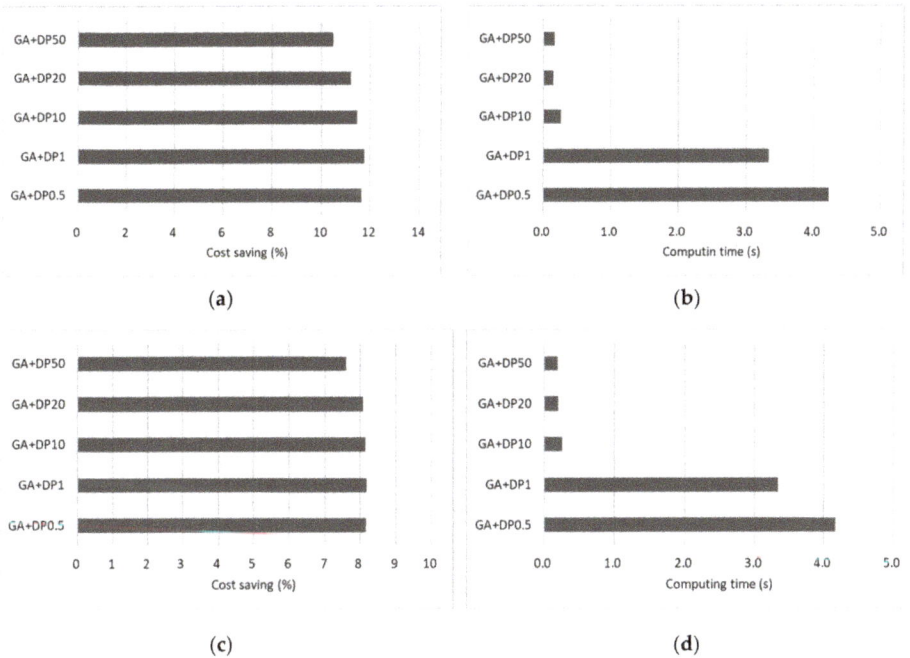

Figure 6. Performance of the combined methods of a genetic algorithm (GA) and dynamic programming (DP) with various sizes of base unit. (**a**) Average cost saving over all cases without demand charge; (**b**) average computing time over all cases without demand charge; (**c**) average cost saving over all cases with demand charge; (**d**) average computing time over all cases with demand charge.

For the case without demand charge (Figure 6a,b), GA+DP1 yields the best cost saving. GA+DP1 is even better than GA+DP0.5 though the results of the two methods have a significance level of 3.0 e^{-1} in *t*-test. Usually, in the design of DP, the smaller the size of the base unit, the better the performance. However, when DP is combined with GA, it is not definitely true. In our experiments, if the base unit is smaller than 1 kWh, the combined method could not show the improved performance. So, if we consider only cost saving, the use of 1 kWh as base unit will be a reasonable choice. However, as mentioned in Section 4.2, the results of GA+DP1 and GA+DP10 have a significance level of $1.2\,e^{-1}$ in t-test, which shows that the performance of GA+DP10 is significantly similar to that of GA+DP1, and GA+DP10 is much faster than GA+DP1. Hence, GA+DP10 is preferable for practical application in terms of both cost saving and time.

In the results with demand charge (Figure 6c,d), GA+DP0.5 is slightly better than GA+DP1, and GA+DP1 is also slightly better than GA+DP10. However, the differences are very small. In this case with demand charge, the effect of GA is larger than that of DP, so the quality of solutions is not so sensitive of the size of base unit. Nevertheless, the cost saving of GA+DP50 is surely worse than that of GA+DP20. From these observations, we could conclude that 10 kWh or 20 kWh base unit size is the most practical when we apply the combined method of GA and DP in the case with demand charge, because they spend much less time than GA+DP1 obtaining similar quality of solutions to GA+DP1.

4.5. Experiments with Combined Methods of Improved GA and DP

In the experiments in Sections 4.1 and 4.2, the population size and the number of generations of the proposed GA have been set to be a fixed number, 100 and 100,000. However, in those experiments, the average running times of GA ($\approx$0.3 s) and GA+DP1 ($\approx$3.3 s) have no significant difference in terms of 24 h. Even though we spend more time to running of GA, there will be no problem in temporal performance. So, we did additional experiments with an improved GA by increasing the number of population and generations. For this improved GA, the population size is set to be 200 and the number of generations is set to be 1,000,000. Increasing those numbers further is meaningless because there is no improvement in the quality of the solution.

Figure 7 shows the results with this improved GA (IGA). For the case without demand charge (Figure 7a,b), the performance of IGA was improved compared with GA and it spends more time than GA about 10 times. However, IGA is not better than GA+DP10 even though it spends more time. So in this case without demand charge, we could conclude that the combination of GA and DP is more effective than using a standalone GA. For the case with demand charge (Figure 7c,d), the results showed a different pattern. The performance of IGA was also better than that of GA and, moreover, it is even better than that of GA+DP1 while IGA consumed less time than GA+DP1. That is, the effect of DP was not so much in this case when compared to that of GA. However, IGA+DP1 and IGA+DP10 showed better performances than IGA. IGA still can be more improved by combining DP. In particular, IGA+DP10 spent not so much time when compared to IGA, so we can conclude that IGA+DP10 is the most practical among the compared methods in this case.

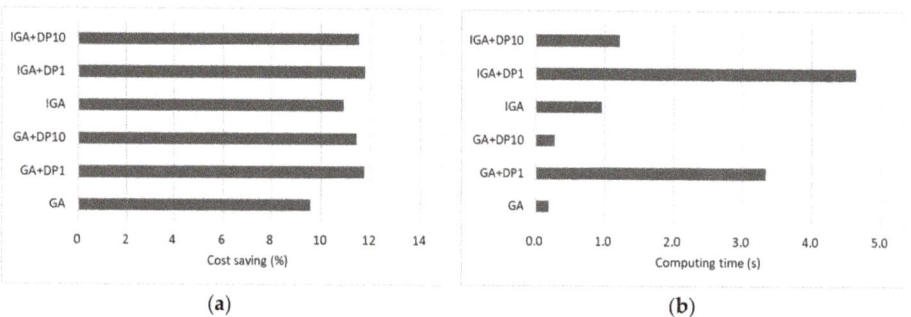

Figure 7. *Cont.*

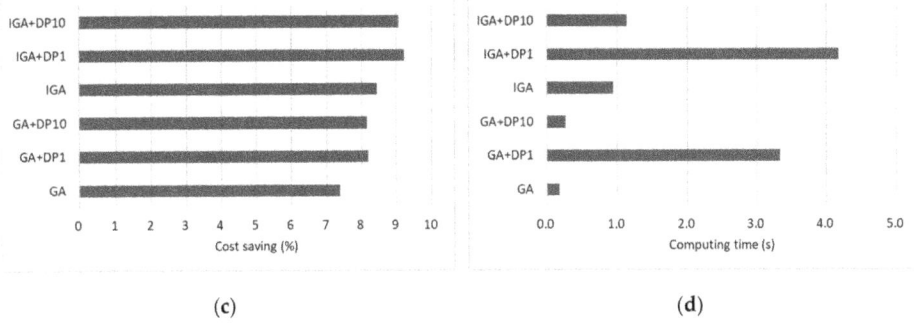

(c) (d)

Figure 7. Comparison between the combined methods with GA and those with improved genetic algorithm (IGA). (**a**) Average cost saving over all cases without demand charge; (**b**) average computing time over all cases without demand charge; (**c**) average cost saving over all cases with demand charge; (**d**) average computing time over all cases with demand charge.

5. Conclusions

In this study, we proposed a combined method of GA and DP for ESS scheduling problems with and without demand charge. Because the temporal performance of DP degrades as the ESS capacity increases, we improved the performance of the combined method by adjusting the size of the base unit in DP.

Without demand charge, DP with a small base unit (1 kWh) showed good economic performance, but its temporal performance was lower than that of DP with a large base unit (10 kWh). The temporal performance of DP with a large base unit was excellent, but its economic performance did not match DP with a small base unit. GA also has a disadvantage of not producing good solutions. Therefore, we improved both the temporal and economic performances by combining GA and DP with a large base unit. Through experiments with various sizes of base units, we could observe that the combined method could not show further improved performance if the base unit is smaller than 1 kWh. Hence, the combined method of GA and DP with the base unit smaller than 1 kWh is not efficient. Moreover, the experimental results showed that GA+DP1 and GA+DP10 had significantly similar economic performances through t-test, and GA+DP10 is much faster than GA+DP1. Therefore, we could conclude that GA+DP10 was the most practical among the compared methods in terms of both economic and temporal performances in the case without demand charge.

When demand charge was included, we could observe that the economic performance of DP significantly degraded through experiments. By combining GA and DP, the advantages of both methods can be utilized, that is, that of GA, which optimizes the demand charge, and that of DP, which optimizes the energy charge. In this case, the combined method of GA and DP with a small base unit showed better economic performance than that with a large base unit. However, the p-value obtained from t-test between the two methods was larger than 0.05, which means that it is not sure that their performances differ. Actually, we could observe that the quality of solutions is not so sensitive of the size of base unit in the case with demand charge through experiments with various sizes of base units. The experimental results showed that the proposed combined method of GA and DP with a base unit 10 kWh or 20 kWh could not only save computing time but also find good solutions when compared with the same method with other base units.

Our study also has some limitations in that we did not consider the cost of battery cycling. For more practical applications, the advanced research considering battery cycling might be required as future work.

Author Contributions: Conceptualization, Y.Y.; methodology, S.-J.L. and Y.Y.; software, S.-J.L.; validation, Y.Y.; formal analysis, Y.Y.; investigation, S.-J.L. and Y.Y.; resources, S.-J.L. and Y.Y.; data curation, S.-J.L. and Y.Y.;

writing—original draft preparation, S.-J.L.; writing—review and editing, Y.Y.; visualization, S.-J.L.; supervision, Y.Y.; project administration, Y.Y.; funding acquisition, Y.Y. All authors have read and agreed to the published version of the manuscript.

Funding: This work was supported by the National Research Foundation of Korea (NRF) grant funded by the Korea government (Ministry of Science, ICT and Future Planning), grant number 2017R1C1B1010768.

Acknowledgments: The authors thank Yong-Hyuk Kim for his valuable suggestions, which greatly improved this paper.

Conflicts of Interest: The authors declare no conflict of interest.

Nomenclature

ESS	Energy storage system
TOU	Time-of-use
DP	Dynamic programming
DP1	DP with a base unit of 1 kWh
DP10	DP with a base unit of 10 kWh
GA	Genetic algorithm
GA+DP1	The combined method of GA and DP with a base unit 1 kWh
GA+DP10	The combined method of GA and DP with a base unit 10 kWh
HS	Harmony search
HS+DP1	The combined method of HS and DP with a base unit 1 kWh
HS+DP10	The combined method of GA and DP with a base unit 10 kWh

References

1. Roberts, B.P.; Sandberg, C. The role of energy storage in development of smart grids. *Proc. IEEE* **2011**, *99*, 1139–1144. [CrossRef]
2. Wang, H.; Huang, J. Joint investment and operation of microgrid. *IEEE Trans. Smart Grid* **2017**, *8*, 833–845. [CrossRef]
3. Mozafari, B.; Mohammadi, S. Optimal sizing of energy storage system for microgrids. *Sadhana* **2014**, *39*, 819–841. [CrossRef]
4. Li, Y.; Yang, Z.; Li, G.; Zhao, D.; Tian, W. Optimal scheduling of an isolated microgrid with battery storage considering load and renewable generation uncertainties. *IEEE Trans. Ind. Electron.* **2018**, *66*, 1565–1575. [CrossRef]
5. Tushar, M.H.K.; Zeineddine, A.W.; Assi, C. Demand-side management by regulating charging and discharging of the EV, ESS, and utilizing renewable energy. *IEEE Trans. Ind. Inform.* **2017**, *14*, 117–126. [CrossRef]
6. Barton, J.P.; Infield, D.G. Energy storage and its use with intermittent renewable energy. *IEEE Trans. Energy Convers.* **2004**, *19*, 441–448. [CrossRef]
7. Smith, S.C.; Sen, P.K.; Kroposki, B. Advancement of energy storage devices and applications in electrical power system. In Proceedings of the IEEE Power and Energy Society General Meeting: Conversion and Delivery of Electrical Energy in the 21st Century, Pittsburgh, PA, USA, 20–24 July 2008; pp. 1–8.
8. Sanghvi, A.P. Flexible strategies for load/demand management using dynamic pricing. *IEEE Trans. Power Syst.* **1989**, *4*, 83–93. [CrossRef]
9. Grillo, S.; Marinelli, M.; Massucco, S.; Silvestro, F. Optimal management strategy of a battery-based storage system to improve renewable energy integration in distribution networks. *IEEE Trans. Smart Grid* **2012**, *3*, 950–958. [CrossRef]
10. PG&E's TOU Rate Plan. Available online: https://www.pge.com/en_US/residential/rate-plans/rate-plan-options/time-of-use-base-plan/tou-everyday.page (accessed on 30 August 2020).
11. Time-Of-Use|Hydro Ottawa. Available online: https://hydroottawa.com/accounts-services/accounts/time-use (accessed on 30 August 2020).
12. Electric Rates Table|KEPCO. Available online: https://home.kepco.co.kr/kepco/EN/F/htmlView/ENFBHP00103.do?menuCd=EN060201 (accessed on 30 August 2020).
13. Sheen, J.-N.; Chen, C.-S.; Yang, J.-K. Time-of-use pricing for load management programs in Taiwan power company. *IEEE Trans. Power Syst.* **1994**, *9*, 388–396. [CrossRef]

14. Hwangbo, S.W.; Kim, B.J.; Kim, J.H. Application of economic operation strategy on battery energy storage system at Jeju. In Proceedings of the IEEE PES Conference on Innovative Smart Grid Technologies, Vancouver, BC, Canada, 1–6 April 2013; pp. 1–8.
15. Nguyen, M.Y.; Yoon, Y.T. Optimal scheduling and operation of battery/wind generation system in response to real-time market prices. *IEEE Trans. Electr. Electron. Eng.* **2014**, *9*, 129–135. [CrossRef]
16. Nottrott, A.; Kleissl, J.; Washom, B. Energy dispatch schedule optimization and cost benefit analysis for grid-connected, photovoltaic-battery storage systems. *Renew. Energy* **2013**, *55*, 230–240. [CrossRef]
17. Squartini, S.; Boaro, M.; de Angelis, F.; Fuselli, D.; Piazza, F. Optimization algorithms for home energy resource scheduling in presence of data uncertainty. In Proceedings of the 4th International Conference on Intelligent Control and Information Processing, Nanyang, China, 16–17 November 2019; pp. 323–328.
18. Ionescu, L.M.; Bizon, N.; Mazare, A.G.; Belu, N. Reducing the cost of electricity by optimizing real-time consumer planning using a new genetic algorithm-based strategy. *Mathematics* **2020**, *8*, 1144. [CrossRef]
19. Maly, D.K.; Kwan, K.S. Optimal battery energy storage system (BESS) charge scheduling with dynamic programming. *IEEE Proc. Sci. Meas. Technol.* **1995**, *142*, 453–458. [CrossRef]
20. Van de Ven, P.; Hegde, N.; Massoulie, L.; Salonidis, T. Optimal control of residential energy storage under price fluctuations. In Proceedings of the 1st International Conference on Smart Grids, Green Communications and IT Energy-Aware Technologies, Athens, Greece, 2–6 June 2019; pp. 159–162.
21. Koutsopoulos, I.; Hatzi, V.; Tassiulas, L. Optimal energy storage control policies for the smart power grid. In Proceedings of the 2nd IEEE International Conference on Smart Grid Communications, Brussels, Belgium, 17–20 October 2011; pp. 475–480.
22. Romaus, C.; Gathmann, K.; Bocker, J. Optimal energy management for a hybrid energy storage system for electric vehicles based on stochastic dynamic programming. In Proceedings of the IEEE Vehicle Power and Propulsion Conference, Lille, France, 1–3 September 2010; pp. 1–6.
23. Neufeld, J.L. Price discrimination and the adoption of the electricity demand charge. *J. Econ. Hist.* **1987**, *47*, 693–709. [CrossRef]
24. Taylor, T.N.; Schwarz, P.M. A residential demand charge: Evidence from the Duke power time-of-day pricing experiment. *Energy J.* **1986**, *7*, 135–151. [CrossRef]
25. Yoon, Y.; Kim, Y.-H. Effective scheduling of residential energy storage systems under dynamic pricing. *Renew. Energy* **2016**, *87*, 936–945. [CrossRef]
26. Zhang, J.; Zhang, C.; Wu, S.; Zhang, X.; Li, C.; Xue, C.; Cheng, B. High-columbic-efficiency lithium battery based on silicon particle materials. *Nanoscale Res. Lett.* **2015**, *10*, 1–5. [CrossRef]
27. Zhang, Y.; Duan, X.; Wang, J.; Wang, C.; Wang, J.; Wang, J.; Wang, J. Natural graphene microsheets/sulfur as Li-S battery cathode towards >99% coulombic efficiency of long cycles. *J. Power Sources* **2018**, *376*, 131–137. [CrossRef]
28. Zhou, X.L.; Zhao, T.S.; An, L.; Wei, L.; Zhang, C. The use of polybenzimidazole membranes in vanadium redox flow batteries leading to increased coulombic efficiency and cycling performance. *Electrochim. Acta* **2015**, *153*, 492–498. [CrossRef]
29. Yoon, Y.; Kim, Y.-H. Charge scheduling of an energy storage system under time-of-use pricing and a demand charge. *Sci. World J.* **2014**, *2014*, 937329. [CrossRef]
30. Yang, X. *Introduction to Mathematical Optimization*; Cambridge International Science Publishing: Cambridge, UK, 2008; pp. 84–85.
31. Bianchi, L.; Dorigo, M.; Gambardella, L.M.; Gutjahr, W.J. A survey on metaheuristics for stochastic combinatorial optimization. *Nat. Comput.* **2009**, *8*, 239–287. [CrossRef]
32. Choi, Y.; Kim, H. Optimal scheduling of energy storage system for self-sustainable base station operation considering battery wear-out cost. *Energies* **2016**, *9*, 462. [CrossRef]
33. Larson, D.P.; Nonnenmacher, L.; Coimbra, C.F.M. Day-ahead forecasting of solar power output from photovoltaic plants in the American Southwest. *Renew. Energy* **2016**, *91*, 11–20. [CrossRef]
34. Yang, H.-T.; Huang, C.-M.; Huang, Y.-C.; Pai, Y.-S. A weather-based hybrid method for 1-day ahead hourly forecasting of PV power output. *IEEE Trans. Sustain. Energy* **2014**, *5*, 917–926. [CrossRef]
35. Foley, A.M.; Leahy, P.G.; Marvuglia, A.; McKeogh, E. Current methods and advances in forecasting of wind power generation. *Renew. Energy* **2012**, *37*, 1–8. [CrossRef]
36. Panapakidis, I.P. Clustering based day-ahead and hour-ahead bus load forecasting models. *Int. J. Electr. Power Energy Syst.* **2016**, *80*, 171–178. [CrossRef]

37. Sadaei, H.J.; Enayatifar, R.; Abdullah, A.H.; Gani, A. Short-term load forecasting using a hybrid model with a refined exponentially weighted fuzzy time series and an improved harmony search. *Int. J. Electr. Power Energy Syst.* **2014**, *62*, 118–129. [CrossRef]
38. Shayeghi, H.; Ghasemi, A.; Moradzadeh, M.; Nooshyar, M. Simultaneous day-ahead forecasting of electricity price and load in smart grids. *Energy Convers. Manag.* **2015**, *95*, 371–384. [CrossRef]
39. Sudheer, G.; Suseelatha, A. Short term load forecasting using wavelet transform combined with Holt–Winters and weighted nearest neighbor models. *Int. J. Electr. Power Energy Syst.* **2015**, *64*, 340–346. [CrossRef]
40. Hodge, B.; Lew, D.; Milligan, M. Short-term load forecast error distributions and implications for renewable integration studies. In Proceedings of the IEEE Green Technologies Conference (GreenTech), Denver, CO, USA, 4–5 April 2013; pp. 435–442.
41. Ortega-Vazquez, M.A.; Kirschen, D.S. Economic impact assessment of load forecast errors considering the cost of interruptions. In Proceedings of the IEEE Power Engineering Society General Meeting, Montreal, QC, Canada, 18–22 June 2006; pp. 1–8.
42. Herrera, F.; Lozano, M.; Verdegay, J.L. Tackling real-coded genetic algorithms: Operators and tools for behavioral analysis. *Artif. Intell. Rev.* **1998**, *12*, 265–319. [CrossRef]
43. Kim, Y.-H.; Yoon, Y.; Moon, B.-R. A Lagrangian approach for multiple personalized campaigns. *IEEE Trans. Knowl. Data Eng.* **2008**, *20*, 383–395. [CrossRef]
44. Yoon, Y.; Kim, Y.-H. An efficient genetic algorithm for maximum coverage deployment in wireless sensor networks. *IEEE Trans. Cybern.* **2013**, *43*, 1473–1483. [CrossRef] [PubMed]
45. Yoon, Y.; Kim, Y.-H. Geometricity of genetic operators for real-coded representation. *Appl. Math. Comput.* **2013**, *219*, 10915–10927. [CrossRef]
46. Yoon, Y.; Kim, Y.-H.; Moraglio, A.; Moon, B.-R. A theoretical and empirical study on unbiased boundary-extended crossover for real-valued representation. *Inf. Sci.* **2012**, *183*, 48–65. [CrossRef]
47. Commercial and Residential Hourly Load Profiles for All TMY3 Locations in the United States. Available online: https://openei.org/datasets/dataset/commercial-and-residential-hourly-load-profiles-for-all-tmy3-locations-in-the-united-states (accessed on 31 July 2020).
48. The Photovoltaic (PV) Watts Calculator. Available online: https://pvwatts.nrel.gov/ (accessed on 31 July 2020).
49. Geem, Z.W.; Yoon, Y. Harmony search optimization of renewable energy charging with energy storage system. *Int. J. Electr. Power Energy Syst.* **2017**, *86*, 120–126. [CrossRef]

© 2020 by the authors. Licensee MDPI, Basel, Switzerland. This article is an open access article distributed under the terms and conditions of the Creative Commons Attribution (CC BY) license (http://creativecommons.org/licenses/by/4.0/).

Article

Reducing the Cost of Electricity by Optimizing Real-Time Consumer Planning Using a New Genetic Algorithm-Based Strategy

Laurentiu-Mihai Ionescu [1,*], Nicu Bizon [1,*], Alin-Gheorghita Mazare [1] and Nadia Belu [2]

[1] Faculty of Electronics, Communications and Computers, University of Pitesti, Targul din Vale 1, 110040 Pitesti, Romania; alin.mazare@upit.ro

[2] Faculty of Mechanics and Technology, University of Pitesti, Targul din Vale 1, 110040 Pitesti, Romania; nadia.belu@upit.ro

* Correspondence: laurentiu.ionescu@upit.ro (L.-M.I.); nicu.bizon@upit.ro (N.B.)

Received: 3 June 2020; Accepted: 10 July 2020; Published: 13 July 2020

Abstract: To ensure the use of energy produced from renewable energy sources, this paper presents a method for consumer planning in the consumer–producer–distributor structure. The proposed planning method is based on the genetic algorithm approach, which solves a cost minimization problem by considering several input parameters. These input parameters are: the consumption for each unit, the time interval in which the unit operates, the maximum value of the electricity produced from renewable sources, and the distribution of energy production per unit of time. A consumer can use the equipment without any planning, in which case he will consume energy supplied by a distributor or energy produced from renewable sources, if it is available at the time he operates the equipment. A consumer who plans his operating interval can use more energy from renewable sources, because the planning is done in the time interval in which the energy produced from renewable sources is available. The effect is that the total cost of energy to the consumer without any planning will be higher than the cost of energy to the consumer with planning, because the energy produced from renewable sources is cheaper than that provided from conventional sources. To be validated, the proposed approach was run on a simulator, and then tested in two real-world case studies targeting domestic and industrial consumers. In both situations, the solution proposed led to a reduction in the total cost of electricity of up to 25%.

Keywords: renewable energy; consumer planning; real-time strategy; consumption monitoring

1. Introduction

Through the range of energy producers, from small (photovoltaic panels at the level of households) and medium (fields of photovoltaic panels, wind power plants) producers, to large traditional electricity producers (hydroelectric power stations, thermal power plants, nuclear power plants), the need to expand power networks has emerged. The tendency is to implement on-demand electricity production and distribution techniques according to consumer's needs. On the one hand, these consumers are increasing in number and have steadily increasing needs. On the other hand, producers of electricity from renewable sources are less likely to adapt their production capacity according to the consumers' requirements.

The use of smart grids and the concepts derived from it, such as dynamic prices and demand management, have had a significant impact in many areas.

An alternative to production management and the distribution of electricity on demand is to apply dynamic tariff plans. Such plans motivate consumers to save electricity during some periods by re-scheduling household activities (chores) to other time periods. Thus, the distribution of total electricity consumption becomes uniform over time by eliminating the consumption peaks in some short time intervals. Dynamic tariff schemes can be changed according to area and consumption time interval. Several studies [1] show that a dynamic price of electricity reduces electricity costs at the level of consumers by as much as 10%. Dynamic pricing favors the consumption of electricity during periods when energy producers from renewable sources can supply this energy.

The main problem related to the production and distribution of electricity on demand and the dynamic pricing of electricity is the real-time monitoring of consumers—it is necessary to determine the energy consumption for short periods of time in order to apply differential tariffing. The existing meters in the electricity distributors' infrastructure have two limitations: they only have a local, short-range communication interface to an operator, and they do not transmit the data acquired in real-time. A third limitation is related to the fixed location of the meter, i.e., at the level of a consumer or at the level of a grid node. Thus, it can either provide a small amount of data regarding the status of a single consumer, which cannot be used in a statistical analysis without violating the consumer's right to privacy, or too large amounts of data regarding the status of an entire grid. In the second case, the status of different branches, in terms of new consumers attaching to a node, modifications of the electricity grid, appearance of new grid nodes, etc., cannot be seen.

The real-time monitoring of electricity consumption (current and voltage) at different points of the electricity distribution grid is therefore an essential element in dynamic pricing. It is important to use non-invasive solutions that monitor electricity consumption, which can be easily placed in the existing electricity distribution infrastructure. The main problem when measuring energy consumption is the measurement of current consumption. This involves mounting the ammeter in series on the power cable. There are several technologies that allow for electricity measurement without the need for infrastructure intervention.

The current consumption sensor is the main component of a monitoring system for consumer planning and dynamic pricing. For an autonomous solution, the main requirement is that the sensor allows for energy harvesting.

Our paper proposes a consumer planning solution, using a genetic algorithm (GA), to reduce the consumer electricity cost by using as much electricity as possible from renewable sources by local micro-producers. The novel elements presented in this paper compared to previous publications and research projects are as follows:

- The use of a genetic algorithm for the re-planning of consumers to reduce the electricity cost using the energy of local producers through renewable means. The use of a genetic algorithm for consumer planning and the way in which the coding solution was realized (i.e., obtaining the chromosome) are the original components of the paper.
- The implementation of a simulator that allows the generation of numerous consumer–producer configurations (i.e., hundreds), and the study of the impact of the consumer planning algorithm.
- A complete platform implemented for real-time consumption monitoring, analysis, and intelligent planning, using the genetic algorithm, and consumer communication/information. Unlike other works in the field, this paper presents a complete solution that allows for the implementation of a consumer planning algorithm that consists of sensor cells for the acquisition of current consumption with self-harvesting, data collectors, a server for analysis and planning, and a client application for informing consumers. The system, developed as part of a research project, was used in two case studies that highlight the efficiency of the consumer planning algorithm.

There are several solutions proposed for consumer planning. They have different approaches to re-planning consumers in order to reduce the total cost of electricity. In contrast, the solution proposed in this paper addresses the issue of consumer planning from multiple perspectives: the ability to produce electricity through renewable means at different times of the day, for certain regions and using different energy sources; the cost of electricity produced from renewable sources; the cost of electricity produced by traditional methods; and the characteristics of some consumers to be able to be reprogrammed at longer or shorter time intervals. All of these elements represent criteria that must be taken into account when determining the optimal planning scheme of consumption units (equipment) during a day, to finally obtain the minimum cost of electricity. The genetic algorithm is an algorithm specialized in solving multi-criteria optimization problems: this is the reason why this type of algorithm was used here.

The objectives of this study are as follows:

- proposing a new strategy to reduce the electricity cost in a nano- or micro-grid through optimal consumer planning;
- implementing the proposed strategy in a simulator (using Python libraries, such as Numpy and Matplotlib);
- validation of simulation results in two experiments, using an innovative genetic algorithm proposed and analyzed in this paper;
- highlighting the advantages for all participants (micro-producers, regular consumers, or prosumers) based on case studies where this strategy was implemented.

The next section will present strategies from the specialized literature that are designed to reduce the electricity cost in a smart grid through optimal consumer planning.

Compared to the previous publications and research projects presented below, the novelty of this paper is highlighted as follows: (1) a new real-time optimization strategy based on an objective function to minimize the total cost by planning consumers to use available electricity produced from renewable sources is proposed; (2) a new chromosome coding solution is proposed and used by the genetic algorithm applied to reduce the electricity cost.

The validation of the reduction of the energy cost for a consumer who applies this new strategy was conducted both in simulation and in experiment. For this, a specific simulator and a complete experimental platform were implemented to evaluate the performance of the proposed strategy, based on hundreds of consumer–producer configurations and case studies emulating the functioning of a smart nano-grid. The experiment was developed as part of a research project (using sensor cells for the acquisition of current consumption with self-harvesting, data collectors, a server for analysis and planning, a client application for informing consumers, etc.), in order to emulate the functioning of a smart nano-grid as closely as possible to reality. For both implementations, the genetic algorithm optimizes the cost function for the consumer taking into account the requirements and constraints imposed (for example, times when devices have to be used, electricity available from renewable sources), which are variable over time. The only difference for the performed experiments is that the input data (such as the possible operating interval and the actual operating interval for consumers, and the cost of energy produced, the available production capacity, and the daily variation in production capacity for producers) are acquired from installed sensors at consumers and producers, rather than simulated data. Thus, the originality of the paper is related to the development and implementation of the consumer planning algorithm on the central station server, as well as to the implementation plan to easily apply the platform in various experimental case studies.

The remainder of this paper is structured as follows: Section 2 presents a literature review. The subsequent section presents the model and coding schema used for the genetic algorithm, and is divided into four subsections: Section 3.1 The Consumer–Producer–Distributor Mode, Section 3.2 Genetic Algorithm, Section 3.3 Simulation Platform and Data Collection, and Section 3.4 Methods. Section 4.1 presents the simulation and the results obtained from the simulation, and Section 4.2 presents the system implementation and its operation in two real case studies. The paper concludes with Section 5 Conclusions and Future Trends.

2. Literature Review

In reference [2], a comprehensive study of the use of the smart grid in communication infrastructure to reduce energy consumption is offered. The study begins by addressing the concept of the smart grid, and continues with the presentation of smart grid techniques for increasing energy efficiency in communications, and minimizing energy costs and emissions of data centers and cloud infrastructures. This is the also the purpose pursued in our work. The objective of the current study is to carry out a re-planning of consumers in order to reduce the cost of electricity. This is possible using the proposed real-time monitoring platform, as presented in this paper. Alahakoon et al. [3] present a comprehensive study of smart grids and their use, technologies used in the metering process, and solutions to satisfy the interests of stakeholders. Moreover, the paper highlights challenges and opportunities that arise using big data and the cloud in smart grids. Problems relating to data privacy in smart grids (SGs) and advanced measurement infrastructure (AMI) are treated in reference [4]. The use of SGs and AMI allows the implementation of consumer planning solutions through the possibility of real-time monitoring on different nodes of the distribution network. However, at present, the monitoring infrastructure is not always available for real-time monitoring. This is the reason why, in our paper, a flexible monitoring solution is proposed, which allows real-time monitoring for the application of dynamic tariffs, and to favor the consumption of electricity from renewable sources. After data acquisition, it is necessary to analyze them, and to develop consumption planning. Deng et al. [5] refer to energy consumption planning using a game model. As a result, consumption peaks are reduced. The issue of consumer planning following the reprogramming of processes with the implementation of a simulator is addressed by Kliazovich et al. [6]. Tang et al. [7] address the reprogramming of tasks (at a certain manufacturing process) in order to reduce energy consumption. Starting from the characteristics of the production process, the consumption pattern is determined by a model developed in reference [8]. A genetic algorithm is then used to determine which of the elaborated schemes is optimal. The key element in the consumer planning solution proposed in the current paper is also a genetic algorithm. Unlike the mentioned works, consumers' planning is realized to reduce the total cost of electricity, with a flexible configuration scheme that applies to both domestic and industrial consumers. This is done by the coding schema performed on the genetic algorithm.

Previous studies have dealt with the problem of consumer planning and have proposed different algorithms for solving it. Cionca et al. [9] refer to the planning of the energy produced and consumed to accumulate surpluses and to reduce losses (which occur when there is a surplus of produced energy). Derakhshan et al. [10] present algorithms for energy consumption: Teaching and Learning-based Optimization (TLBO) and Shuffled Frog Leaping (SFL). These are applied to a case study for a real environment. As with our approach, in the above-mentioned work, the total energy cost is reduced by reducing the costs for certain periods using the electricity produced from unconventional sources. Ghasemi et al. [11] propose a new algorithm for forecasting the electricity price and its demand, which uses data preprocessing techniques, forecasting, and adjustment algorithms. The three components of the forecasting algorithm are described, including an artificial intelligence (AI) component of the artificial bee colony (ABC) type used in the learning process. This generates coefficients for the other modules. The proposed hybrid forecast algorithm is evaluated on several real markets, illustrating its high accuracy in the forecast of the price and demand of electricity. Moreover, the impact of its use on demand management techniques is investigated, emphasizing the improvements made.

Ma et al. [12] present a new concept of a residential consumption planning framework based on cost efficiency. Cost efficiency means the ratio between the total benefit and the total payment of electricity over a certain period. In this work, a consumption planning algorithm based on daily costs is developed using a fractional programming approach. The proposed planning algorithm can reflect and affect consumer behavior, and ultimately generate an optimal cost-effective energy consumption profile. Similarly, in the current paper, the planning of consumption units takes place, but the cost is related to the method of obtaining the electricity.

The aforementioned paper presents a simulation that confirms that the proposed planning algorithm significantly improves cost efficiency for consumers compared to conventional planning solutions. In a similar way, a simulator is also created for validating the efficiency of the proposed consumer planning algorithm. Moreover, the algorithm is implemented using a real-time monitoring platform and results for two real-world case studies are presented.

During the implementation of the platform, there are several challenges related, in particular, to connecting the current sensors to the electricity transmission lines and their power supply. On this topic, several studies have been conducted of the construction of the monitoring platform and the implementation of the algorithm. Along with solutions with neural networks and GAs, other algorithms have been used to reduce consumption and plan hybrid energy sources. Thus, the paper [13] presents an algorithm-based chaotic search and harmony search, which is used to plan the storage of energy produced from five renewable sources (photovoltaic and wind). A system composed of a chemical cell based on hydrogen and a battery is used for energy storage and the objective is to extend the life of the storage cells and reduce the cost by optimally planning the storage sequence (charges/discharges). In our case, a GA is used to plan the operating time interval in order to use as much energy as possible from renewable sources. A novel particle swarm optimization algorithm based on the Hill function is presented in paper [14] for reducing energy consumption in an industrial process by dynamically planning its component tasks. We propose a similar task in our paper, but we address a much wider category of consumers. Our goal is not to reduce energy consumption but to reduce the cost of energy by using renewable sources. An energy management system produced from three different sources (renewable with photovoltaic panels, batteries, and generator with a diesel engine) based on a modified gravitational search algorithm is presented in [15]. Here, a solution is provided by which to choose one of the three sources to meet the needs of the consumer while reducing fuel consumption as much as possible, and extending battery life. The solution we propose is aimed at consumer-level programming to reduce the cost of energy using energy from different sources. Our approach is somewhat more general—it aims to provide solutions for a wide range of consumers and producers of energy from renewable sources. The solution we propose is based on GA but can be extended with other algorithms, such as the whale optimization algorithm (WOA).

The use of genetic algorithms for consumer planning to reduce the cost of electricity is a subject that has also been addressed in very recent works and research (2019). Xu et al. [16] used a genetic algorithm to plan the operation of an irrigation system so that it is used when the cost of energy is lower; thus, the goal is to reduce the cost of electricity. The solution proposed in the current paper is also the application of an optimization problem to reduce the cost of electricity. However, our approach involves, more generally, the planning of different consumers, in order to reduce the cost of electricity. Another GA-based approach, applied in an industrial environment, is presented in reference [17]. Here, it was used in the planning of an automatic guided vehicle (AGV) inside a factory, to make electricity consumption more efficient. Chang et al. [18] illustrate a solution for planning the operation of heating, ventilation, and air conditioning (HVAC) equipment through a neuro-genetic hybrid solution, to reduce annual energy consumption. The work addresses an optimization problem that aims to reduce cost and increase satisfaction among the tenants in the building connected to the air conditioning system. Farhadiana [19] presents a solution for planning how to allocate virtual machines on physical equipment considering the operation and the resources used by each one. The objective is to reduce energy consumption—allocation planning is undertaken through a genetic algorithm solution. Our approach brings, as a novelty element, the coding scheme proposed in the genetic algorithm that allows the planning of different consumers. The objective function is to minimize the total cost of planning consumers, to use electricity produced from renewable sources when it is available.

The involvement of consumers in the process of planning the operation of electrical appliances to reduce the cost of electricity is a modern trend, as shown in Gram-Hanssen et al. [20]. The authors state that such policies have a direct impact on the environment by reducing carbon emissions and the use of renewable energy sources. It is shown that, for a sample of about 2000 people surveyed, 75% stated that they are interested in re-planning consumers to use more renewable energy. This idea is taken up in the present paper. Moreover, its authors propose a solution to help the consumer plan in the most efficient way. Along with the desire to protect the environment, the consumer must also be materially motivated. For this reason, a consumer profile must be determined and then associated with a tariff scheme. Some consumer profiles use more energy from renewable sources and lead to a lower total cost. Determining a consumer's profile is not easy. It depends on the type of consumption units (equipment) but also on its needs. That is why different algorithms are used to identify the profile. In reference [21], the authors propose the use of an AI algorithm with neural networks to identify the consumer profile and its association with a tariff scheme. Several patterns are identified that are used in training. In this way, consumer profile classifications and associations with tariff schemes can be made. What the authors of the present paper propose is not only the identification of a profile, but the determination of the profile that leads to the lowest total cost. For this, both the consumer and the producer of energy from renewable sources will be considered. The identification of the connection between the producer capacity and the consumer's requirements that lead to the lowest cost is treated in [22], through a deep recurrent neural network. The solution allows a prognosis of what the consumer must do to minimize the cost. This is also the target of the present paper. Nonetheless, the approach is different in the current work: it starts from the minimization of the cost (objective function), and identifies a consumer profile that corresponds to the minimum cost. The search considers the needs of the consumer and the capacity of the producer. The solution is more flexible and more general: it addresses any type of consumer and any type of producer. The reason for performing tests in a real operating field with several types of consumers (domestic, industrial) and with several producers (photovoltaic panels, wind generators, energy recovery from motors) was to validate the proposed solution. The results obtained are presented in the next sections. Minimizing the cost starting from completely uncorrelated criteria (consumer need, producer capacity) is a problem solved through genetic algorithms. These algorithms specialize in solving these types of problems. They have been successfully used in multi-criteria optimization problems in various fields, ranging from finding the optimal configuration (minimum cost, minimum weight) for the realization of composite beams and frames taking into account a set of input criteria (including were related to the climate) [23], to making financial statistics and predictions starting from several economic indicators, data sets, or moods, as presented in reference [24]. In addition to the papers presented above, other studies have addressed the issue of optimizing energy cost (or energy consumption) through genetic algorithms. Pillai et al. [25] present a method to reduce power consumption in a multi-processor system. Here, task planning (program sequences) must be performed, to keep processors at the point of operation that involves the lowest power consumption for as long as possible. These aspects must be taken into account for the tasks and characteristics of the processors. The present paper does not aim to reduce energy consumption but to reduce energy costs. However, the scheme is similar: the processor is the producer of energy from renewable sources and the task is the consumer. Another paper that deals with reducing both energy consumption and maintenance costs using GA is [26]. A GA-based model for planning the operating mode of some pumping stations is described here. It takes into account criteria related to the nature of the fluid and the process performed and results in a reduction in energy consumption [27] and the number of switches [28], thereby involving less wear and tear and, ultimately, resulting in a reduction of maintenance costs [29,30]. All of the solutions presented demonstrate the power of genetic algorithms in solving cost minimization (or consumption) problems, using multiple uncorrelated input criteria that must be adjusted [31,32]. In the current paper, the requirements of the consumer and the capacity of the producer are criteria that are considered for determining the profile of the consumer that leads to a minimum cost of electricity. The coding scheme that consists of transforming the criteria into input data for the genetic algorithm and the way in which the results are transformed into a 'green' consumer profile represents an innovative element. In addition, unlike many of the solutions previously presented, in addition to a simulator that validates the algorithm and the proposed coding scheme, a pilot for data collection and display of results is proposed. Thus, the solution is tested in a real operating field. In this way, the challenges that appear in the implementation of the algorithm and the way in which they are solved were highlighted. The solution we propose involves consumers who have a maximum number of devices 10. This involves the use of 10 individuals/generation. For a large number of individuals per generation (over 100), there are problems with the convergence of the algorithm, as shown in the paper [33]. However, there are practical solutions that can improve the performance of the algorithm.

3. Proposed Approach

3.1. The Consumer–Producer–Distributor Mode

Figure 1 represents a consumer–producer–distributor structure from the point of view of the electricity circuit.

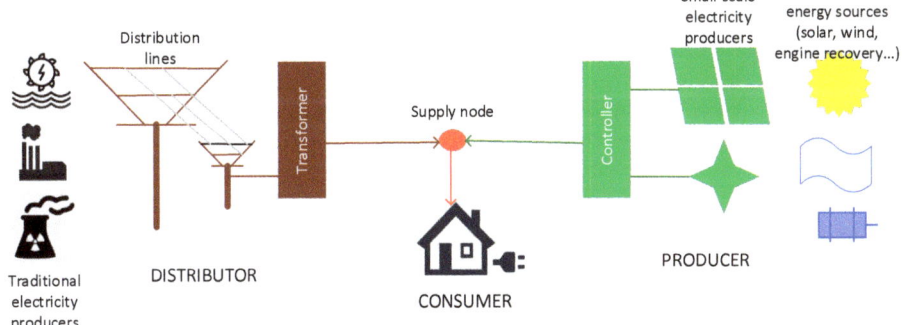

Figure 1. Consumer–producer–distributor structure.

The electricity transmission and distribution operator (TSO-DSO) takes electricity from traditional producers—thermo-power plants, hydro-power stations, and nuclear power plants—and supplies it to domestic and industrial consumers. This is the classic circuit of electricity which has existed for more than a century. The most recent on the market are the local electricity producers (or micro-producers). These have more limited capacity to produce electricity from renewable sources—such as by using photovoltaic panels that capture solar energy and convert it into electricity, or using horizontal and vertical windmills—thus, by converting wind energy into electricity or, as is the case of industrial producers, by converting mechanical energy of machines into electricity or recovering the current from power engines. Unlike a distributor that has a very high-power capacity (SUPPLIER_PEAK), in the case of local producers, the maximum capacity of electricity production, hereinafter referred to as PRODUCER_PEAK, is more limited, and is dependent on the season, time of day, technological process, etc.

In our model, $p(t)$ represents the amount of energy produced by a micro-producer at time t; the total energy produced for a time interval is defined as:

$$\text{PT}_{[a,b]} = \sum_{t \in [a,b]} p(t) \qquad (1)$$

$p(t)$ cannot be represented (or is difficult to represent) using a mathematical relation, because it depends on many factors. In the simulator derived and presented in this paper, uniform random distributions are used to represent $p(t)$ for a period of time. An input data set could be easily generated to simulate the production behavior through photovoltaic electricity cells (maximum ceiling between 11:00–17:00, increase 6:00–11:00, decrease 17:00–21:00, zero 21:00–6:00 on a summer day, with variations depending on the season). The idea is that the algorithm is used to find solutions for any energy sources. For example, one of the case studies of the real operating environment presented in the paper, case study number 2, presents an example of the use of electricity recovered from power motors in an industrial plant. This installation works as follows: when it is necessary to transport (it is a belt for transporting a sheet for rolling), the engines enter an operating mode. When the engines are under load (so the sheet is located exactly on the axles when they move), then the energy recovered is low; conversely, when the engines are not under load, the energy recovered from them is at a maximum. Thus, in this situation we do not have a predictable behavior; energy production is dependent on the operating range and how the weight of the sheet is distributed on the belt. This is why it is preferred to use a source of inputs that is as general as possible, thus simulating the behavior of any electricity generation system. On a practical level, each producer can provide a table of values for $p(t)$ measured at discrete moments of time.

An electricity distributor has an amount of energy produced at a time t, called $s(t)$ but, unlike $p(t)$, it has a constant value (with very few exceptions) and is equal to the maximum production capacity per unit of time:

$$s(t) \approx SUPPLIER\_PEAK \tag{2}$$

The electricity transmission and distribution operator informs the consumer about the value of SUPPLIER_PEAK.

The consumer has several consumption units (home appliances, industrial equipment). A consumption unit has a consumption value given by the function $u(t)$, which depends on the operating mode of the equipment. The total consumption for an operating interval $[t_1, t_2]$ has a formula similar to PT from Equation (1):

$$UT_{[t_1,t_2]} = \sum_{t \in [t_1,t_2]} u(t) \tag{3}$$

For a consumption unit, the possible operating interval $[a, b]$ may include or coincide with the operating interval $[t_1, t_2]$. The possible operating interval is the interval in which the equipment—the consumption unit—can perform its task. There are devices that have a possible operating interval equal to the operating interval or others that have a possible operating interval greater than the operating interval. On the one hand, a TV has a 3-h operating interval that coincides with the possible operating interval. On the other hand, a washing machine that has an operating interval of 1 h can be scheduled to work during the day—between 10 a.m. and 8 p.m. for example. Thus, the 1-h operating interval can be programmed anytime within the 10-h interval of time, as long as there is a possible operating interval.

There are two possible sources of energy that can cover consumption $u(t)$: locally produced energy from renewable sources $p_u(t)$ and energy supplied from conventional sources $s_u(t)$, as follows:

$$u(t) = p_u(t) + s_u(t), \; p_u(t) \leq p(t), \; s_u(t) = \begin{cases} 0, \; u(t) \leq p(t) \\ |p(t) - u(t)|, \; u(t) > p(t) \end{cases} \tag{4}$$

The cost of energy to the consumer is defined in the equation:

$$c(t) = p_u(t) \times pc + s_u(t) \times sc \tag{5}$$

The electricity distribution node allows the consumer to use primarily $p_u(t)$. It is possible that the amount of produced energy covers the consumer's requirements; that is, in this case, $p_u(t) < p(t)$. If not, the consumer uses the entire $p(t)$ and the energy surplus required for the consumer is provided by $s_u(t)$: this is the meaning of the difference in Equation (4). Parameters sc and pc from Equation (5) represent the costs of energy supplied and produced at time t. The cost of energy is usually expressed in monetary units/kW.

For a possible operating interval $[a, b]$ of a device (consumption unit) the total cost is given by:

$$CT_{[a,b]} = \sum_{t \in [a, \; b]} c(t) \tag{6}$$

For a consumer who has several consumption units, the total consumption for a selected time interval [A, B] is defined as:

$$CT([A,B],n) = \sum_{i=1}^{n} CT^i_{[a_i,b_i]}, \quad A \leq a_i < b_i \leq B \tag{7}$$

where [A, B] represents the selected time interval and n represents the number of consumption units.

3.2. The Genetic Algorithm

Genetic algorithms are bio-inspired methods for finding solutions for multi-criteria optimization problems. Their goal is to minimize the cost described by means of a function called an "objective function". They belong to the class of artificial intelligence methods that are used to solve problems that are difficult or impossible to solve by conventional deterministic or heuristic methods. They work with a set of potential solutions that will evolve, following the objective function, until finding the solution or solutions sought.

The genetic algorithm works with the following components:

- The individual, which has one or more chromosomes and represents a potential solution. The main challenge when working with genetic algorithms is to find the method to represent the chromosomes—this is known as the coding schema. To be solved with a genetic algorithm, the problem must be transposed into chromosomes.
- The population represents a set of individuals, that is, a set of potential solutions. A generation represents the state of a population at an iteration of the algorithm.
- The objective function is the criterion after which a generation is evaluated. After the evaluation, each individual will receive a rate called fitness. There may be individuals who have good fitness, so they are closer to the solutions sought, or have poor fitness, i.e., further away from the solutions. The algorithm stops when fitness has reached an acceptable level to one or more individuals; that is, when the solutions sought have been found.
- During each generation, individuals are subjected to so-called genetic operators: crossover and mutation. The results of these operations are offspring (new individuals) or mutants (existing individuals that are changed). These will be added to the population and represent what the algorithm changes for a generation.

In addition to establishing the objective function and determining the coding scheme for the use of an evolutionary algorithm to solve a problem, it is necessary to configure its parameters: population size and type, maximum number of generations, selection method, probability of crossing, probability of mutation, crossing points (number and position), mutation points (number and position), and the replacement policy for offspring. Genetic algorithms are used in applications for determining the optimal route, reducing consumption and finding an optimal consumption scheme, finding patterns that determine the optimization of a process, and generally in solving any search problem known for its objective function. Problems solved with genetic algorithms have several criteria that need to be matched, and which, most often, cannot be correlated.

For the solution proposed by us, the genetic algorithm was used to reduce the cost to the consumer, taking into account its requirements and the constraints it may have (the moments of time at which devices must be used), and the electricity that can be supplied by a local producer from renewable sources (for which production capacity varies depending on external factors over short periods of time).

In our case, the objective function of the proposed genetic algorithm for consumption planning is to minimize the total cost to a consumer who has n units for a selected time interval (in our case a month for domestic consumers, a day for industrial consumers) by planning the operating interval:

$$Obj([A,B],n) = Min(CT([A,B],n)) \tag{8}$$

In order to implement the genetic algorithm, it is necessary, first, to elaborate a coding schema of the problem. The problem is one of determining the minimum cost by re-planning the operating interval of the consumption units within the limits of the possible operating interval.

A chromosome is, in this case, the operating intervals for all consumption units. A gene is an operating interval for a single consumption unit.

For n consumption units (devices) of a consumer, the chromosome structure (individual) is as follows:

$$Chromosome = \bigcup_{i=1}^{n} [t_{1i}, t_{2i}] \in [a_i, b_i], \ a_i \leq t_{1i} < t_{2i} \leq b_i \quad (9)$$

The range $[t_{1i}, t_{2i}]$ that represents a chromosome gene does not necessarily represent a successive set of t values but rather a set of t values. For example, a valid scheduling scheme is at 1:00, 3:00, and 4:00, i.e., the equipment operates at 1:00–2:00, 3:00–4:00, and 4:00–5:00, with a break between 2 and 3. Thus, flexibility in planning (where possible) is maximized. The restriction is that the operating interval be included (or coincide) with the possible operating interval $[a_i, b_i]$.

The genetic crossover operator is implemented in the form of the following function (Relation 10), which receives as parameters the two parent chromosomes (*P1* and *P2*) and the crossover point (*CrP*):

$$Offspring(P1, P2, CrP) = \cup(\cup_{iP1=1}^{CrP}[t_{1iP1}, t_{2iP1}] \in [a_i, b_i], \cup_{iCrP+1}^{n}[t_{1iP2}, t_{2iP2}] \in [a_i, b_i],) \quad (10)$$

Like all individuals of the population, the offspring, presented in Equation (10), has the same number of consumption units, n, as its parents. Until the crossover point, the operating intervals will be taken for the first *CrP* consumer units from the first parent (this is the meaning of the *iP1* index from the first member in Equation (10)), and then the following operating intervals for the following *CrP*+1 consumption units up to n of the second parent (represented by the *iP2* index of the second member). In its chromosome structure, the offspring will contain the two groups of operating intervals. As with parents and, indeed, with all other individuals in the population, the possible operation intervals $[a_i, b_i]$ remain the same; this is the meaning of the index i at a_i and b_i in Equation (10).

Mutation is a function that acts on a single gene. The mutation function is described as:

$$Mutation(I, G) = [t_{1IG}, t_{2IG}] \rightarrow [t'_{1IG}, t'_{2IG}], \ [t_{1IG}, t_{2IG}]\epsilon[a_{IG}, b_{IG}], \ [t'_{1IG}, t'_{2IG}]\epsilon[a_{IG}, b_{IG}] \quad (11)$$

where *I* represents the number of the individual and *G* the number of the gene to which the mutation is applied. Thus, mutation means determining another operating interval for a gene with the restriction of the possible operating interval.

The coding scheme is a binary one: the gene represents a set of bits that indicate the presence of instantaneous consumption on a device at a certain time (e.g., operation with a device for a certain time interval). We have represented this as an interval $[a, b]$, in which the equipment is operated (discrete interval, composed of hours), and which cannot be greater than $[t_1, t_2]$, that is, the maximum allowed interval where the equipment can be programmed respectively to operate. At this interval $[a, b]$, in every hour the device can operate (1) or not (0). After operating for one hour, instantaneous consumption results, and the sum of the instantaneous consumptions is the total consumption of the device. The crossover is the exchange of intervals during a day. Thus, two parents will have certain operating intervals proposed for each device; for example, to the offspring, operation intervals for the first device are from the first parent, and for the second device, they are from the second parent. The mutation represents the reconfiguration at the level of the gene, that is, the change in the way the instantaneous consumption is distributed over an interval to a device. Due to the proposed binary scheme, a crossover at more than one point increases the probability of obtaining a "clone" offspring, which copies a parent without a positive impact on evolution, so crossover is desired at one point. The use of two mutation points per individual/generation can sometimes have positive effects, but leads to the excessive spreading of results within a generation; hence, we discarded the use of two mutations per individual/generation and returned to one mutation per individual/generation, with a probability rate of 0.5, that is, one mutation in two generations.

The characteristics of the genetic algorithm used in this paper are presented in Table 1.

Table 1. Parameters of genetic algorithm.

Parameter	Value
Population type	Fixed size
Population size	10 individuals
Number of generations	800
Selection	Roulette rule
Crossover rate	1/generation
Crossover points	1
Mutation rate	0.5/generation
Mutation points	1
Replacement	Offspring from crossover replace individual with weakest evaluation (fitness)
Evaluation	CT—see Relation (7)
Objective	See Relation (8)
Ending criteria	After 800 generations

The results were initially tested with a larger population (50–100 individuals) but we reduced the population and observed very good results up to a number of 10 individuals/generation; successive tests were performed to reach this number of individuals per generation. Interestingly, under 10 individuals per generation, the convergence time greatly increased. Thus, it appears that 10 individuals/generation is optimal for this kind of problem, in which the number of devices for a consumer is up to 10. All simulations with producer–consumer configurations generated showed a decrease in cost in the first 500–600 generations, then a cost maintenance from generation 600 onwards. Thus, 800 generations were chosen as the fixed number of generations imposed. For each manufacturer–consumer configuration, the algorithm ran for 800 generations and the minimum cost solution obtained after the 800th generation was considered the optimal solution.

The key point in the efficiency of the algorithm is that the time units (t) should be as small as possible, so that the re-planning can include the peaks of local electricity production. These may vary over short periods of time, depending on the oscillation of energy sources that are dependent on uncontrollable external factors. Existing energy consumption measurement systems at the consumer do not have the capacity to make short-term measurements; therefore, an alternative measuring system is needed that can determine real-time energy consumption. Later in the paper, a physical measurement system used to implement the planning algorithm is presented.

The second essential element in the efficiency of the algorithm relates to a lower electricity cost in the case of micro-producers more than in the case of the distributor. Otherwise, the cost optimization algorithm will propose solutions that are close to the classic distributor–consumer arrangement.

3.3. Simulation Platform and Data Collection

The algorithm was implemented on a PC using the Python libraries Numpy and Matplotlib. First, a simulator was designed to validate the proposed method. The simulator was also implemented in Numpy and Matplotlib. Numpy is the Python library used to implement the algorithm. Components of this were used for the generation of chromosome vectors, operations with vectors (i.e., two-dimensional matrices containing chromosomes and fitness), traversal, initialization with pseudo-random sequences, etc. In addition, Matplotlib was used to draw the graphics that appear in the paper. The library was used both to create the simulator and to implement the algorithm in the system that was used in the real operating environment.

Then, the monitoring platform pilot was used to implement and test the algorithm in a real environment. This was composed of sensors connected to a collector used to monitor a branch from a power supply network (sensor cells), a node transceiver used to collect data from sensor cells, packing and transmission, and a central station, which ran a server component based on a complex event processing module connected to the data analyzer engine, which implemented the consumer planning algorithm; see Figure 2.

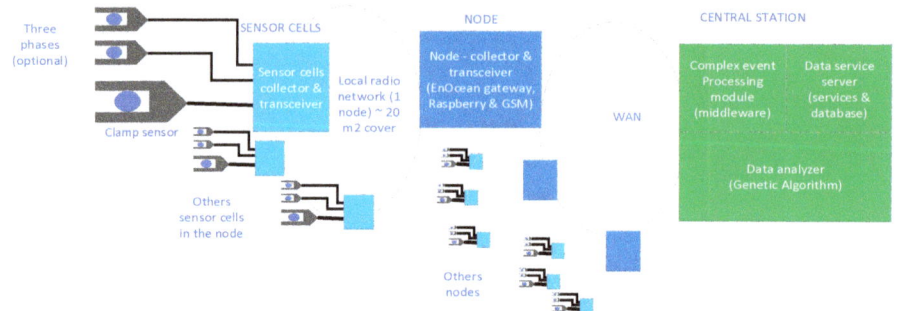

Figure 2. Block diagram of the monitoring platform.

The monitored platform was designed and built within the research project mentioned in the Funding section. This paper adds (i) the development and implementation of the consumer planning algorithm on the central station server; and (ii) the deployment of the platform to be used for the two case studies in a real environment.

Figure 2 shows a monitoring platform based on a sensor cell (mono or three-phase) type clamp that allows for connection to the isolated cable and current measurement. The project in which the monitoring system was implemented aimed only at measuring the current and approximating the voltage, so measured only the active power.

The sensor cell has self-harvesting capabilities and also feeds on the same energy of the cable field that determines the current. The monitoring system allows for easy electricity grid connection for one or more consumers. In addition, the monitoring system allows for the acquisition and transmission of energy consumption for short periods of time (15 min). The data from the sensor cells on a node reaches a data collector and reaches the server through the wide area network (WAN) network (via the Global System for Mobile Communications (GSM)). At the server level, these are analyzed, and the consumer profile is generated with the lowest cost through the presented algorithm.

As parameters, the monitoring system is based on inductive current sensors and the EnOcean circuit, which performs data acquisition and transmission on a low-power wireless local area network in the 933 MHz band. The network supports up to 15 cells (each can be mono and three-phase) and the coverage area is approximately 20 square meters. With self-harvesting technology, the cell does not require additional power, so no maintenance is required.

Due to the low coverage area for the low-power wireless network, a data collector is required to be connected to the network to retrieve the data from it and transmit it to the wide area network (WAN) via Global System for Mobile Communications (GSM). The data collector, also fully created within the project, is based on an embedded PC Raspberry Pi version 3.B+. It has an EnOcean USB gateway that can be connected to the low power Wi-Fi network, and a GSM SIM800 modem on a GSM RPI (Raspberry PI) shield that allows connection to the WAN network via secure HyperText Transfer Protocol (HTTP) protocol. Sensor cells transmit data packages to the local network every 30 s, and the packages are picked up by the data collector, assembled, and transmitted to the server every 15 min.

3.4. Method

In order to validate the proposed solution, a simulation was performed, and the solution was implemented in a real operation using the pilot monitoring platform.

The simulator allows for the generation of different profiles of consumers and producers of electricity. For each consumer–producer–distributor profile generated by the simulator (which represents a simulated case study), the proposed planning algorithm was applied, and the results obtained (consumption distribution and total cost for the consumer) with and without the algorithm were compared. The maximum electricity production capacity (PRODUCER_PEAK) of local producers was varied between 2 and 8 kW and the cost of energy to the producer (PRODUCER_COST) was varied from 0.1 to 0.2 monetary units/1 kW. For each value, between 30 and 40 producer–consumer profiles were generated. The obtained results presented in the next section show a decrease in the cost of electricity to the consumer using the planning solution with the proposed genetic algorithm.

Next, the solution was implemented in a real environment, using the monitoring platform, in two case studies: domestic consumers and industrial consumers.

The monitoring platform was built to operate under outdoor conditions (mounted in IP67 casings). For the two case studies presented, the pilot platform was deployed with seven sensory cells (mono and three-phase) and seven collectors—the system collected data from seven consumers and associations of consumers. It was installed in three localities and operated (and still operates) both in the warm season and in winter, at the latitude of Romania. Figure 3 shows images during the system deployment and the system integration in the IP67 box casings.

Figure 3. The pilot device assembly (**left**) and the mounting (**right**). The mounting was carried out by the beneficiary partner within the project.

In addition to the monitoring system, there are display interfaces (web client) to indicate to the client how to undertake consumer programming. Thus, the flow of planning for household consumers is presented in Figure 4.

In the first phase, the consumer must introduce the consumer units and their respective energy consumption values (read from the product labels), as well as the programming possibilities: the time frame during which a consumer can be programmed and the operating time/day. Further, the data entered by the consumer regarding consumption are stored in the server. In addition, the server stores data from the electricity producers (located near the consumer, and which may make a contribution to the consumer) in terms of maximum production capacity and energy cost. From the producers, the amount of energy produced is purchased at short time intervals (also every 15 min).

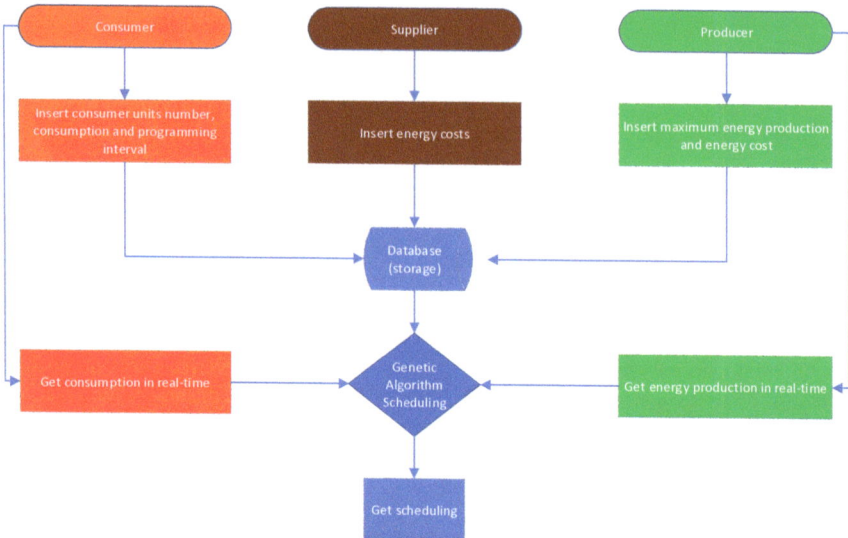

Figure 4. Planning flow for consumer, producer, and supplier. The supplier must enter only the cost of electricity (which can be taken from electricity bills). The producer must enter its maximum production capacity and cost of energy produced (which can be taken from the specifications of the energy source) and electricity produced at some point (which can be given by the controller). The consumer must enter the number of consumption units, the programming interval, and the consumption at a given time using a current consumption monitoring system.

With this data available, the server provides a schedule to the consumer, by means of the genetic algorithm presented and implemented as shown in the previous section, to reduce the cost of electricity using energy produced from renewable sources. This schedule is calculated for each consumer daily. It is then displayed on the client web interface. The consumer will know permanently whether or not his profile is approaching the optimum. Thus, a consumer is able to program his/her home appliances in such a way as to reach the optimum profile, i.e., where the cost is minimal.

In practice, the main reluctance of consumers relates to the need to introduce consumption data and the programming of the devices. However, the introduction of consumption data is made from a common client web graphical interface, with easily accessible controls. For a consumer with minimal IT knowledge, the introduction of consumption data for 10 devices took 10 min. Entering time intervals for each consumer unit requires some effort—this task took about 30 min. This programming is not needed daily—the characteristics of the local electricity producer, as well as those of the consumers are maintained within certain limits for several days. The consumer is also informed that this programming will result in an immediate reduction in the cost of electricity. The number of reprogrammed devices is decided by the consumer, who is continually informed through the display interface the value of the costs at a given moment.

The genetic algorithm implemented in the pilot is the same as that used in the simulator, performed in the same environment (i.e., Numpy). The only difference is the input data: the number of consumers, their possible operating interval, the operating interval, the cost of energy at the producer, the capacity of energy production, and the way the energy produced per day is distributed are information acquired from real consumers and producers.

4. Results

4.1. Simulation

First, a simulator was used for the consumer–producer–distributor structure. Thus, numerous consumer–producer–distributor cases can be generated, which are then planned using the genetic algorithm.

The simulator allows the generation of a set number of consumers (CONSUMERS_NO). Maximum consumption and minimum consumption levels are set and the simulator generates the number of consumers with random values between the two consumption limits; e.g., if the consumption limits are set between 100 and 3000 W (limits within which regular household consumers fall), 10 consumption units will have the consumption generated by the simulator shown in Figure 5.

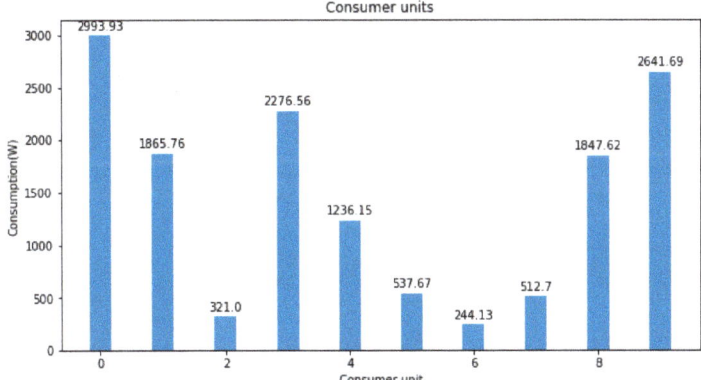

Figure 5. Example of consumers units generated by simulator.

For each consumer, the following are defined: a time interval in which a device can operate (it is given as a start time, 0–23, and an end time, 0–23) and a number of operating time units (hours) from this time interval. The consumer can only work within the time frame, with a number of hours allocated from that time frame.

An example is presented in Table 2.

Table 2. Consumer schedule for a profile (case study) generated by simulator.

Consumer Unit No.	0	1	2	3	4	5	6	7	8	9
Start time	2:00	3:00	19:00	4:00	13:00	14:00	20:00	15:00	18:00	22:00
Stop time	13:00	20:00	20:00	9:00	20:00	19:00	21:00	22:00	18:00	22:00
Number of operations hours per interval	9	4	1	9	1	2	1	2	0	0

The first consumer (with 2993.93 W consumption/hour) can operate between 2 and 13 h with 9 h of operation during this interval. This means that, of the total of 11 h, operation will occur for 9 h and no operation will occur for 2 h. The generation of the interval, as well as the number of operating time units, is derived randomly. In this way, as can be seen in Table 2, consumers are generated that have a large time frame in which they can be programmed, but also a large number of operating units; for example, the case of the first consumer shown in Figure 5 and in Table 2 has a 12-h operating range that operates for 9 h. Consumers may also have a large time interval in which they can be programmed and a low number of operating time units; for example, the second consumer has a range of 17 h in which it can be programmed with only 4 operating hours. Finally, consumers may also have a small programming time interval; for example, in Table 2 and Figure 5, the seventh consumer has only an hour (8 p.m.) at which it can operate.

Thus, among home consumers, a TV can only be programmed when watching it, while a washing machine can be programmed to operate over a longer period of time. In addition, in the industrial environment, a process from a production flow is more flexible and can be programmed over a longer period, while another process cannot be programmed: the production flow requires it to be executed exactly at one point of time. Thus, the simulator allows for the analysis of all types of consumer units.

A micro-producer exists that can be a producer of electricity from renewable sources—solar (through photovoltaic panels), wind, etc.—and is able to provide hybrid sources of electricity. Such a producer is characterized by a maximum nominal value of energy that it can produce (PRODUCER_PEAK), but which, of course, cannot always be achieved. For example, a producer that has eight photovoltaic panels of 250 W each could theoretically produce 2000 W. In practice, its production capacity varies depending on several factors: season, weather, panel status, controller performance, etc. Therefore, the distribution of energy produced can have fluctuations. In the simulator, there is a uniform random distribution of electricity production per day with a range from 0 to the maximum nominal value; Figure 6.

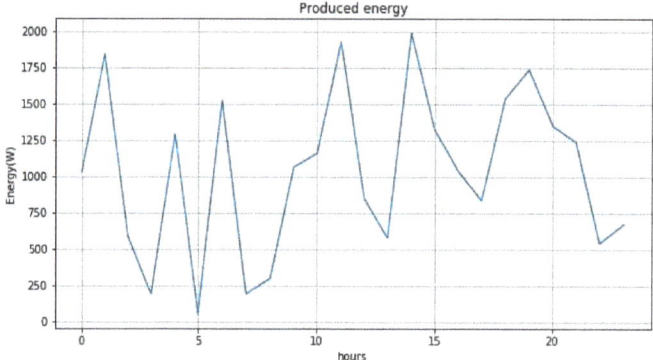

Figure 6. Example of daily distribution of energy produced by a micro-producer generated by the simulator.

The distribution of energy produced in Figure 6 does not necessarily reflect the way electricity is produced by a system with photovoltaic panels. Rather, it is a generic distribution in which fluctuations are present (in which production sometimes reaches 0) in the production process of electricity, using different types of renewable energy generators. As presented in the next section, generators utilize wind energy or mechanical recovery of energy from engines.

The electricity producer has an associated cost of electricity (PRODUCER_COST), which is measured in monetary units/W. Even if the energy from renewable sources is free, the cost of the energy delivered by the producer includes taxes for the amortization of the infrastructure and other fees relating to the electricity grid and energy transport.

An electricity distributor exists that has, at least theoretically, a large capacity to produce energy that comes from conventional sources (the maximum nominal value SUPPLIER_PEAK), so that it can cover the needs of consumers. The electricity distributor also has a cost of distributed energy: SUPPLIER_COST.

The purpose of including a micro-energy producer in the overall consumer–micro-producer–distributor assembly is that the energy produced by it through renewable means must be cheaper than the energy provided by the distributor, so:

PRODUCER_COST < SUPPLIER_COST.

This is made possible by the fact that the raw material of the renewable energy generators is free (sunlight, wind, transforming currents, inertial motion of motors, etc.), and because several states encourage the production of renewable energy by financing the infrastructures.

Given the fact that the consumer planning algorithm proposed in this paper aims to reduce costs, it will work even if at a certain time, under certain conditions, the cost of energy at the micro-producer is higher or equal to the cost of energy from the distributor. The result provided in this case is a trivial one: the electricity from the conventional distribution grid will be used, no matter how the planning is done.

Thus, the algorithm and the system proposed in this paper is motivated if there is at least one local micro-producer of electricity that can supply through an on-grid system in the electricity grid at a lower cost than that of the energy from the distributor. At the simulator level, the energy consumption/day is generated when no planning is made. Thus, for each consumer, the operating time units are placed at the beginning of the time period in which the consumer can operate; in this way, it is a fixed schedule. An example of such planning is shown in Figure 7, for a fixed distribution of consumers over the time intervals presented in Table 2.

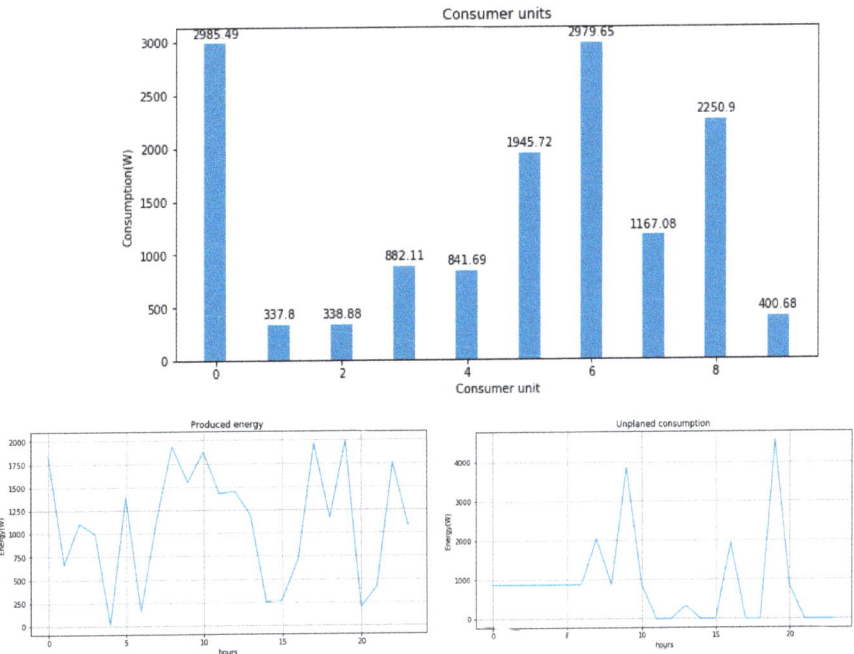

Figure 7. Simulated case study results. **Top**: consumers, **bottom**: distribution of energy produced and unplanned consumption distribution.

The top of Figure 7 shows the consumers units and the energy produced. The bottom left of Figure 7 shows the energy peaks that are typical of a photovoltaic panel power generator at maximum sunlight during the day, while the bottom right of Figure 7 illustrates the unplanned consumption distribution throughout the day. As shown, at the simulator level the time intervals planned for each consumer are randomly generated, but Figure 7 captures a situation close to the real one, where the maximum consumption of a domestic consumer is obtained at certain time intervals, usually in the morning and in the evening.

As shown, the algorithm will run for 800 generations with a population of 10 individuals/generation. The convergence time is very short (a few seconds). The cost evolution for the profile generated by the simulator and illustrated in Figure 7 is shown in Figure 8.

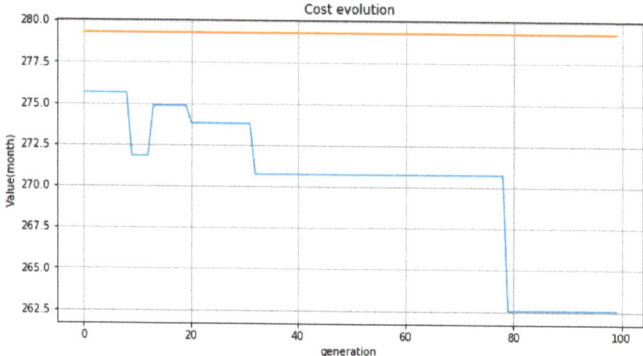

Figure 8. The evolution of the cost during the running of the algorithm: in orange color, the cost for an unplanned scheme, and in blue color, the cost for the best individual of the generation (in the figure the evolution is stopped at 100 generations).

Figure 7 shows a set of simulator-generated configurations. The figure displays the consumption for the 10 devices; the bottom-left shows the fluctuation of the energy produced by renewable means during a full day; and the bottom-right shows the variation of the total consumption (from the 10 devices) during a day, without any planning to reduce the cost. Like the other configurations, this is taken over by the GA module, and a re-planning of consumers is determined, in order to reduce the total energy cost. Figure 8 shows the evolution of the cost during the running of the planning algorithm: the initial cost, as shown in the graph in Figure 7 (bottom, right), is over 275 MU (monetary units)/month, and after planning, it reaches 262.5 MU/month.

In Figure 8, the orange line represents the cost resulting from the fixed planning (presented at the bottom of Figure 7) and the blue line represents the lowest cost for a population generated by the genetic algorithm (the result from the "best" individual). The cost is expressed in monetary units (RON, where 1 RON~0.2 EUR) per month. A lower cost can be observed from the beginning (generation 0). The explanation is that generation 0 contains a total of 10 individuals. Of the 10 individuals, at least one has a lower cost solution than the result of fixed planning. There is the possibility that all generated individuals will have a higher cost than the fixed planning cost at generation 0.

The idea is that the algorithm evolves so that the cost decreases over generations to the lowest values. In the case presented in Figure 7, a decreasing evolution of the cost from a value of approximately 276 to 262 can be observed. The simulation was performed for the values presented in Table 3.

Table 3. Set of parameters from the simulator used for measurements presented below.

CONSUMERS NO	consMin [W]	consMax [W]	PRODUCER PEAK [W]	SUPPLIER PEAK [W]	SUPPLIER COST RON/W	PRODUCER COST RON/W
10	100	3000	2000	6000	0.6/1000	0.3/1000

A typical feature of genetic algorithms is what happens, in our case, between generations 10 and 20 (Figure 8). A decrease in cost can be observed somewhere around the 10th generation and thereafter an increase. Local growth of this kind is not necessarily harmful to convergence. A local optimum is reached at the cost of approximately 272. Then, the algorithm escapes from this, the cost increases, and finally the algorithm reaches the minimum (global optimal) cost of 262.5. The individual with the lowest cost from generation 100 generated the following plan (Figure 9).

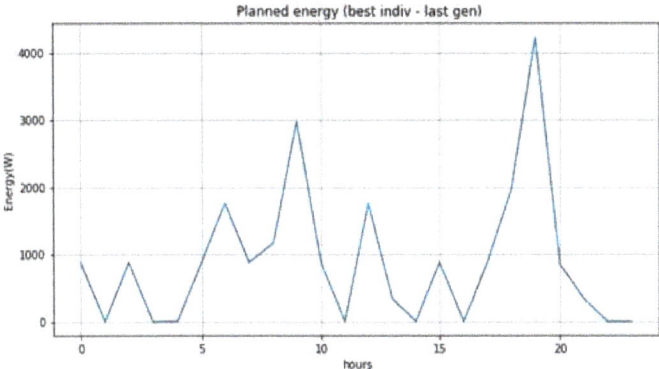

Figure 9. The individual from the 100th generation obtained using the genetic algorithm (GA), in which the planned distribution resulted in the lowest cost for profile generated by the simulator and is illustrated in Figure 7.

Consumption is redistributed from the two high peaks of the day to the areas where more electricity is produced. It is natural that the general form of the evolution of the optimally distributed daily consumption to reduce the cost will be similar to that of the evolution of the daily consumption with fixed planning; this is caused by consumers who cannot be re-planned.

Several sets of measurements made with the simulator led to interesting conclusions. First of all, all of the measurements led to a lower cost from re-planning the consumers than when the consumers were not planned. It should be noted that the comparison is made between two consumption schemes, both of which use electricity from the distributor and the micro-producer. In the unplanned consumption scheme, the consumption units are used at the beginning of the period when they must be activated, without any planning. In the programmed consumption scheme, the consumption units are planned, where possible, with the idea of bringing them into the time window where the energy producer produces more and, therefore, cheaper energy. Table 4 presents the results for the following sets of measurements.

Table 4. Results from a set of profiles generated by the simulator for PRODUCER_COST = 0.3 RON/1000 W.

PRODUCER_PEAK [W]	Measure No.	Range of Cost Decrease between Unplanned and Planned Consumption Schema in Favor of the Planned Consumption [%]	Average Cost Decrease between Unplanned and Planned Consumption Schema in Favor of the Planned Consumption [%]
2000	30	3.25–7.14	4.81
3000	40	5.48–11.59	8.95
4000	40	2.97–11.24	6.55
5000	40	7.14–16.11	11.48
6000	40	5.79–19.51	11.90
7000	40	8–21.43	14.23
8000	30	15.38–22.5	18.48

Thus, as can be seen in Table 4, a maximum decrease in the cost of electricity of almost 18.5% can be observed, influenced by the micro-producer's capacity to produce electricity.

Interestingly, at a decrease of 0.1 monetary units per kW of electricity produced (that is, a decrease in the kW cost of energy by 2 euro cents), the results presented in Table 5 were generated.

Table 5. Results from a set of measurements in which PRODUCER_COST = 0.2 RON/1 kW.

PRODUCER_PEAK [W]	Measure No.	Range of Decreasing Cost between Unplanned and Planned Consumption Schema [%]	Medium Cost Decrease between Unplanned and Planned Consumption Schema [%]
2000	40	3.06–36.54	14.15
3000	40	2.43–25.78	11.14
4000	40	7.52–25.39	16.45
5000	40	6.33–24.02	14.21
6000	40	19.07–35.71	25.33
7000	40	5.60–28.83	19.50
8000	40	16.41–36.49	25.65

It can be observed that in the case of a price decrease per 1000 W of only 0.1 RON (~0.02 EUR), higher cost reductions and a lower sensitivity to the production capacity are obtained using the proposed planning algorithm. A large variation of the cost reduction percentages between the unplanned and the planned version is shown in Table 5. The explanation is related to how the "production" of electricity was modeled: it is a random uniform distribution that varies throughout the day between 0 and the maximum production capacity. If the unplanned consumption profile is in areas with maximum production then the cost difference is smaller. More relevant is the percentage decrease in average cost (column 4, Tables 4 and 5)—this reflects the trend for each peak of production and the cost of the electricity produced. In Table 4, it can be seen that the increase of this percentage is more accentuated by the peak of production. In Table 5, a higher cost reduction is obtained than in Table 4, because a lower producer cost was used. The conclusion would be for micro-electricity producers, rather, to focus their attention on reducing the cost of energy produced by using cheaper modern technologies, planning the distribution of the infrastructure cost over a longer term, obtaining subsidies for infrastructure construction, etc. In addition, the percentage variation is also caused by the type of consumers. If the consumer units (devices) do not allow reprogramming, then the costs will not be significantly reduced. From this point of view, consumers must focus on consumption units (devices) that allow for the programming of the processes that execute them and to focus on processes that can be programmed, although it is acknowledged that this is not always possible.

Moreover, increasing the efficiency of electricity production has a major impact in reducing the cost to the customer by using our re-planning algorithm. The results obtained in the tables above and the graphs shown in the figures were generated with the created simulator.

Validation in a real environment of the results obtained using the genetic algorithm for re-planning was conducting by testing it on two case studies.

The practical requirements that emerged from the case studies discussed in this paper led us to work with a maximum of 10 pieces of equipment/consumer. However, the simulator was used to perform an analysis for several consumers (from 20 to 80 consumers). As can be seen in Figure 10, the algorithm finds a plan with a lower cost. However, it was found that, for 50 individuals or more, it does not always find a better solution within 800 generations: it was necessary to increase the number of generations several times. Therefore, for a larger number of devices, the parameterization of the algorithm must be reconsidered, that is, by increasing both the number of individuals per generation and the number of generations, as well as testing with different mutation and crossover coefficients.

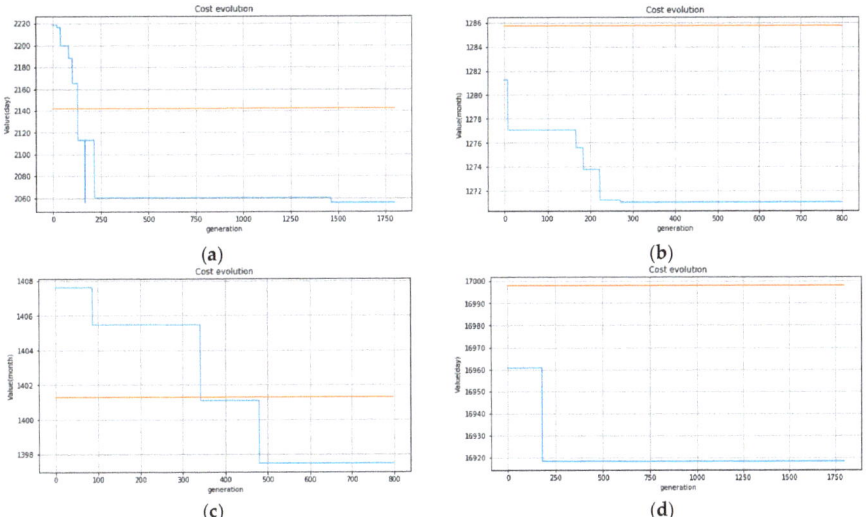

Figure 10. Cost evolution by running the GA for consumers with different numbers of devices: 20 (**a**), 40 (**b**), 60 (**c**), and 80 (**d**).

4.2. Testing in Real Operation

Both of the case studies presented below involved the existence of an alternative measuring system that can be placed in different nodes of the electricity grid, as well as consoles that indicate to the consumer the profile to be followed to reduce the cost of electricity.

The two case studies involve eight consumers (four individuals, three blocks of flat associations and one industrial manufacturing line) located in three different localities, where there are also three local electricity producers. The producers comprise one network of producers with photovoltaic panels with a production capacity of 6 kW, one producer with a field of 8 × 8 panels each with a capacity of 250 W (i.e., a total of 16 kW), one producer with a field of 10 × 10 panels each of 250 W (i.e., a total of 25 kW), and one manufacturing line with a system of recovery energy from eight power engines.

It is clear that lowering the monthly cost of electricity encourages consumers to use a certain consumption scheme, and, thus, to program their electric devices and appliances.

A. Case study 1: domestic consumers and micro-producer with 2000 W PV (Photovoltaic Panels)

Here, we demonstrate the efficiency of the algorithm in two real cases. The first is that of domestic consumers (14 households) and a micro-producer of electricity using a field of photovoltaic panels that leads to a maximum electricity output of 2 kW for each household.

Consumers have typical household appliances: TV, fridge, lighting, PC, microwave oven, air conditioning, electric oven, vacuum cleaner, washing machine, and a water pump with hydrophore. The distribution of consumption per consumer units, distribution of the energy produced (from photovoltaic panels), and an unplanned distribution in which a maximum of consumption is concentrated during the morning and evening can be observed in Figure 11. In addition, in the figure, at the bottom, it is shown how the electricity consumption is distributed for a solution generated by the genetic algorithm. One can observe the preservation of a similar pattern to the unplanned use—this is because of the consumption units that do not allow for re-planning (for example TV, PC, illuminator), but is also due to a distribution of consumption during the day when the production of electricity through photovoltaic panels is at maximum level. The main re-planned unit of consumption is the hydrophore pump. Thus, in the unplanned solution, the pump comes into operation every morning. A basin with a 200-L volume is filled, which is sufficient for the daily needs of the household. In contrast, the water filling of the basin was re-planned by the intelligent algorithm starting at 1 pm, at the peak of electricity production from the photovoltaic panels. Although the solution may be intuitive, the fact that it was found by the intelligent algorithm shows its feasibility. Another producer–consumer assembly may have less obvious solutions, as shown in the following case study.

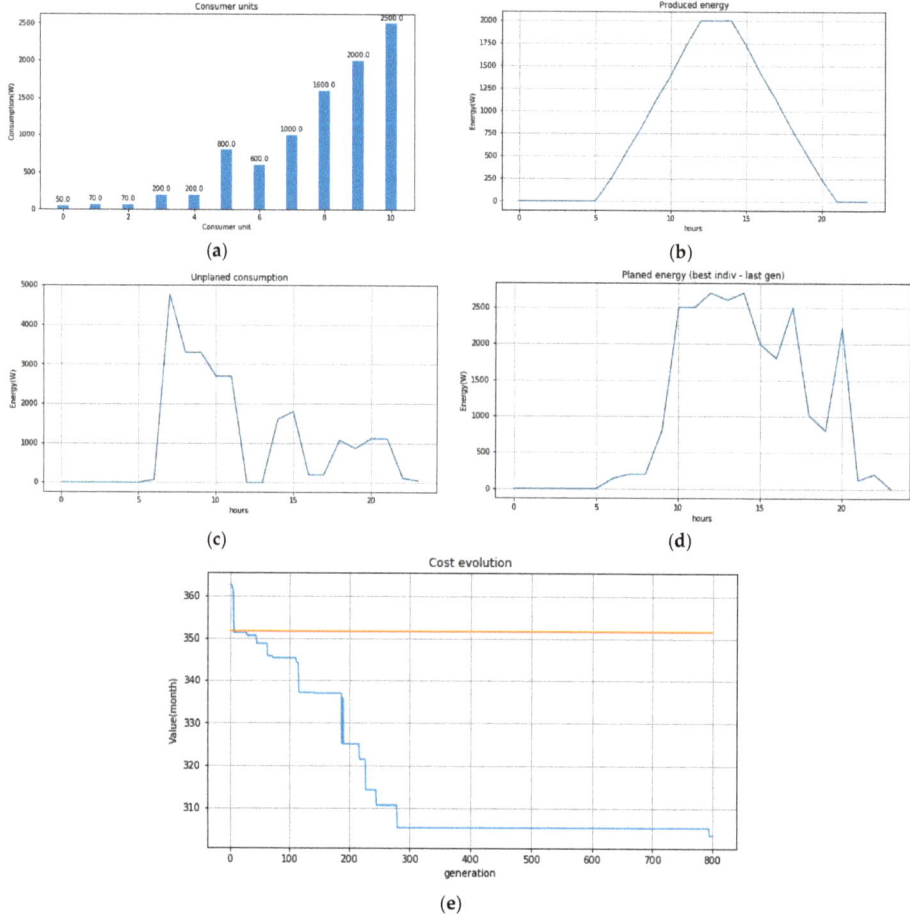

Figure 11. Results for the domestic consumer case study presented in this section. (**a**) Consumer Units; (**b**) Produced Energy; (**c**) Unplaned Consumption; (**d**) Planed Energy; (**e**) Cost Evolution.

B. Case study 2—industrial consumer using electricity recovered from an installation with electric engines of 5 kW

This system was also tested on a production process (production of chassis in the automotive industry). The process involves five consumers (one automatic welding system, two vacuum pumps, and two electric motors).

Figure 11 represents the consumer–producer characteristics for case study number 1 (domestic consumer). The top left of the figure shows the consumption of the 10 devices (household appliances), the top right shows the energy produced by the photovoltaic panels, the middle left shows the total consumption without any planning, and the middle right shows the consumption with planning. It can be seen that consumption somewhat follows the pattern of the producer of energy from renewable sources, which leads to a reduction in cost. The pulse at 8 pm is due to the impossibility of planning some devices that are used at that time. The bottom of the graph represents the evolution of the cost from the initial cost, for the unplanned solution, to the final cost obtained by the algorithm with the planned solution, for which consumption is illustrated in the figure in the middle right.

From another section of the factory there is a system of electricity recovery from eight power engines. This system can produce electricity of up to 5 kW (variable depending on the engine operating regime).

The task performed by the first vacuum pump cannot be re-planned (it must necessarily take place at 10 o'clock on that day) and the tasks performed by the second electric motor and the automatic welding system can only be re-planned with a deviation of +/−1 h. In contrast, the second vacuum pump and the first electric motor have great flexibility in planning. The objective is to use consumer planning to reduce the cost of electricity by using the energy generated by recovery from the engines in the neighboring section as much as possible.

Figure 12 represents the consumer–producer characteristics for case study number 2 (industrial consumer). The top left of the figure shows the five industrial devices with their consumption, the right shows the energy produced (recovery from power engines), the middle left shows the total consumption for the unplanned solution, and the total consumption for the solution with planning is shown at the middle right. The lower chart shows the evolution of the cost during the running of the optimization algorithm from the unplanned cost to the planned cost.

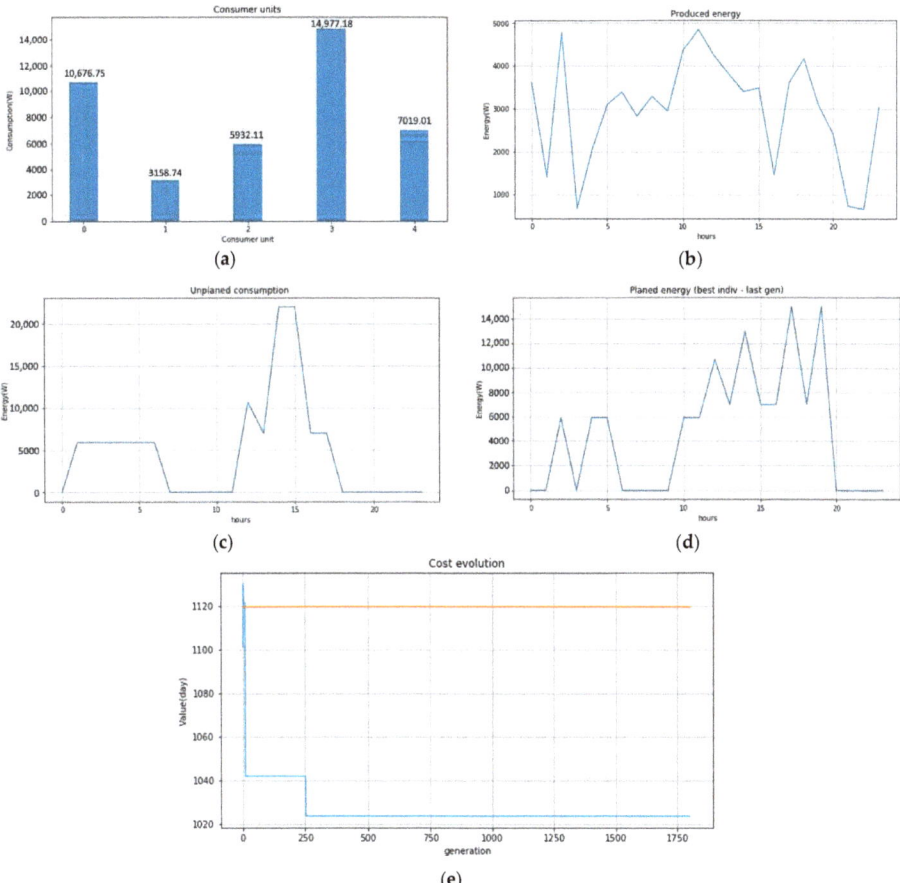

Figure 12. Results for industry consumer case study. (**a**) Consumer Units; (**b**) Produced Energy; (**c**) Unplaned Consumption; (**d**) Planed Energy; (**e**) Cost Evolution.

Figure 12 shows the distribution of the produced electricity, which, this time, depends on the tasks that are performed in the neighboring section. The algorithm was applied and it can be seen how the energy consumption was re-planned and how the cost "evolved" over 800 generations. As can be seen, this was achieved using the presented pilot to achieve a cost reduction of 90 monetary units/day (~EUR 20). This occurred under the conditions in which the cost of energy—at the producer—was taken to be 0.2 monetary units/1000 W for the amortization of the investment in the energy recovery systems, and the implementation was made only at the level of two sections. The figures show that as the system was used, the cost was clearly reduced.

The solution proposed by us involves running the algorithm for each customer. The time in which the algorithm finds the solution (reaches the limit generation) varies between 3–5 s when running on a server PC with 2.59 GHZ processor and 32 GB RAM on a 64-bit architecture. For a larger number of customers, the time to find the solution would increase linearly. Using a dedicated server and parallelizing the evolution process by using parallel computing technologies (e.g., CUDA from graphics accelerators) would reduce computation time. Future studies will be done on these.

5. Conclusions and Future Trends

The central point of the current paper was the presentation of an algorithm, based on a GA, for planning the consumption units in a household or company to reduce the cost of electricity, by using energy produced from renewable sources as much as possible. As seen from the presented simulations, in addition to the two real operation case studies (one for domestic consumers and one for industrial consumers), the solution with the GA for the planning of the operating intervals for each device (where possible) proved its efficiency; by the intelligent re-planning of the operating intervals the cost of electricity was decreased. In some cases, this reduction was only a few percent; however, it should be kept in mind that only a moderate effort from the consumer, to program their equipment according to the plan generated by the algorithm, was necessary to reduce the cost. Advantages apply to all participants: consumers pay less, producers are encouraged to sell their produced energy, and pressure on distributors decreases, especially during peak hours (evening and morning). Globally, such a system will encourage consumer planning, reduce the consumption of raw materials at the level of large electricity energy producers and, ultimately, reduce pollution. This paper presented not only a method with the stated advantages, but also described how it can be implemented in an interactive consumer planning system. Its impact was presented in two case studies.

Future development will involve the implementation of automatic planning methods. On the one hand, the system could integrate with the Internet of Things (IoT) consumption units (devices) that are increasingly present in both domestic and industrial activities. On the other hand, IoT sources can be used to connect to non-IoT equipment. Even when devices are older and do not have programming capabilities, they can be connected to IoT power supplies, and, thus, be automatically programmed. Automatic programming will take over the consumer programming task, which practically would reduce the cost of electricity. Observing the potential of a metaheuristic algorithm in this field, we plan to conduct research on the use of other similar algorithms in determining the minimum cost to a consumer using renewable energy. Our current research involves a whale optimization algorithm (WOA) used for the same purpose, including both the performance obtained using the algorithm and the possibility to implement it in a system. Furthermore, other solutions, such as particle swarm optimization (PSO) or Gravitational Search Algorithm (GSA), can be tested and implemented.

Author Contributions: Conceptualization, Methodology, and Writing—Original Draft Preparation: L.-M.I.; Validation and Supervision: N.B. (Nicu Bizon); Formal analysis: N.B. (Nadia Belu); Writing—Review & Editing: A.-G.M. All authors have read and agreed to the published version of the manuscript.

Funding: The research that led to the results shown here has received funding from the project "Cost-Efficient Data Collection for Smart Grid and Revenue Assurance (CERA-SG)", ID: 77594, 2016-19, ERA-Net Smart Grids Plus.

Acknowledgments: This work was also partially supported by Research Center "Modeling and Simulation of the Systems and Processes" based on grants of the Ministry of National Education and Ministry of Research and Innovation, CNCS/CCCDI-UEFISCDI within PNCDI III "Increasing the institutional capacity of bioeconomic research for the innovative exploitation of the indigenous vegetal resources in order to obtain horticultural products with high added value" PN-III P1-1.2-PCCDI2017-0332.

Conflicts of Interest: The authors declare no conflict of interest.

Nomenclature

ABC	Artificial Bee Colony
AI	Artificial Intelligence
AMI	Advanced Measurement Infrastructure
AGV	Automatic Guided Vehicle
BSN	Body Sensor Network
CMOS	Complementary metal–oxide–semiconductor
HTTP	HyperText Transfer Protocol
HVAC	Heating, Ventilation, and Air Conditioning
MPP	Maximum Power Point
RF	Radio-Frequencies
GSM	Global System for Mobile Communications
SFL	Shuffled Frog Leaping
SG	Smart Grids
SMM	Social Media Marketing
TLBO	Teaching & Learning based Optimization
VLSI	Very-large-scale integration
WAN	Wide Area Network
GA	Genetic Algorithm

References

1. Dutta, G.; Mitra, K. A literature review on dynamic pricing of electricity. *J. Oper. Res. Soc.* **2017**, *68*, 1131–1145. [CrossRef]
2. Erol-Kantarci, M.; Mouftah, H.T. Energy-Efficient Information and Communication Infrastructures in the Smart Grid: A Survey on Interactions and Open Issues. *IEEE Commun. Surv. Tutor.* **2015**, *17*, 179–197. [CrossRef]
3. Alahakoon, D.; Yu, X. Smart Electricity Meter Data Intelligence for Future Energy Systems: A Survey. *IEEE Trans. Ind. Inform.* **2016**, *12*, 425–436. [CrossRef]
4. Finster, S.; Baumgart, I. Privacy-Aware Smart Metering: A Survey. *IEEE Commun. Surv. Tutor.* **2015**, *17*, 1088–1101. [CrossRef]
5. Deng, R.; Yang, Z.; Chen, J.; Asr, N.R.; Chow, M. Residential Energy Consumption Scheduling: A Coupled-Constraint Game Approach. *IEEE Trans. Smart Grid* **2014**, *5*, 1340–1350. [CrossRef]
6. Kliazovich, D.; Bouvry, P.; Khan, S.U. DENS: Data center energy-efficient network-aware scheduling. *Cluster Comput.* **2013**, *16*, 65–75. [CrossRef]
7. Tang, Z.; Qi, L.; Cheng, Z.; Li, K.; Khan, S.U.; Li, K. An Energy-Efficient Task Scheduling Algorithm in DVFS-enabled Cloud Environment. *J. Grid Comput.* **2016**, *14*, 55–74. [CrossRef]
8. Zhang, Z.; Tang, R.; Peng, T.; Tao, L.; Jia, S. A method for minimizing the energy consumption of machining system: Integration of process planning and scheduling. *J. Clean. Prod.* **2016**, *137*, 1647–1662. [CrossRef]
9. Cionca, V.; McGibney, A.; Rea, S. MALLEC: Fast and optimal scheduling of energy consumption for energy harvesting devices. *IEEE Internet Things J.* **2018**, *5*, 5132–5140. [CrossRef]
10. Derakhshan, G.; Shayanfar, H.A.; Kazemi, A. The optimization of demand response programs in smart grids. *Energy Policy* **2016**, *94*, 295–306. [CrossRef]
11. Ghasemi, A.; Shayeghi, H.; Moradzadeh, M.; Nooshyar, M. A novel hybrid algorithm for electricity price and load forecasting in smart grids with demand-side management. *Appl. Energy* **2016**, *177*, 40–59. [CrossRef]
12. Ma, J.; Chen, H.H.; Song, L.; Li, Y. Residential Load Scheduling in Smart Grid: A Cost Efficiency Perspective. *IEEE Trans. Smart Grid* **2016**, *7*, 771–784. [CrossRef]

13. Zhang, W.; Maleki, A.; Rosen, M.A.; Liu, J. Optimization with a simulated annealing algorithm of a hybrid system for renewable energy including battery and hydrogen storage. *Energy* **2018**, *163*, 191–207. [CrossRef]
14. Tang, D.; Dai, M.; Salido, M.A.; Giret, A. Energy-efficient dynamic scheduling for a flexible flow shop using an improved particle swarm optimization. *Comput. Ind.* **2016**, *81*, 82–95. [CrossRef]
15. Ghavidel, S.; Aghaei, J.; Muttaqi, K.M.; Heidari, A. Renewable energy management in a remote area using Modified Gravitational Search Algorithm. *Energy* **2016**, *97*, 391–399. [CrossRef]
16. Yuan, S.; Huang, Y.; Zhou, J.; Xu, Q.; Song, C.; Thompson, P. Magnetic Field Energy Harvesting Under Overhead Power Lines. *IEEE Trans. Power Electron.* **2015**, *30*, 6191–6202. [CrossRef]
17. Yang, L.; Chen, X.; Zhang, J.; Poor, H.V. Cost-Effective and Privacy-Preserving Energy Management for Smart Meters. *IEEE Trans. Smart Grid* **2015**, *6*, 486–495. [CrossRef]
18. Chang, Y.; Choi, G.; Kim, J.; Byeon, S. Energy Cost Optimization for Water Distribution Networks Using Demand Pattern and Storage Facilities. *Sustainability* **2018**, *10*, 1118. [CrossRef]
19. Xu, W.; Guo, S.A. A Multi-Objective and Multi-Dimensional Optimization Scheduling Method Using a Hybrid Evolutionary Algorithms with a Sectional Encoding Mode. *Sustainability* **2019**, *11*, 1329. [CrossRef]
20. Satrio, P.; Mahlia, T.M.I.; Giannetti, N.; Saito, K. Optimization of HVAC system energy consumption in a building using artificial neural network and multi-objective genetic algorithm. *Sustain. Energy Technol. Assess.* **2019**, *35*, 48–57.
21. Farhadian, F.; Kashani, M.M.R.; Rezazadeh, J.; Farahbakhsh, R.; Sandrasegaran, K. An Efficient IoT Cloud Energy Consumption Based on Genetic Algorithm. *Digit. Commun. Netw.* **2019**, *150*, 1–8. [CrossRef]
22. Gram-Hanssen, K.; Hansen, A.R.; Mechlenborg, M. Danish PV Prosumers' Time-Shifting of Energy-Consuming Everyday Practices. *Sustainability* **2020**, *12*, 4121. [CrossRef]
23. Kontogiannis, D.; Bargiotas, D.; Daskalopulu, A. Minutely Active Power Forecasting Models Using Neural Networks. *Sustainability* **2020**, *12*, 3177. [CrossRef]
24. Yaprakdal, F.; Yılmaz, M.B.; Baysal, M.; Anvari-Moghaddam, A. A Deep Neural Network-Assisted Approach to Enhance Short-Term Optimal Operational Scheduling of a Microgrid. *Sustainability* **2020**, *12*, 1653. [CrossRef]
25. Whitworth, A.H.; Tsavdaridis, K.D. Genetic Algorithm for Embodied Energy Optimisation of Steel-Concrete Composite Beams. *Sustainability* **2020**, *12*, 3102. [CrossRef]
26. Ahmed, S.A.M. Performance Analysis of Power Quality Improvement for Standard IEEE 14-Bus Power System based on FACTS Controller. *JEEECCS* **2020**, *5*, 11.
27. Mihaescu, M. Applications of multiport converters. *JEEECCS* **2016**, *2*, 13.
28. Ahn, W.; Lee, H.S.; Ryou, H.; Oh, K.J. Asset Allocation Model for a Robo-Advisor Using the Financial Market Instability Index and Genetic Algorithms. *Sustainability* **2020**, *12*, 849. [CrossRef]
29. Pillai, A.S.; Singh, K.; Saravanan, V.; Anpalagan, A.; Woungang, I.; Barolli, L. A genetic algorithm-based method for optimizing the energy consumption and performance of multiprocessor systems. *Soft Comput.* **2018**, *22*, 3271. [CrossRef]
30. Karami, J.; Moghaddam, A.; Faridhosseini, A.; Ziaei, A.N.; Rouholamini, M.; Moghbeli, M. Using Fast Messy Genetic Algorithm to Optimally Schedule Pump Operation. In *Frontiers in Water-Energy-Nexus—Nature-Based Solutions, Advanced Technologies and Best Practices for Environmental Sustainability*; Springer: Cham, Switzerland, 2020; Volume 1, pp. 509–512.
31. Shahzad, U. Significance of Smart Grids in Electric Power Systems: A Brief Overview. *JEEECCS* **2020**, *6*, 7.
32. Rai, K.; Jandhyala, A.; Shantharama, C. Power Intelligence and Asset Monitoring for System. *JEEECCS* **2019**, *5*, 31.
33. Beg, A.H.; Islam, M.Z. Advantages and limitations of genetic algorithms for clustering records. In Proceedings of the IEEE 11th Conference on Industrial Electronics and Applications (ICIEA), Hefei, China, 5–7 June 2016; Volume 11, p. 1.

© 2020 by the authors. Licensee MDPI, Basel, Switzerland. This article is an open access article distributed under the terms and conditions of the Creative Commons Attribution (CC BY) license (http://creativecommons.org/licenses/by/4.0/).

Enhanced IoV Security Network by Using Blockchain Governance Game

Song-Kyoo (Amang) Kim

School of Applied Sciences, Macao Polytechnic Institute, R. de Luis Gonzaga Gomes, Macao; amang@ipm.edu.mo; Tel.: +853-8599-6455

Abstract: This paper deals with the design of the secure network in an Enhanced Internet of Vehicles by using the Blockchain Governance Game (BGG). The BGG is a system model of a stochastic game to find best strategies towards preparation of preventing a network malfunction by an attacker and the paper applies this game model into the connected vehicle security. Analytically tractable results for decision-making parameters enable to predict the moment for safety operations and to deliver the optimal combination of the number of reserved nodes with the acceptance probability of backup nodes to protect a connected car. This research helps for whom considers the enhanced secure IoV architecture with the BGG within a decentralized network.

Keywords: IoT security; Internet of Vehicles; IoV; connected car; Blockchain Governance Game; mixed game; stochastic model; fluctuation theory; 51 percent attack

1. Introduction

A connected car transfers data to others based on vehicle-to-vehicle (V2V) communication technologies and it typically means that cars are equipped for Internet access usually with wireless local area networks (WLANs). Cars have been evolved to support enhanced driving aids for full autonomous driving by using the artificial intelligence (AI) and its maneuvers [1]. The Internet of Vehicles (IoV) is a superset of a connected car which contains sensors, GPS, entertainment systems, brakes and throttles. The IoV is a moving network which is made up of IoT enabled cars through the usage of modern electronics and the integrated information to maintain traffic flow. Cars have evolved from mechanical transportation to the smart vehicles with varieties of communication and sensing capabilities. The IoV has developed over time from the conventional vehicular networks that connect the smart vehicles to the smart city with the development of Internet of Things (IoT) [2]. The IoV is designed to perform more effective fleet management and accident avoidance [3,4]. An ad-hoc network is applied to connect IoT components as nodes in a connected car [4]. Adapting the Blockchain technologies into IoV networks brings huge attentions from researchers and developers because of decentralization, anonymity, and trust characteristics [5–7]. Additionally, data sharing among vehicles is critical to improve driving safety and to enhance vehicular services in IoV networks. The studies in the security and the tractability of data sharing indicate that utilizing consensus schemes are as hard as establishing Blockchain enabled IoV (BIoV) [6]. Some other studies have proposed a decentralized trust management system for vehicle data credibility assessment using Blockchain with joint Proof-of-Work (PoW) and Proof-of-Stake (PoS) consensus schemes [7–9]. Vehicle manufacturers Volkswagen and Ford have filed the patents that enable secure inter-vehicle communication through Blockchain technologies [10,11]. In the view point of the IoT securities, several studies are dealing with similar topics regarding the Blockchain based IoT securities [6,12–14]. Most studies are surveys [12] but none of them are mathematically approached for developing a network architecture for a Blockchain based IoT [13,14].

The Enhanced BIoV (EBIoV) network, which has been conceptually introduced on public [15], is reloaded in this paper. The EBIoV is an IoV network architecture based on the Edge computing [16,17] with enabling the Blockchain Governance Game (BGG) [18,19] for improving network securities [15]. Although a public Blockchain is designed for empowering their decentralization [20], current public Blockchain based services including cybercurrencies are not fully safe from attacks especially based on the mining computation [18]. Hence, a private Blockchain which is a permission-based Blockchain [20] has been proposed for business or government usage [21,22]. Many peoples are interested in a private Blockchain technology even in a consortium Blockchain technology [20] but they are not comfortable with a level of control compared to the offered control level in a public decentralized network. More importantly, the control levels in a Blockchain network should be balanced to retain the strengths of a decentralized network to avoid all security matters what atypical centralized network contains [18]. By adapting the BGG, we do not need to concern about the computation power for mining (i.e., generating ledgers) without losing the strength from fully decentralized networks. A conventional BIoV model could be applied to trace the provenance of spare parts back through every step of the supply chain to its original manufacture date and location [15,23]. Car manufacturers concern that service centers and garages are knowingly fitting counterfeit spare parts to their vehicles of customers. Counterfeit parts shall damage a brand reputation when the parts become causes of accidents. Identifying genuinity of car parts by using the EBIoV has been studied and the network within car components shall be considered as an Edge network [16,17,24].

The Fog computing pushes information to the neighborhood area network amount of community knowledge at an IoT gateway and it could be combined with the Blockchain technology for enhancing security matters [17]. In the EBIoV network architecture, diagnosis equipment in car service centers is in a Fog network and the database in a headquarter (HQ) is positioned the level of the cloud network (see Figure 1). This paper provides the mathematical functional of the EBIoV network architecture for enhancing security particularly from counterfeits of car parts. The EBIoV security regarding counterfeits of car parts has already been studied [15] and this research is focused on avoiding conventional IoV attack in a decentralized network. The BGG is the game model that an attacker and a defender compete each other by building blocks in private and public chains as a sequence of stages to generate ledgers [18,19]. The historical strategies and the probabilistic stage transitions can be observed by both an attacker and a defender. Hence, the interaction between an attacker and a defender can be modeled as a stochastic game [18,19,25]. This joint functional between two players of the predicted time of the first observed threshold to cross the half of the total nodes along with values of each component upon this time. The defender (a car company) could take a preliminary action (i.e., request to add honest nodes as a safety mode) for protecting the Blockchain in a vehicle.

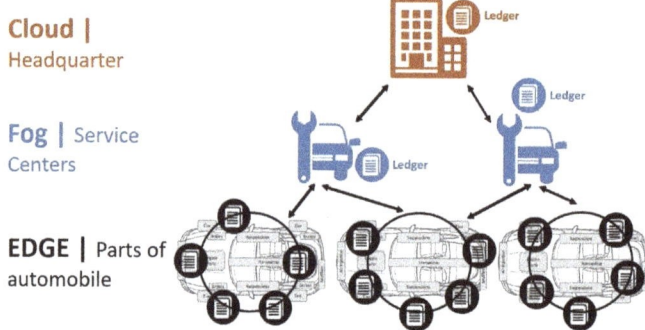

Figure 1. Adapting Fog Computing in Auto Services in Blockchain [15].

This paper is organized as follows: Section 2 presents the Enhanced BIoV (EBIoV) network and it describes how to construct EBIoV architecture by using the BGG model which is a stochastic game between an attacker and a defender. Once a dishonest blocks are generated, the model predicts how many blocks will be generated and finds the moment when more than the half of nodes are covered by an attacker. The framework for setting up the mixed strategic game is provided in Section 3. The optimal values of an EBIoV network for the memoryless case are analytically calculated in Section 4. The memoryless property implies that a defender does not spend additional resources to store past information. Lastly, Section 5 provides the conclusion of the paper.

2. Stochastic Game for BIOV Network Security

The Blockchain Governance Game (BGG) [18,19] has been adapted the BIoV network architecture to defend against the attacks [15]. This system model consists of one attacker (i.e., the miner which intends to fork a private chain) and one defender (the miner which honestly mines on the public chain) [25]. This explicit function (Theorem BGG-1) from the BGG (Blockchain Governance Game) gives the predicted moment of one step prior to an attack [18].

2.1. Enhanced IoV Network Structure

The proposed BIoV network structure [15] is explored in detail in this section. The components in a vehicle, the equipment of a service center and a HQ database are hooked up as one Blockchain network (see Figure 2). Each smart components in a connected car could mechanically or electronically generate random values and share these values with other smart components. Tires, brakes, an engine, a transmitter in a car could be the smart components which shall be capable to communicate with other components and to construct ledgers. Connected car components beside a CPU (smart controller) generate values based on their mechanical actions and these generated values are sharing with all other components including assigned service centers and a company headquarter. Each values from car components is unique and randomly generated. And sharing these generated number is a transaction in a conventional Blockchain network. A service center has a database which contains information from cars which are served by service centers. The unique value based on a registered car database are generated and sharing with other nodes including a company headquarter.

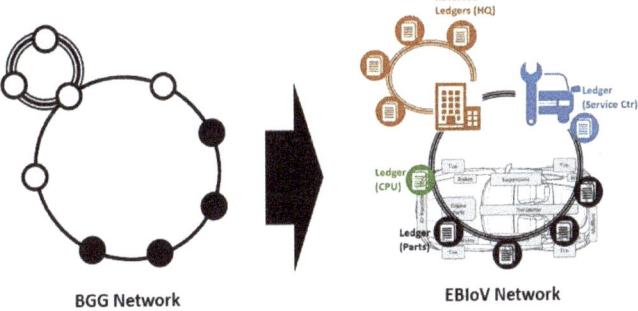

Figure 2. Adapting BGG for the EBIoV architecture [18].

Unlike conventional Blockchain networks, an EBIoV network does not have any reward system which requires a heavy computational power for being a miner to generate ledgers. All nodes even in a service center and a HQ have same contribution power and the equal chance to be a miner. For instance, a node of a CPU is same as nodes of other car parts although a CPU controls other car parts. The verifiable random function (VRF) which maps inputs to verifiable pseudorandom outputs is applied to select a node for generating ledgers [26,27]. The VRF has been applied to perform secret cryptographic solution to select committees to run the consensus protocol [28,29]. By applying the VRF, all nodes in the EBIoV network shall have the equal chance to become a miner who generates ledgers without requiring heavy computational powers. The mechanism for protecting an EBIoV network is exactly same as the mechanism of the BGG. The governance in a Blockchain network is followed by the decision making parameters which include the prior time before catching more than half of total nodes by an attacker. Any action shall not be taken until one step prior to the time when it passes the first passage time. It still has the chance that all nodes are governed by an attacker although an attacker catches less than the half of nodes.

2.2. BGG Model for Enhanced BIoV Network

To apply the BGG model into the BIoV network structure, the antagonistic game of two players (called "A" and "H") are introduced to describe the Blockchain network in a connected car as a defender and an attacker. Both players compete to build the blocks either for honest or false nodes. Let $(\Omega, \mathcal{F}(\Omega), P)$ be probability space $\mathcal{F}_A, \mathcal{F}_H, \mathcal{F}_\tau \subseteq \mathcal{F}(\Omega)$ be independent σ-subalgebras. Suppose:

$$\mathcal{A} := \sum_{k \geq 0} X_k \varepsilon_{s_k}, \; s_0(=0) < s_1 < s_2 < \cdots, \text{a.s.} \tag{1}$$

$$\mathcal{H} := \sum_{j \geq 0} Y_j \varepsilon_{t_j}, \; t_0(=0) < t_1 < t_2 < \cdots, \text{a.s.} \tag{2}$$

are $\mathcal{F}_A$-measurable and $\mathcal{F}_H$-measurable marked Poisson processes (ε_w is a point mass at w) with respective intensities λ_a and λ_h. These two values are related with the computing performance for generating blocks for attackers and honest nodes in the blockchain network. They will represent the actions of player A (an attacker) and H (a defender). Player A builds the blocks with fake information and sustain respective build the blocks of magnitudes X_1, $X_2, \ldots$ formalized by the process. Similarly, player H generates the blocks with authorized information with the blocks of magnitudes $Y_1, Y_2, \ldots$ Both players compete each other to build their blocks (either genuine or fake). The processes $\mathcal{A}$ and $\mathcal{H}$ are specified by their transforms

$$\mathbb{E}\left[g^{\mathcal{A}(s)}\right] = e^{\lambda_a s(g-1)}, \mathbb{E}\left[z^{\mathcal{H}(t)}\right] = e^{\lambda_h t(z-1)}. \tag{3}$$

The game is observed at random times in accordance with the point process and it is equivalent with the duration of the PoW (Proof-of-Work) completion in a Blockchain based network:

$$\mathcal{T} := \sum_{i \geq 0} \varepsilon_{\tau_i}, \; \tau_0(>0)), \tau_1, \ldots, \tag{4}$$

which is assumed to be delayed renewal process. The observation process could be formalized as

$$\mathcal{A}_\tau \otimes \mathcal{H}_\tau := \sum_{k \geq 0} (X_k, Y_k) \varepsilon_{\tau_k}, \tag{5}$$

and it is with position dependent marking and with X_k and Y_k being dependent with the notation

$$\Delta_k := \tau_k - \tau_{k-1}, \; k = 0, 1, \ldots, \tau_{-1} = 0, \tag{6}$$

and

$$\gamma(g, z) = \mathbb{E}\left[g^{X_k} \cdot z^{Y_k}\right], \|g\| \leq 1, \|z\| \leq 1. \tag{7}$$

By using the double expectation,

$$\gamma(g, z) = \delta(\lambda_A(1-g) + \lambda_H(1-z)), \tag{8}$$

$$\gamma_0(g, z) = \delta_0(\lambda_A(1-g) + \lambda_H(1-z)), \tag{9}$$

where

$$\delta(\theta) = \mathbb{E}\left[e^{-\theta \Delta_1}\right], \delta_0(\theta) = \mathbb{E}\left[e^{-\theta \tau_0}\right], \quad (10)$$

are the magical transform of increments $\Delta_1, \Delta_2, \ldots$. This game contains a stochastic process $\mathcal{A}_\tau \otimes \mathcal{H}_\tau$ which describes the evolution of a conflict between players A and H known to an observation process $\mathcal{T} = \{\tau_0, \tau_1, \ldots\}$. The process ends when on the k-th observation epoch τ_k, the collateral building blocks to player H (or A) exceeds more than the half of the total nodes M in the regular operation or player A exceeds more than $\left(\frac{M}{2}\right) + B$ nodes under the safety mode. To further formalize the game, the exit indexes [18] are as follows:

$$\nu := \inf\left\{k : A_k = A_0 + X_1 + \cdots + X_k \geq \left(\frac{M}{2}\right) + B\right\}, \quad (11)$$

$$\mu := \inf\left\{j : H_j = H_0 + Y_1 + \cdots + Y_j \geq \left(\frac{M}{2}\right)\right\} \quad (12)$$

where B is the number of the reserved honest nodes from a headquarter (HQ) which is depends on the availability from the HQ. Since, an attacker is win at time τ_ν, otherwise an honest node generates the correct blocks to share with others nodes. We are targeting the confined game in the view point of player A. The joint functional of the BIoV network model is as follows:

$$\Phi\left(\zeta; \left\lceil\frac{M}{2}\right\rceil + B, \left\lceil\frac{M}{2}\right\rceil\right) = \mathbb{E}\left[\mathbb{E}\left[\zeta^\nu \cdot g_0^{A_{\nu-1}} \cdot g_1^{A_\nu} \cdot z_0^{H_{\nu-1}} \cdot z_1^{H_\nu} \mathbf{1}_{\{\nu < \mu\}} \Big| B\right]\right], \quad (13)$$

$$\|\zeta\| \leq 1, \|g_0\| \leq 1, \|g_1\| \leq 1, \|z_0\| \leq 1, \|z_1\| \leq 1,$$

where M indicates the total number of nodes (or ledgers) in the BIoV network for each car (see Figure 2). The BGG-1 Theorem [18] establishes an explicit formula $\Phi\left(\xi, \left\lceil\frac{M}{2}\right\rceil + B, \left\lceil\frac{M}{2}\right\rceil\right)$ and the functional (13) satisfies following expression:

$$\Phi\left(\zeta; \left\lceil\frac{M}{2}\right\rceil + B, \left\lceil\frac{M}{2}\right\rceil\right) = \mathfrak{D}_{(u,v)}^{(\lceil\frac{M}{2}\rceil+B,\lceil\frac{M}{2}\rceil)}\left[\Gamma_0^1 - \Gamma_0 + \frac{\xi \cdot \gamma_0}{1-\xi\gamma}\left(\Gamma^1 - \Gamma\right)\right] \quad (14)$$

$$\gamma := \gamma(g_0 g_1 u, z_0 z_1 v), \gamma_0 := \gamma_0(g_0 g_1 u, z_0 z_1 v), \quad (15)$$

$$\Gamma := \gamma(g_1 u, z_1 v), \Gamma_0 := \gamma_0(g_1 u, z_1 v), \quad (16)$$

$$\Gamma^1 := \gamma(g_1, z_1 v), \Gamma_0^1 := \gamma_0(g_1, z_1 v). \quad (17)$$

Additionally, the operator $\mathfrak{D}_{(x,y)}^{(m,n)}$ in (14) is defined as follows [18]:

$$\mathfrak{D}_{(x,y)}^{(m,n)}(\bullet) = \begin{cases} \left(\frac{1}{m! \cdot n!}\right) \lim_{(x,y) \to 0} \frac{\partial^m \partial^n}{\partial x^m \partial y^n} \frac{1}{(1-x)(1-y)}(\bullet), & m, n \geq 0, \\ 0, & \text{otherwise,} \end{cases} \quad (18)$$

$$\|x\| < 1, \|y\| < 1,$$

then we can find

$$g(m, n) = \mathfrak{D}_{(x,y)}^{(m,n)}\left[\mathcal{D}_{(m,n)}\{g(m,n)\}\right], \quad (19)$$

where

$$\mathcal{D}_{(m,n)}[g(m,n)](x, y) := (1-x)(1-y) \sum_{m \geq 0} \sum_{n \geq 0} g(m,n) x^m y^n, \|x\| < 1, \|y\| < 1. \quad (20)$$

It is noted that both operators $\mathfrak{D}$, $\mathcal{D}$ are originated from the first exceed theory [30,31]. We can find the PGFs (probability generating functions) of the exit index ν from (14):

$$\mathbb{E}[\zeta^\nu] = \mathbb{E}\left[\Phi\left(\zeta; \left\lceil \frac{M}{2} \right\rceil + B, \left\lceil \frac{M}{2} \right\rceil\right)\bigg| B\right]\bigg|_{(g_0,g_1,z_0,z_1)\to 1} \quad (21)$$

3. Strategies in Blockchain Governance Game

Let us consider a two-person mixed strategy game which is played by player H as a defender and player A as an attacker. Player H, who is mostly a car company in an EBIoV network, has two strategies at the observation moment when one step prior to complete for generating alternative chains with fake information. Basically, player H has the following strategies (i.e., operation modes): (1) Regular—regular operations which implicates that the BIoV network in a connected car are running as usual, and (2) Safety—the network is running under the safety mode for avoiding attacks by adding honest nodes from a HQ. Alternatively, player A (an attacker) might succeed to catch the blocks or fail to catch the honest nodes. Therefore, the response of player A would be either "Not Burst" or "Burst." Let us assume that the cost for reserving additional honest nodes is c_b where b is a set of the factors that related with the reserved nodes from a HQ and these related factors could be one or multiple values.

The headquarter of a car company reserves a certain portion of nodes for protecting the BIoV network integrity. If an attacker succeeds to generate alternative blocks within car parts, the network in a car is burst and the whole car value V is lost. It still has a chance to burst a car network although a defender (or a car company) adds honest nodes before catching blocks of an attacker. In this case, the lost cost includes not only a full car value but also a cost for additional reserved honest nodes. The normal form of a game is as follows:

$$\begin{aligned} \cdot \text{ Players:} &\quad N = \{A, H\}, \\ \cdot \text{ Strategy sets:} &\\ &\quad s_a = \{\text{"NotBurst"}, \text{"Burst"}\}, \\ &\quad s_h = \{\text{"Regular"}, \text{"Safety"}\} \end{aligned} \quad (22)$$

Based on the above conditions, the general cost matrix at $\tau_{\nu-1}$ when is the prior time just before bursting could be composed as shown in Table 1 where $q(s_h)$ is the probability of bursting a blockchain network (i.e., an attacker wins a game) and it depends on a strategic choice of player H:

$$q(s_h) = \begin{cases} \mathbb{E}\left[\mathbf{1}_{\{A_\nu \geq \frac{M}{2}\}}\right], & s_h = \{\text{"Regular"}\}, \\ \mathbb{E}\left[\mathbb{E}\left[\mathbf{1}_{\{A_\nu \geq \frac{M}{2}+B\}}\big|B\right]\right], & s_h = \{\text{"Safety"}\}. \end{cases} \quad (23)$$

Table 1. Cost matrix.

	NotBurst $(1-q(s_h))$	Burst $(q(s_h))$
Regular	0	V
Safety	c_b	$c_b + V$

It is noted that the cost for reserved nodes (i.e., the cost of "Safety" operation strategy by player H) should be smaller than the whole cost of the other strategy. Additionally, the number of reserved honest nodes from the HQ is random and this variable B shall have a certain probability distribution. Let us consider the number of reserved honest nodes has the binomial distribution with the success probability ρ and the number of trial n. The PGF of the binomial distribution is as follows:

$$\sigma_n = \mathbb{E}\left[b^B\right] = (\rho b - (1-\rho))^n. \tag{24}$$

The optimal number of reserved nodes n^* which supported by the HQ depends on the cost function and the optimal value ρ^* is the acceptance rate when the reserved honest nodes are requested to a HQ. The best combination (n^*, ρ^*) could be found as follows:

$$(n^*, \rho^*) = \inf\left\{(n,\rho) \geq 0 : \mathfrak{S}_{\text{Reg}}\left(q^0\right) \geq \mathfrak{S}_{\text{Safe}}(n,\rho)\right\}, \tag{25}$$

where, at the moment $\tau_{\nu-1}$,

$$\mathfrak{S}_{\text{Reg}}\left(q^0\right) = V \cdot q^0, \tag{26}$$

$$\mathfrak{S}_{\text{Safe}}(n,\rho) = c_{(n,\rho)}\left(1 - q_\eta^1\right) + \left(c_{(n,\rho)} + V\right)q_{(n,\rho)}^1, \tag{27}$$

$$q^0 = \mathbb{E}\left[\mathbf{1}_{\{A_\nu \geq \lceil \frac{N}{2} \rceil\}}\right], \tag{28}$$

$$q_{(n,\rho)}^1 = \mathbb{E}\left[\mathbb{E}\left[\mathbf{1}_{\{A_\nu \geq \lceil \frac{N}{2} \rceil + B\}} \Big| B\right]\right]. \tag{29}$$

We would like to design the BGG adapted BIoV network that is capable to take the safety operation at the decision making moment $\tau_{\nu-1}$. The governance of the Blockchain network is driven by the BGG decision making parameters. It is noted that no safety actions are required until the time $\tau_{\nu-1}$. Additionally, it still has the chance that all nodes are governed by an attacker if the attacker catches more than the half of nodes at $\tau_{\nu-1}$ (i.e., $\left\{A_{\nu-1} \geq \frac{M}{2}\right\}$). If the attacker catches less than half of all nodes at $\tau_{\nu-1}$ (i.e., $\left\{A_{\nu-1} < \frac{M}{2}\right\}$), then the defender could run the safety mode to avoid the burst at τ_ν. The total cost for developing the enhanced BIoV network is as follows:

$$\mathfrak{S}(q^0; n, \rho)_{\text{Total}} = \mathbb{E}\left[\mathfrak{S}_{\text{Safe}}(n,\rho) \cdot \mathbf{1}_{\{A_{\nu-1} < \frac{M}{2}\}} + \mathfrak{S}_{\text{Reg}}\left(q^0\right) \cdot \mathbf{1}_{\{A_{\nu-1} \geq \frac{M}{2}\}}\right]$$
$$= \left\{c_{(n,\rho)}\left(1 - q_{(n,\rho)}^1\right) + \left(c_{(n,\rho)} + n\rho\right)q_{(n,\rho)}^1\right\}p_{A_{-1}} + B \cdot q^0(1 - p_{A_{-1}}) \tag{30}$$

where

$$p_{A_{-1}} = \mathbf{P}\left\{A_{\nu-1} < \frac{M}{2}\right\} = \sum_{k=0}^{\lfloor \frac{M}{2} \rfloor} \mathbf{P}\{A_{\nu-1} = k\}. \tag{31}$$

3.1. Memoryless BGG Observation Process for EBIoV Networks

It is assumed that the observation process has the memoryless properties. This might be a special condition but very practical for actual implementation of a Blockchain Governance Game [18]. It implies that a defender does not spend additional cost for storing past information. After building a cost function of a BGG network, we could find explicit solutions of q^0, $p_{A_{-1}}$ and the moment of decision making after finding the closed form (PGF) of the first exceed index ν, each probability (generating function) of the number of blocks at the moment τ_ν (i.e., $\mathbb{E}\left[g_1^{A_\nu}\right]$) and $\tau_{\nu-1}$ (i.e., $\mathbb{E}\left[g_0^{A_{\nu-1}}\right]$). The functional $\mathfrak{D}$ is defined on the space of all analytic functions at 0. Recalling from (7)–(10), we have:

$$\gamma(g,z) = \delta(\lambda_a(1-g) + \lambda_h(1-z)) = \gamma_a(g) \cdot \gamma_h(z), \tag{32}$$

$$\gamma_a(g) = \delta(\lambda_a(1-g)), \gamma_h(z) = \delta(\lambda_h(1-z)), \tag{33}$$

and

$$\gamma_0(g,z) = \delta_0(\lambda_a(1-g) + \lambda_h(1-z)) = \gamma_a^0(g) \cdot \gamma_h^0(z), \tag{34}$$

$$\gamma_a^0(g) = \mathbb{E}\left[g^{A_0}\right] = \delta_0(\lambda_a(1-g)), \tag{35}$$

$$\gamma_h^0(z) = \mathbb{E}\left[z^{H_0}\right] = \delta_0(\lambda_h(1-z)), \tag{36}$$

from (15)–(17),

$$\gamma = \gamma_a \cdot \gamma_h := \gamma_a(g_0 g_1 u)\gamma_h(z_0 z_1 v), \tag{37}$$

$$\gamma_0 = \gamma_a^0 \cdot \gamma_h^0 := \gamma_a^0(g_0 g_1 u)\gamma_h^0(z_0 z_1 v), \tag{38}$$

$$\Gamma := \gamma_a(g_1 u)\gamma_h(z_1 v), \Gamma_0 := \gamma_a^0(g_1 u)\gamma_h^0(z_1 v), \tag{39}$$

$$\Gamma^1 := \gamma_a(g_1)\gamma_h(z_1 v), \Gamma_0^1 := \gamma_a^0(g_1)\gamma_h^0(z_1 v). \tag{40}$$

The exit index (aka, the first exceed level index) is the most important factor to be fully analyzed because the decision making parameters including the marginal mean of $\tau_{\nu-1}$, A_ν and $A_{\nu-1}$ could be calculated easily once the exit index is explicitly determined from (18) and (37)–(40):

$$\mathbb{E}[\zeta^\nu] := L^1 + L^2 - L^3 \tag{41}$$

where

$$L^1 = \mathfrak{D}_{(u,v)}^{\left(\frac{M}{2}+B,\frac{M}{2}\right)}\left[\gamma_h^0(v) - \gamma_a^0(u)\gamma_h^0(v)\right], \tag{42}$$

$$L^2 = \mathfrak{D}_{(u,v)}^{\left(\frac{M}{2}+B,\frac{M}{2}\right)}\left[\frac{\zeta \cdot \gamma_a^0(u)\gamma_h^0(v)\gamma_h(v)}{1 - \zeta\gamma_a(u)\gamma_h(v)}\right], \tag{43}$$

$$L^3 = \mathfrak{D}_{(u,v)}^{\left(\frac{M}{2}+B,\frac{M}{2}\right)}\left[\frac{\zeta \cdot \gamma_a^0(u)\gamma_h^0(v)\gamma_a(u)\gamma_h(v)}{1 - \zeta\gamma_a(u)\gamma_h(v)}\right]. \tag{44}$$

Since the observation process has the memoryless properties, the inter-arrival time for observation is exponentially distributed and the functionals (37)–(40) could be reconstructed as follows:

$$\gamma_a^0(u) = \frac{\beta_a^0}{1-\alpha_a^0 \cdot u}, \gamma_a(u) = \frac{\beta_a}{1-\alpha_a \cdot u}, \gamma_h^0(v) = \frac{\beta_h^0}{1-\alpha_h^0 \cdot v}, \gamma_h(v) = \frac{\beta_h}{1-\alpha_h \cdot v}, \tag{45}$$

$$\beta_a^0 = \frac{1}{\left(1+\widetilde{\delta}_0 \lambda_a\right)}, \alpha_a^0 = \frac{\widetilde{\delta}_0 \cdot \lambda_a}{\left(1+\widetilde{\delta}_0 \cdot \lambda_a\right)}, \beta_a = \frac{1}{\left(1+\widetilde{\delta} \cdot \lambda_a\right)}, \alpha_a = \frac{\widetilde{\delta}_0 \cdot \lambda_a}{\left(1+\widetilde{\delta} \cdot \lambda_a\right)}, \tag{46}$$

$$\beta_h^0 = \frac{1}{\left(1+\widetilde{\delta}_0 \cdot \lambda_h\right)}, \alpha_h^0 = \frac{\widetilde{\delta}_0 \cdot \lambda_h}{\left(1+\widetilde{\delta}_0 \cdot \lambda_h\right)}, \beta_h = \frac{1}{\left(1+\widetilde{\delta} \cdot \lambda_h\right)}, \alpha_h = \frac{\widetilde{\delta}_0 \cdot \lambda_h}{\left(1+\widetilde{\delta} \cdot \lambda_h\right)}, \tag{47}$$

where

$$\widetilde{\delta}_0 = \mathbb{E}[\tau_0], \widetilde{\delta} = \mathbb{E}[\Delta_k]. \tag{48}$$

From (42),

$$L^1 = \mathfrak{D}_{(u,v)}^{\left(\frac{M}{2}+B,\frac{M}{2}\right)}\left[\gamma_h^0(v)\right] - \mathfrak{D}_{(u,v)}^{\left(\frac{M}{2}+B,\frac{M}{2}\right)}\left[\gamma_a^0(u)\gamma_h^0(v)\right]$$

$$= \beta_h^0\left[\frac{1-\left(\alpha_h^0\right)^{\frac{M}{2}+1}}{1-\left(\alpha_h^0\right)}\right]\left(1-\beta_A^0\left[\frac{1-\left(\alpha_a^0\right)^{\frac{M}{2}+B+1}}{1-\left(\alpha_a^0\right)}\right]\right) \tag{49}$$

and, from (43),

$$L^2 = \mathfrak{D}_{(u,v)}^{\left(\frac{M}{2}+B,\frac{M}{2}\right)}\left[\frac{\zeta \cdot \gamma_a^0(u)\gamma_h^0(v)\gamma_h(v)}{1-\zeta\gamma_a(u)\gamma_h(v)}\right]$$

$$= \sum_{n\geq 0} \zeta^{n+1}\left\{\left(\beta_a^0 \cdot (\beta a)^n\right) \cdot \sum_{j=0}^{\frac{M}{2}+B}\left\{(\alpha_a)^j \psi_{n-1}^a(j)\right\}\right\} \cdot \left\{\left(\beta_h^0 \cdot (\beta_h)^{n+1}\right) \cdot \sum_{k=0}^{\frac{M}{2}}\left\{(\alpha_h)^k \psi_n^h(k)\right\}\right\} \tag{50}$$

and, from (44),

$$L^3 = (\zeta \beta_a^0 \beta_h^0 \beta_a \beta_h) \left[\sum_{n\geq 0}(\zeta \beta_a \beta_h)^n \Xi_n^a\left(\frac{M}{2}+B\right) \cdot \Xi_n^h\left(\frac{M}{2}\right)\right]$$

$$= \sum_{n\geq 0} \zeta^{n+1} \left\{ \left(\beta_a^0 \cdot (\beta a)^{n+1}\right) \cdot \sum_{j=0}^{\frac{M}{2}+B} \left\{(\alpha_a)^j \psi_n^a(j)\right\} \right\} \cdot \left\{ \left(\beta_h^0 \cdot (\beta_h)^{n+1}\right) \cdot \sum_{k=0}^{\frac{M}{2}} \left\{(\alpha_h)^k \psi_n^h(k)\right\} \right\} \quad (51)$$

where

$$\Xi_n^a(m) = \sum_{j=0}^m \left\{(\alpha_a)^j \psi_n^a(j)\right\}, \Xi_n^h(m) = \sum_{k=0}^m \left\{(\alpha_h)^k \psi_n^h(k)\right\}, \quad (52)$$

$$\psi_n^a(j) = \left(\sum_{i=0}^j \binom{n+i}{i}\left(\frac{\alpha_a^0}{\alpha_a}\right)^i\right), \psi_n^h(k) = \left(\sum_{i=0}^k \binom{n+i}{i}\left(\frac{\alpha_h^0}{\alpha_h}\right)^i\right). \quad (53)$$

From (49)–(53), the PGF of the exit index ν satisfies the following formula from the lemma in the BGG [18]:

$$\mathbb{E}[\zeta^\nu] = \beta_h^0 \left[\frac{1-(\alpha_h^0)^{\frac{M}{2}+1}}{1-(\alpha_h^0)}\right]\left(1-\beta_A^0 \cdot \mathbb{E}\left[\frac{1-(\alpha_a^0)^{\frac{M}{2}+B+1}}{1-(\alpha_a^0)}\right]\right)$$
$$+ \sum_{n\geq 0} \zeta^{n+1} \left[(\beta_a^0 \beta_h^0 \beta_h)(\beta_a \beta_h)^n \Xi_n^h\left(\frac{M}{2}\right) \cdot \mathbb{E}\left[\Xi_{n-1}^a\left(\frac{M}{2}+B\right) - \Xi_n^a\left(\frac{M}{2}+B\right)\beta_a\right]\right] \quad (54)$$

and

$$\mathbb{E}[\nu] = \left(\beta_a^0 \beta_h^0 \beta_h\right) \sum_{n\geq 1} n \left\{(\beta_a \beta_h)^n \Xi_{n-1}^h\left(\frac{M}{2}\right) \mathbb{E}\left[\Xi_{n-2}^a\left(\frac{M}{2}+B\right) - \Xi_{n-1}^a\left(\frac{M}{2}+B\right)\beta_a\right]\right\} \quad (55)$$

where

$$\Xi_{-1}^a(m) = 0, \Xi_{-2}^a(m) = 0, \Xi_{-1}^h(m) = 0. \quad (56)$$

3.2. Marginal Means of EBIoV Decision Making Parameters

In the EBIoV network, conventional decision making parameters are $\nu, \tau_{\nu-1}, A_\nu$ and $A_{\nu-1}$. Although all decision making parameters could be fully analyzed, using a marginal mean of each parameter is occasionally more efficient than finding the explicit PGFs of parameters. The marginal means of EBIoV decision making parameters could be found as follows:

$$\mathbb{E}[\nu] = \frac{\partial}{\partial \zeta} \mathbb{E}\left[\Phi\left(\zeta; \left[\frac{M}{2}\right]+B, \left[\frac{M}{2}\right]\right)\Big|B\right]\Big|_{(\zeta,g_0,g_1,z_0,z_1))\to 1}, \quad (57)$$

$$\mathbb{E}[\tau_{\nu-1}] = \mathbb{E}[\tau_0] + \mathbb{E}[\Delta_1](\mathbb{E}[\nu]-1), \quad (58)$$

$$\mathbb{E}[A_\nu] = \mathbb{E}[A_0] + \mathbb{E}[\nu-1]\mathbb{E}[X_k], \quad (59)$$

$$\mathbb{E}[A_{\nu-1}] = \mathbb{E}[A_0] + \mathbb{E}[\nu-2]\mathbb{E}[X_k]. \quad (60)$$

Recalling from (23), the probability of bursting a Blockchain network (i.e., an attacker wins a game) under the memoryless properties becomes a Poisson compound process:

$$q(s_h) = \sum_{k>\frac{M}{2}} \mathbb{E}\left[\mathbf{1}_{\{A_\nu=k\}}\right], s_h = \{\text{"Regular"}\}, \quad (61)$$

or

$$q(s_h) = \mathbb{E}\left[\sum_{k>\frac{M}{2}+B} \mathbb{E}\left[\mathbf{1}_{\{A_\nu=k\}}\right]\Big|B\right], s_h = \{\text{"Safety"}\}, \quad (62)$$

where

$$\mathbb{E}\left[\mathbf{1}_{\{A_\nu=k\}}\right] = \mathbb{E}\left[\frac{\lambda_a \tau_\nu}{k!} \cdot e^{-\lambda_a \tau_\nu}\right]. \quad (63)$$

4. The EBIoV Optimization Practice

This section deals with the BIoV network security optimization practice in a connected car. The strategy for protecting the EBIoV network is supporting additional nodes to give the less chance that an attacker catches blocks with false control requests. The example in this paper is targeting a connected car which consists 16 IoV components in a BIoV network (15 nodes from car parts and 1 node from a service center) and the estimated car value is around 50,000 USD (see Table 2).

Table 2. Initial conditions for the cost function.

Name	Value	Description
M	16 [Component]	Total number of the nodes in each BIoV network
V	50,000 [USD]	Average value of a BIoV enabled connected car
$c(n, \rho)$	$= 25 \cdot n \cdot \rho$ [USD]	Cost for reserving nodes to avoid attacks per each car
$\mathbb{E}[A_0]$	2 [Blocks]	Total number of blocks that changed by an attacker at $\tau_0 (=0)$
B_M	32 [Nodes]	Maximum number of honest nodes supported from the HQ

It is noted that the values on Table 2 are artificially made up for demonstration purposes only. Since the BGG adapted IoV network (EBIoV) has been analytically solved, finding optimal values of a cost function and calculating a probability distribution of a model are straight forward. However, software implementation by using a programming language is still required for solving a LP (Linear Programming) problem. Based on the above conditions, a LP model could be described as follows from (25) and (28):

Objective:
$$G = min \ \mathfrak{S}(n, \rho)_{Total} \tag{64}$$

Subject to:
$$n \geq \frac{c_{(n,\rho)}}{V \cdot q^0 - c(n,\rho)}; \tag{65}$$

From (28), the total cost $\mathfrak{S}(n, \rho)_{Total}$ is as follows:

$$\mathfrak{S}(n, \rho)_{Total} = \left\{ c(n,\rho)\left(1 - q^1_{(n,\rho)}\right) + \left(c(n,\rho) + V\right)q^1_{(n,\rho)} \right\} p_{A_{-1}} + Vq^0 \cdot (1 - p_{A_{-1}}) \tag{66}$$

where

$$p_{A_{-1}} = P\left\{ A_{\nu-1} < \tfrac{M}{2} \right\} \simeq P\left\{ A_\nu < \tfrac{M}{2} - \lambda_a \widetilde{\delta} \right\}$$
$$= \sum_{k=0}^{\left\{\tfrac{M}{2} - \lambda_a \widetilde{\delta}\right\}} \left(\frac{\left\{\lambda_a\left(\widetilde{\delta}_0 + \mathbb{E}[\nu-1]\widetilde{\delta}\right)\right\}^k}{k!} \cdot e^{-\lambda_a\left(\widetilde{\delta}_0 + \mathbb{E}[\nu-1]\widetilde{\delta}\right)} \right), \tag{67}$$

$$q^0 \simeq 1 - \sum_{k=0}^{\tfrac{M}{2}} \left(\frac{\left\{\lambda_a\left(\widetilde{\delta}_0 + \mathbb{E}[\nu-1]\widetilde{\delta}\right)\right\}^k}{k!} \cdot e^{-\lambda_a\left(\widetilde{\delta}_0 + \mathbb{E}[\nu-1]\widetilde{\delta}\right)} \right), \tag{68}$$

$$q^1_{(n,\rho)} = \sum_{j=0}^{n} \sum_{\{k \geq \tfrac{M}{2} + B + j\}} \left(\frac{\lambda_a\left(\widetilde{\delta}_0 + \mathbb{E}[\nu-1]\widetilde{\delta}\right)}{k!} \cdot e^{-\lambda_a\left(\widetilde{\delta} + \mathbb{E}[\nu-1]\widetilde{\delta}\right)} \right) P_j, \tag{69}$$

$$P_j = \binom{n}{j} \rho^j (1-\rho)^{n-j}. \tag{70}$$

The total cost $\mathfrak{S}(n, \rho)_{Total}$ could be minimized by the given parameter set (n, ρ) and the parameter set (n^*, ρ^*) is the optimal combination of an acceptance rate and the number of total backup nodes which are supported from the HQ. The below illustration in Figure 3 is the conventional graph that shows the optimal result of the BGG based BIoV (EBIoV) network based on the given initial conditions in Table 2.

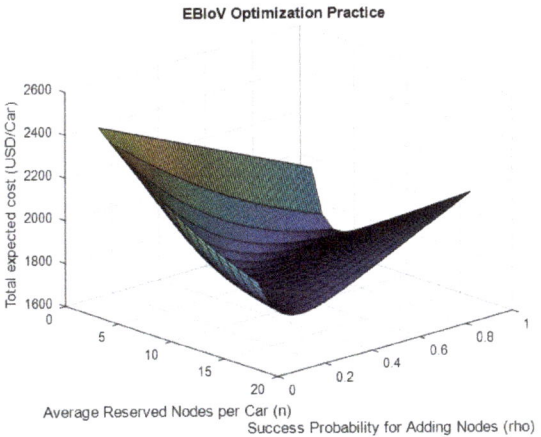

Figure 3. Optimization Example for the EBIoV.

It is noted that the total cost for reserving backup nodes should be more than the car value when an EBIoV network is burst. Hence, the limitation of the success probability for adding nodes is as follows:

$$\frac{V \cdot q^0}{c(n, \rho) \cdot (B_M - 1)} \leq 1, \tag{71}$$

where B_M is the maximum available reserved nodes per a connected car. According to this demonstration (based on Table 2), the optimal cost is 1700 USD (per a car) when the defender reserves 4 additional nodes per each car with the 82.6% acceptance rate for managing the risk from attackers (i.e., $n^* = 4$, $\rho^* = 0.826$). The moment of requesting the additional nodes will be the time τ_{v-1} when is one step prior to the moment when an attacker catches more than the half of whole blocks.

5. Conclusions

This paper establishes the enhanced Blockchain based IoV network architecture by bringing a theoretical model in stochastic modeling. The Enhanced Blockchain enabled Internet of Vehicles (EBIoV) is an advanced secure IoT network architecture for protecting a connected car from attackers. This new architecture has been designed for a decentralized network by adapting the Blockchain Governance Game (BGG) to improve the connected car security. The BGG is a mathematically proven game model to develop optimal defense strategies to protect systems from attackers. The practical case in the paper demonstrates how an EBIoV network could be implemented for connected car securities. The EBIoV network is the first research that applies a BGG model into the IoV security domain. The BGG model shall be extended to various Blockchain based cybersecurity areas including IoT security and secured decentralized service network design.

Funding: This research received no external funding.

Institutional Review Board Statement: Not applicable.

Informed Consent Statement: Not applicable.

Data Availability Statement: There are no available data to be stated.

Acknowledgments: Special thanks to the Guest Editor, Nicu Bizon, who guides the author to submit the proper topic of the journal and also thankful to the referees whose comments are very constructive.

Conflicts of Interest: The author declares no conflict of interest.

References

1. Rouse M. Internet of Vehicles. 2018. Available: https://whatis.techtarget.com/definition/Internet-of-Vehicles (accessed on 1 May 2019).
2. Kim S.; Shrestha, R. Internet of Vehicles, Vehicular Social Networks, and Cybersecurity. In *Automotive Cyber Security*; Springer: Singapore, 2020; pp. 149–181.
3. Dandala, T.T.; Krishnamurthy, V.; Alwan, R. Internet of Vehicles (IoV) for traffic management. In Proceedings of the 2017 International Conference on Computer, Communication and Signal Processing (ICCCSP), Chennai, India, 10–11 January 2017; pp. 1–4.
4. Hamid, U.Z.A.; Zamzuri, H.; Limbu, D.K. Internet of Vehicle (IoV) Applications in Expediting the Implementation of Smart Highway of Autonomous Vehicle: A Survey. In *Performability in Internet of Things*; Springer: Cham, Germany, 2018; pp. 137–157.
5. Dorri, A.; Steger, M.; Kanhere, S.S.; Jurdak, R. BlockChain: A Distributed Solution to Automotive Security and Privacy. *IEEE Commun. Mag.* **2017**, *55*, 119–125. [CrossRef]
6. Kang, J.; Xiong, Z.; Niyato, D.; Ye, D.; In Kim, D.; Zhao, J. Towards Secure Blockchain-enabled Internet of Vehicles: Optimizing Consensus Management Using Reputation and Contract Theory. *IEEE Trans. Veh. Technol.* **2019**, *68*, 2906–2920. [CrossRef]
7. Yang, Z.; Yang, K.; Lei, L.; Zheng, K.; Leung, V.C. Blockchain-based decentralized trust management in vehicular networks. *IEEE Internet Things J.* **2018**, *6*, 1495–1505. [CrossRef]
8. Steger, M.; Dorri, A.; Kanhere, S.S.; Römer, K.; Jurdak, R.; Karner, M. Secure wireless automotive software updates using blockchains: A proof of concept. In *Advanced Microsystems for Automotive Applications*; Springer: Cham, Germany, 2017; pp. 137–149.
9. Cebe, M.; Erdin, E.; Akkaya, K.; Aksu, H.; Uluagac, S. Block4forensic: An integrated lightweight blockchain framework for forensics applications of connected vehicles. *IEEE Commun. Mag.* **2018**, *56*, 50–57. [CrossRef]
10. Blockchain News. Available online: https://www.ccn.com/volkswagen-seeks-patent-for-inter-vehicular-blockchain-communications-system/ (accessed on 1 May 2019).
11. Haig, S. Ford to Use Cryptocurrency for Inter-Vehicle Communication System. 2018. Available online: https://news.bitcoin.com/ford-cryptocurrency-inter-vehicle-communication-system/ (accessed on 1 May 2019).
12. Conoscenti, M.; Vetrò, A.; De Martin, J.C. Blockchain for the Internet of Things: A systematic literature review. In Proceedings of the 13th IEEE/ACS International Conference of Computer Systems and Applications, AICCSA 2016, Agadir, Morocco, 29 November–2 December 2016.
13. Restuccia, F. Blockchain for the Internet of Things: Present and Future. Available online: https://arxiv.org/abs/1903.07448 (accessed on 1 May 2019).
14. Jesus, E.F.; Chicarino, V.R.L.; de Albuquerque, C.V.N.; Rocha, A.A.D.A. Survey of How to Use Blockchain to Secure Internet of Things and the Stalker Attack. *Secur. Commun. Netw.* **2018**, *2018*, 9675050. [CrossRef]
15. Kim, S.-K.; Yeun, C.Y.; Damiani, E.; Al-Hammadi, Y.; Lo, N.-W., New Blockchain Adoption For Automotive Security by Using Systematic Innovation. In Proceedings of the 2019 IEEE Transportation Electrification Conference and Expo Asia-Pacific, Jeju, Korea, 8–10 May 2019; pp. 1–4.
16. Baker, J.; Edge Computing—The New Frontier of the Web. Available online: https://hackernoon.com/edge-computing-a-beginners-guide-8976b6886481 (accessed on 1 May 2019).
17. ERPINNEW. Fog Computing vs Edge Computing. 2017. Available online: https://erpinnews.com/fog-computing-vs-edge-computing (accessed on 1 May 2019).
18. Kim, S.-K. Blockchain Governance Game. *Comput. Ind. Eng.* **2019**, *136*, 373–380. [CrossRef]
19. Kim, S.-K. Strategic Alliance for Blockchain Governance Game. *Probab. Eng. Inf. Sci.* **2020**, 1–17. doi:10.1017/s0269964820000406. [CrossRef]
20. Hammoud, A.; Sami, H.; Mourad, A.; Otrok, H.; Mizouni, R.; Bentahar, J. AI, Blockchain, and Vehicular Edge Computing for Smart and Secure IoV: Challenges and Directions. *IEEE Internet Things Mag.* **2020**, *3*, 68–73. [CrossRef]
21. Narayanan, A.; Clar, J. Bitcoin's Academic Pedigree. *Mag. Commun. ACM* **2017**, *60*, 36–45. [CrossRef]
22. Weiss, M.; Corsi, E. *Bitfury: Blockchain for Government*; HBP Case 9—818-031; Harvard University: Cambridge, MA, USA, 2018; 29p.
23. Jones, M. Blockchain for Automotive: Spare Parts and Warranty. 2017. Available online: https://www.ibm.com/blogs/internet-of-things/iot-Blockchain-automotive-industry/ (accessed on 1 May 2019)
24. Bonomi, F.; Milito, R. Fog computing and its role in the internet of things. In Proceedings of the First Edition of the MCC Workshop on Mobile Cloud Computing, Helsinki, Finland, 17 August 2012; pp. 13–16.
25. Liu, Z.; Luong, N.C.; A Survey on Applications of Game Theory in Blockchain. *arXiv* **2019**, arXiv:1902.10865.
26. Micali, S.; Rabin, M.; Wang, W.; Niyato, D.; Wang, P.; Liang, Y.C.; In Kim, D. Verifiable random functions. In Proceedings of the 40th IEEE Symposium on Foundations of Computer Science, New York, NY, USA, 17–19 October 1999; pp. 120–130.
27. Dodis, Y.; Yampolskiy, A. A Verifiable Random Function with Short Proofs and Keys. In *Lecture Notes in Computer Science*; Springer: Berlin/Heidelberg, Germany, 2005; Volume 3386, pp. 416–431.
28. Gorbunov, S. Algorand Releases First Open-Source Code: Verifiable Random Function. 2018. Available online: https://medium.com/algorand/ (accessed on 1 May 2019).

29. Zhao, W. MIT Professor's Blockchain Protocol Nets 62 Million in New Funding. 2018. Available online: https://www.coindesk.com/mit-professors-Blockchain-protocol-nets-62-million-in-new-funding (accessed on 1 May 2019).
30. Dshalalow, J.H. *First Excess Level Process, Advances in Queueing*; CRC Press: Boca Raton, FL, USA, 1995; pp. 244–261.
31. Dshalalow, J.H.; Ke, H.-J. Layers of noncooperative games. *Nonlinear Anal.* **2009**, *71*, 283–291. [CrossRef]

Article

An Optimization Model for the Temporary Locations of Mobile Charging Stations

Maria-Simona Răboacă [1,2,†], **Irina Băncescu** [3], **Vasile Preda** [3,4] **and Nicu Bizon** [5,*]

1. Faculty of Electrical Engineering and Computer Science, Stefan cel Mare University of Suceava, 13 Universității Street, 720229 Suceava, Romania; simona.raboaca@icsi.ro
2. Faculty of Electrical Engineering, Technical University of Cluj-Napoca, 26-28 G. Barițiu Street, 400027 Cluj-Napoca, Romania
3. Costin C. Kirițescu, National Institute of Economic Research, 13 Calea 13 Septembrie Street, 050711 Bucharest, Romania; irina_adrianna@yahoo.com (I.B.); vasilepreda0@gmail.com (V.P.)
4. Gheorghe Mihoc-Caius Iacob, Institute of Mathematical Statistics and Applied Mathematics, 13 Calea 13 Septembrie Street, 050711 Bucharest, Romania
5. Faculty of Electronics, Communications and Computers University of Pitești, 1 Târgu din Vale Street, 110040 Pitești, Romania
* Correspondence: nicubizon@yahoo.com
† Current Address: National Research and Development Institute for Cryogenic and Isotopic Technologies-ICSI, 4 Uzinei Street, 240050 Râmnicu Vâlcea, Romania.

Received: 25 January 2020; Accepted: 15 March 2020; Published: 21 March 2020

Abstract: A possible solution with which to alleviate the range anxiety of electric vehicle (EV) drivers could be a mobile charging station which moves in different places to charge EVs, having a charging time of even half an hour. A problem that arises is the impossibility of charging in any location due to heavy traffic or limited space constraints. This paper proposes a new operational mode for the mobile charging station through temporarily stationing it at different places for certain amounts of time. A mathematical model, in the form of an optimization problem, is built by modeling the mobile charging station as a queuing process, the goal of the problem being to place a minimum number of temporary service centers (which may have one or more mobile charging stations) to minimize operating costs and the charger capacity of the mobile charging station so that the service offered is efficient. The temporary locations obtained are in areas with no or few fixed charging stations, making the mobile station infrastructure complementary to the fixed charging station infrastructure. The temporary location operational mode, compared to current moving operational mode, is more efficient, having a small miss ratio, short mean response time and short mean queuing time.

Keywords: mobile charging station; electric vehicle; operational mode; location-allocation problem

1. Introduction

The market for electric vehicle charging stations has grown rapidly in recent years. At present, the most developed markets for electric vehicle (EV) charging stations are in North America and Europe, with prospects for the Asian-Pacific market, in the near future (Table 1). Markets for EV charging stations are dependent on the state of their adoption, and the objectives and legislative targets in the region, which means that the market for charging stations for electric vehicles is higher in regions where electric cars are widely adopted. Some major problems faced by charging station infrastructure include its location and size, these two issues being extensively studied in the literature.

Table 1. Sales of charging stations (thousands), 2012–2017.

Region	2012	2013	2014	2015	2016	2017
Africa/Middle East	<15	<20	<25	<40	<50	~50
Asia Pacific	~200	~350	~650	~800	~950	~1000
Europe	~75	~200	~240	~350	~450	~500
Latin America	<5	<5	<10	<15	<25	<25
North America	~70	~200	~260	~350	~400	~450

Source: Global EV Outlook 2017 [1].

In the European Union, in 2019, it is estimated that approximately 165,064 charging stations were installed [2]. At present, there are no leaders in the charging station market, but the competitive market in the field is emerging through new, specialized companies, so there are opportunities for larger societies to own big market shares. In Europe, the electrical capacity of an electric car battery is approximately 20 kWh, which allows you to have a range of approximately 150 km [3]. Hybrid vehicles have a capacity of about 3–5 kWh for a range between 20 and 40 km. As this range is still limited, the vehicle battery should be recharged periodically. For a normal load (3 kW), vehicle manufacturers have developed a charger inside the car. For fast charging (22 kW, even 43 kW or more), manufacturers have developed two solutions: the use of a car charger designed to charge between 3 and 43 kW at single-phase (230V) or three-phase (400V) voltage (using an external charger that converts the alternating current to DC and charges the vehicle to 50 kW) [4]. Table 2 shows the charging time for an EV, depending on the type of connection.

According to IEC 61851-1 standard, the electric vehicle conduction stations can be made in four ways (Table 2) as follows:

- Mode 1 (L1) is the simplest EV charging solution. In this case, an EV is connected to AC power supply using standard sockets, but it must contain a contractor to disconnect the power supply in case of overload or an electric shock. The EV's charging is done without communication between it and the station, and the maximum current accepted is 16 A.
- Mode 2 (L2), where the EV is charged from the power supply using standardized single-phase or three-phase sockets and charging conductors containing a box of control integrated into the cable with a pilot function command and a protection system against electrical shocks (RCD). The charge current value should not exceed 32 A.
- Mode 3 (L3), where an EV is connected via a specific scheme to the charging station (EAVE) that has installed control and protection functions. The maximum current is 3×63 A.
- Mode 4 (L4), where EVs are charged from stations using direct current.

Table 2. Charging time of an EV according to charger type.

Charging Time	Type of Source (kW)	Tension (V)	Maximum Current (A)	Mode
6–9 h	mono-phase 3.3 kW	230 V AC	16 A	Mode 1
2–3 h	triple-phase 10 kW	400 V AC	16 A	Mode 1
3–4 h	mono-phase 7 kW	230 V AC	32 A	Mode 2
20–30 min	triple-phase 43 kW	400 V AC	63 A	Mode 3
20–30 min	voltage continues 50 kW	400–500 V DC	100–125 A	Mode 4
1–2 h	triple-phase 24 kW	400 V AC	32 A	Mode 2

Source: IEC 61851-1 standard [4].

Another obstacle in the development of EV market is range anxiety. Depending on the style of driving, the outside temperature and the traffic, the distance an EV can travel decreases, requiring more frequent recharging.

In the initial phase of charging infrastructure's development, a solution with which to alleviate range anxiety may be a mobile charging station which can move to different places to charge EVs, having a charging time of even less than half an hour (Table 2). Furthermore, the mobile charging station can also be used as a car attendant service.

Current, the operational mode of a mobile charging station involves moving it, on demand, within a certain service area to charge EVs, only staying in that location for the time of charging. Multiple operating strategies for this type of charging station can be considered; for example, the mobile station may move on the FIFO (first-in, first-out) or the NJN (nearest-job-next) principles [5]. Based on a recent study [5], the last strategy seems to be the most appropriate, a question of interest being, "What would be the optimal charger capacity of the mobile charging station?" A higher charger capacity implies a higher cost for a mobile charging station. Based on data and statistical methods, Huang et al. [5] proposed an optimum battery capacity of 40 kWh and a charge rate between 15 and 30 kW for an urban environment such as Singapore, in order to design a mobile charging station. The cost range per unit of a DC fast charging station is $10,000–$40,000, while its installation cost range per unit is $4000–$51,000 [6,7]. An estimated cost per unit of a mobile charging station is $40,000 [5].

A problem that arises is the impossibility to charge in any location due to heavy traffic or limited space constraints. For example, in 2019, Gong et al. [8] adopted the concept of reasonable location in a location problem for fixed charging stations. A reasonable location is defined as "a location having a insignificant variance in the access frequencies to different charging stations" [8].

This paper proposes a new operational mode of the mobile charging station by temporarily stationing it, on demand, in different locations. A real frame, using data from New York taxis, is considered. A mathematical model, in the form of an optimization problem, is built by modeling the mobile charging station as a queuing process and the charging requests as Poisson processes. Furthermore, this optimization problem can also be used to position fixed charging stations. The purpose of the problem is to place a minimum number of temporary service centers (which may have one or more mobile charging stations) in order to minimize the operating costs and the charger capacity of the mobile charging station so that the service offered is effective, meaning, for example, that upon arrival at a service center, each EV will stand in a queue with no more than b clients with a probability of at least β.

Looking from the perspective of a company that operates the mobile charging stations, the company is interested in each service center having a queue of at least a clients with a probability of at least α. The two queuing conditions mentioned can be combined so that in each service center we have at least a clients and at most b clients with a probability of at least γ.

The mobile charging station is not intended to be the main charging method for electric vehicles but a supplement to the fixed charging station, especially in areas where there are few or no fixed charging stations. Moreover, in countries with old electricity grid infrastructure (for example, Romania) a mobile charging station could take over the increased electricity demand due to the penetration of electric vehicles in the market until the modernization of the electricity distribution network infrastructure and until the installation of sufficient fixed stations. The mobile charging station is complementary to the fixed station in the incipient phase of developing charging station infrastructure.

The paper is structured in the following way. Section 2 is dedicated to a literature review. In Section 3, the proposed new operational mode of the mobile charging station is presented. Additionally, in this section, a mathematical model is discussed. Section 4 deals with estimating the optimization problem parameters. Section 5 discusses the results obtained from simulations using the New York taxi database [9], and in Section 6 we compare two operational modes of a mobile charging station. The paper is concluded in Section 7.

2. Literature Review

The power management of a mobile charging station was investigated in [10], considering Internet-of-Things to manage its supply power. The routing problem that arises with the current operational mode of a mobile charging station was also analyzed in the literature. Two decisions (location and routing) with time-windows were investigated in [11], with the results showing a need for a more efficient service in the sense of improving the charging rate or increasing battery capacity. Similar results were obtained by Cui et al. in 2018 [12] by modeling the routing problem of a mobile charging station as a mixed-integer linear problem with time windows. In [13], heterogeneous networks were used to optimally schedule the EV's charge requests, assuming that EV users send their respective power demands and locations, when requesting battery recharging, to the mobile charging station.

In 2012, Jia et al. [14] considered several factors in the location of fixed charging stations, including demand for charging stations, driver behavior, road structure, fixed construction costs and operating costs, building a mathematical model that minimizes the total cost of investment. Additionally, charging infrastructure planning was discussed in [15] using multiday travel data. Zhu et al. [16] considered charging station location for plug-in EVs by minimizing the total cost of investment and how many chargers a charging station should have. Recently, in 2019, a two-level charging station locating method, which uses weighted multicriteria methods, was considered for locating charging stations [17].

Some studies take into consideration queuing time for the location problems of charging stations (for example, in [18–22]); various studies revealed that queuing time strongly affects the adoption of EVs [23,24]. Fast-charging stations have been proposed to address this issue; however, the installation of this type of charging station has a negative impact on the distribution network, such as high voltage [19,25] and overload distribution system during peak hours [26,27]. User behavior was considered in [28], while in [29] a resource planning scheme was investigated. An equilibrium modeling framework was considered in [30].

Other approaches consider the maximal covering location problem, taking into account the existing charging infrastructure [31] for urban taxi drivers. Frade et al. (2011) [32] also considered a maximal covering model with daytime versus nightime charging demands. Long travel data is considered in [33,34]. A quality of service index is introduced in [35] in a location optimization problem with charging reliability constraints.

3. Mathematical Model

In this section, we describe the operational mode of a mobile charging station which considers the temporary locations of this type of charging station.

Suppose we have some places (depots) where the mobile charging station can stay. In these places, the mobile station charges both its own battery and the EVs' batteries. Instead of moving the mobile station to where the electric vehicles that require the load are located, the mobile charging station stays for a period of time in a place, moving only between the location set, depending on the intensity of the demand in the depot areas within a certain time interval. Thus, a mobile charging station fulfills the functions of a stationary charging station and those of a mobile charging station. Examples of stationary stations of a mobile station can be gas stations and tramway heads. We no longer assume that the mobile station can move and load anywhere, but only in certain places (Figure 1). If there are many requests for a sub-zone, a mobile station stays there for a while, while another moves between places. Depending on the day and time (during the week or weekend), the mobile station stays for a while in a place, after which time it moves to another place. Through a smartphone application, EVs can see where the mobile station is stationed.

A charging station's infrastructural development deals with constructing optimization problems considering budget constraints and capacity optimization [36,37]. Techniques used involve particle swarm optimization [38,39], the hybrid optimization algorithm [40], ant colony optimization [41] and the genetic algorithm [42,43].

This paper proposes a mixed linear optimization location-allocation problem of mobile charging stations based on a simulation.

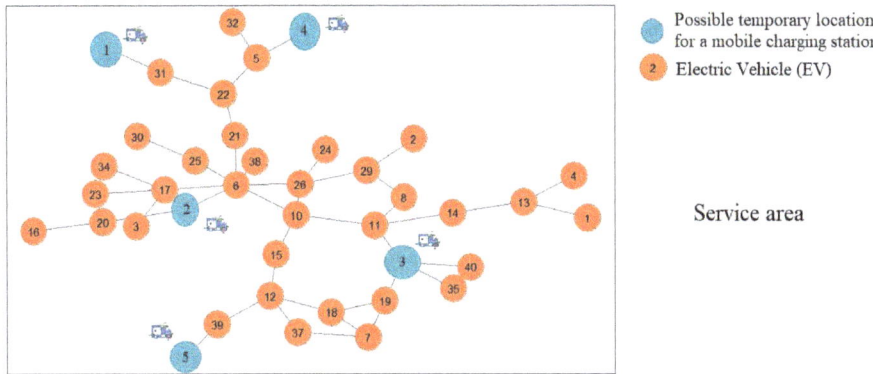

Figure 1. Operational mode of a mobile charging station.

In 1998 and 2002, Marianov and Serra [44,45] proposed a location-allocation problem considering queuing constraints in the form of queue length and waiting time of a customer. Similar constraints are considered in this paper.

The purpose of the optimization problem is to place a minimum number of temporary service centers for EVs' charging, each center having a minimum number of mobile charging stations that can charge the EVs while minimizing the charger capacity of mobile charging stations required for running the service system in some optimal parameters (stated a priori) such that

(i) Each EV will be assigned to a center at the shortest possible distance from its location.
(ii) Upon arrival at the service center, each EV will stand in a queue with no more than b clients with a probability of at least β; or in each center will be at least a clients, with a probability of at least α; or when arriving at the service center, each EV will be in a queue with no more than b clients, and in each center will be at least a clients, with a probability of at least γ.
(iii) The battery capacity of the mobile charging station is enough to meet all requests during the considered temporary location period.

When a mobile charging station becomes idle, having no customers, it can be used in another service area. If a mobile charging station is not used in a service area, for a day or for several hours, it may be temporarily allocated to a service area lacking a mobile charging station; thus, the location and battery state of charge can take different values depending on the day, time and area.

The operating cost associated with a mobile charging station may be composed of the cost of the electricity required to charge the battery of the mobile charging station and maintenance costs.

3.1. Nonlinear Optimization Problem

The problem (P1), given below, minimizes the number of service centers; the number of mobile stations in each service center; the distances between the demand nodes and the locations of the mobile charging stations; and the charger capacity of the mobile charging station battery.

Optimization problem (P1):

$$\min \sum_{j \in J_1} \sum_{m=2}^{C_j} z_{jm} P_{jm} + \sum_{j \in J_1} z_{j1} P_{j1} + \sum_{i \in I} \sum_{j \in J} d_{ij} x_{ij} + \sum_{j \in J} \sum_{m=1}^{C_j} SoC_{jm} z_{jm}, \quad \text{where } J = J_1 \cup J_2 \quad (1)$$

subject to the following constraints:

$$\sum_{j \in J} x_{ij} = 1, \quad \forall i \in I \quad (2)$$

$$z_{jm} \leq z_{j1}, \quad j \in J_1, m = 2, 3, \ldots, C_j \quad (3)$$

$$x_{ij} \leq z_{j1}, \quad \forall i \in I, j \in J \tag{4}$$

$$\sum_{j \in J_1} \sum_{m=1}^{C_j} z_{jm} = B \tag{5}$$

$$\sum_{m=1}^{C_j} SoC_{jm} z_{jm} \geq \sum_{i \in I} g_i x_{ij}, \quad \forall j \in J \tag{6}$$

$$C_j = 1, \quad \forall j \in J_2 \tag{7}$$

$$d_{ij} x_{ij} \leq D, \quad \forall i \in I, j \in J_2 \tag{8}$$

$$P[\text{center } j \text{ has at most } b \text{ EVs in queue}] \geq \beta, \quad \forall j \in J \tag{9}$$

or

$$P[\text{center } j \text{ has at least } a \text{ EVs in queue}] \geq \alpha, \quad \forall j \in J \tag{10}$$

or

$$P[\text{center } j \text{ has at least } a \text{ EVs and at most } b \text{ EVs in queue}] \geq \gamma, \quad \forall j \in J \tag{11}$$

$$x_{ij}, z_{jm} \in \{0,1\}, \quad \forall j \in J, m = 1, 2, \cdots, C_j \tag{12}$$

$$SoC_{jm} \text{ positive integer}, \quad SoC_{jm} \leq \omega_u, \quad \forall j \in J, m = 1, 2, \cdots, C_j \tag{13}$$

$$B \text{ positive integer}, \quad B \leq B_{max} \tag{14}$$

Constraint (2) is allocated to every node a single service center, while constraint (3) limits the number of mobile stations to C_j and indicates that mobile charging station m cannot be placed at node j without first placing $(m-1)^{th}$ mobile station. We force the allocation of a demand node to an open service center by constraint (4); z_{j1}, for $j \in J_2$ takes the value 1 when node i is assigned to fixed charging station from node $j \in J_2$. The maximum number of mobile charging station which can be placed is B (constraint (5)). The total charging capacity of service center j having at most C_j mobile charging stations is greater than or equal to the sum of the average charge load required from node i satisfied by the center j (constraint (6)). Constraint (7) refers to fixed charging stations, assuming the existence of just one fixed charging station in node $j \in J_2$. Additionally, we assume that nodes i cannot be assigned to a fixed charging station located more than D km away (Equation (8)). The decision variables of problem (P1) are z_{jm}, x_{ij}, B and SoC_{jm}.

Upon arrival at the serving center, each EV will stand in queue with no more than b clients with a probability of at least $\beta \in (0,1)$ (Equation (9)); or in each center there will be at least a clients with a probability of at least $\alpha \in (0,1)$ (Equation (10)); or when arriving at the service center, each EV will stand in queue with no more than b clients, and in each center there will be at least a customers, with a probability of at least $\gamma \in (0,1)$ (Equation (11)).

Due to the fact that the optimization problem (P1) is nonlinear, we define the equivalent mixed linear problem to (P1) in the following way.

3.2. Linear Optimization Problem

The equivalent mixed linear problem (P2) is given by

$$\min \sum_{j \in J_1} \sum_{m=2}^{C_j} z_{jm} P_{jm} + \sum_{j \in J_1} z_{j1} P_{j1} + \sum_{i \in I} \sum_{j \in J} d_{ij} x_{ij} + \sum_{j \in J} \sum_{m=1}^{C_j} SoC_{jm} \quad (15)$$

subject to constraints (2), (3), (4), (5), (7), (8), (9) or (10) or (11), (12), (13), (14) and

$$\sum_{m=1}^{C_j} SoC_{jm} \geq \sum_{i \in I} g_i x_{ij}, \quad \forall j \quad (16)$$

$$SoC_{jm} \leq z_{jm} \omega_U, \quad \forall j \in J, m = 1, 2, 3, \ldots, C_j \quad (17)$$

If z_{jm} is 0, then SoC_{jm} is 0. If z_{jm} is 1, then SoC_{jm} cannot be greater than the maximum charger capacity ω_U. The decision variables of problem (P2) are z_{jm}, x_{ij}, B and SoC_{jm}.

Constraint (9) can be rewritten according to the intensity of requests in nodes i and the mean serving rate of the mobile charging stations in nodes j. Assuming that the charge requests from each node i follow a Poisson process of intensity f_i and that the serving time is exponentially distributed with an mean serving rate of μ_j, condition (9) is equivalent to the following linear condition (for mobile charging stations) [44,46]

$$\sum_{i \in I} f_i x_{ij} \leq \mu_j \left[z_{j1} \rho_{\beta j1} + \sum_{m=2}^{C_j} z_{jm} (\rho_{\beta jm} - \rho_{\beta j(m-1)}) \right], \quad (18)$$

$\forall \ j \in J_1$, where the values $\rho_{\beta jm}$ are determined a priori from the following inequality

$$\sum_{k=0}^{m-1} \frac{(m-k)m!m^b}{k!} \frac{1}{\rho^{m+b+1-k}} \geq \frac{1}{1-\beta}, \quad (19)$$

those being the values for which in (19) we have equality, while for fixed charging stations we have the equivalent linear condition

$$\sum_{i \in I} f_i x_{ij} \leq \mu_j z_{j1} \rho_{\beta j1}, \quad j \in J_2 \quad (20)$$

Fixed values of $\rho_{\beta jm}$ are computed for every possible level of probability-chosen β, provided there are m mobile charging stations at node j.

Similarly, we obtain the following new result for constraint (10) which is equivalent to the following linear condition (for mobile charging station)

$$\sum_{i \in I} f_i x_{ij} \geq \mu_j \left[z_{j1} \theta_{\alpha j1} + \sum_{m=2}^{C_j} z_{jm} (\theta_{\alpha jm} - \theta_{\alpha j(m-1)}) \right], \quad (21)$$

$\forall j \in J_1$, where the values $\theta_{\alpha jm}$ are determined a priori from the following inequality:

$$\sum_{k=0}^{m-1} \frac{(m-k)m!m^a}{k!} \frac{1}{\theta^{m+a+1-k}} \leq \frac{1}{\alpha} \quad (22)$$

The equivalent linear condition for fixed charging station infrastructure is

$$\sum_{i \in I} f_i x_{ij} \geq \mu_j z_{j1} \theta_{\alpha j1}, \quad j \in J_2 \quad (23)$$

Fixed values of $\theta_{\alpha jm}$ are computed for every possible level of probability-chosen α provided there are m mobile charging stations at node j. The equivalence of inequality (22) to constraint (10) is given by the next theorem.

Theorem 1. *Constraint $\theta \geq \theta_\alpha$, where θ_α is the value for which in Equation (22) holds equality, is equivalent to constraint (10).*

Inequality (22) is equivalent to (10), i.e., Theorem 1 holds, and because θ can be rewritten as a function of variables x_{ij}, it yields linear deterministic constraints (21) with $\theta_{\alpha jm}$ such that (22) holds as an equality.

A linear form of condition (11) can be obtained similarly to those obtained for (9) and (10). Assuming service times in each service center are exponentially assigned with a mean serving rate $\mu_j > 0$, to have a balance in the system, we have the condition $\mu_j \geq m_j \lambda_j$ where m_j represents the number of mobile stations of center j, and λ_j is the intensity of the Poisson process of charge requests arriving to center j. Marianov and Serra [45] showed that

$$\lambda_j = \sum_{i \in I} f_i x_{ij}$$

Assuming that each service center is represented by a M/M/m queuing process, we can calculate the probability from (11). Assuming that the two events are independent, this probability can be written as

P[center j has at least a EVs and at most b EVs in queue] = P[center j has at least a EVs in queue]$\times$ P[center j has at most b EVs in queue]

The assumption of independence leads to an approximation of reality, but ease of calculus. In a future paper we will address the issue of dependence.

We want the probability of an EV arriving at a center to find a queue with less than b other EVs, and the probability that the queue length at a center is greater than a EVs.

The first probability can be written as $1 - (p_{m+b+1} + p_{m+b+2} + \ldots + p_\infty)$, and the second is $p_{m+a+1} + p_{m+a+2} + \ldots + p_\infty$, while p_{m+1} are the state probabilities of the system, where a state $m+1$ corresponds to the event when we have $m+1$ EVs at the center, one EV being in the queue waiting to be recharged, while the others are recharged; for a state $k \leq m$, we have k EVs being recharged at a service center. In the following, we will drop the indices for a easier reading.

Expressions for the state probabilities are the following [47]:

$$p_k = p_0 \omega^k / k!, \quad k \leq m$$

$$p_k = p_0 \omega^k / m! m^{k-m}, \quad k > m$$

$$p_0 = \left(\frac{\omega^m}{\left(1 - \frac{\omega}{m}\right) m!} + \sum_{j=0}^{m-1} \frac{\omega^j}{j!} \right)^{-1}$$

where $\omega = \lambda / \mu$. With these formulae, Equation (11) becomes

$$\left[\frac{p_0 m^m}{m!} \left(\frac{\omega}{m} \right)^{m+a+1} \frac{1}{1 - \frac{\omega}{m}} \right] \times \left[1 - \frac{p_0 m^m}{m!} \left(\frac{\omega}{m} \right)^{m+b+1} \frac{1}{1 - \frac{\omega}{m}} \right] \geq \gamma_j, \quad \forall j \in J \tag{24}$$

Replacing p_0 and after some calculus, we get

$$\sum_{j=0}^{m-1} \frac{(m-j)m!m^b}{j!} \frac{1}{\omega^{m+b+1-j}} - \gamma_j \sum_{j=0}^{m-1} \sum_{k=0}^{m-1} \frac{(m-j)(m-k)(m!)^2 m^{a+b}}{k!j!} \times \frac{1}{\omega^{2m+2+a+b-k-j}} \geq 1 \quad (25)$$

This equation is equivalent to Equation (11). However, due to it being a nonlinear equation, we require an equivalent linear equation.

We denote by $\varphi_j : [0,1] \to \mathbb{R}$ the left hand side of Equation (25) which depends on ω, for all $j \in J$. We prove that function $\varphi_j(\omega)$ is increasing.

Lemma 1. *Function $\varphi_j(\omega)$ is an increasing function for all $\omega \in [0,1]$ and for all*

$$\gamma_j \geq \left[\sum_{k=0}^{m-1} \frac{(m-k)m!m^a}{k!} \frac{2m+2+a+b-k-j}{m+b+1-j}\right]^{-1}$$

Proof. The derivative of $\varphi_j(\omega)$ is

$$\varphi'_j(\omega) = \sum_{j=0}^{m-1} \left\{ \frac{(m-j)m!m^b(-1)}{j!} \frac{(m+b+1-j)}{\omega^{m+b+2-j}} \left[1 - \gamma_j \sum_{k=0}^{m-1} \frac{(m-k)m!m^a}{k!} \frac{2m+2+a+b-k-j}{\omega^{m+a+1-k}(m+b+1-j)}\right]\right\}$$

We denote by $h_j : [0,1] \to \mathbb{R}$ function

$$h_j(\omega) = 1 - \gamma_j \sum_{k=0}^{m-1} \frac{(m-k)m!m^a}{k!} \frac{2m+2+a+b-k-j}{\omega^{m+a+1-k}(m+b+1-j)}$$

The derivative of function $h_j(\omega)$ is

$$h'_j(\omega) = -\gamma_j \sum_{k=0}^{m-1} \frac{(m-k)m!m^a}{k!} \frac{(-1)(m+a+1-k)(2m+2+a+b-j-k)}{\omega^{m+a+2-k}(m+b+1-j)}$$

Because $m+a+1-k$ and $m+b+1-j$ are positive, like the rest of the terms, we obtain that $h'_j(\omega)$ is positive. Hence, we get $h_j(\omega)$ is increasing. We have $\lim_{\omega \to 0} h_j(\omega) = -\infty$, and

$$\lim_{\omega \to 1} h_j(\omega) = 1 - \gamma_j \sum_{k=0}^{m-1} \frac{(m-k)m!m^a}{k!} \frac{2m+2+a+b-j-k}{m+b+1-j}$$

For

$$\gamma_j \geq \left[\sum_{k=0}^{m-1} \frac{(m-k)m!m^a}{k!} \frac{2m+2+a+b-j-k}{m+b+1-j}\right]^{-1},$$

$j \leq m-1$, we have that $\lim_{\omega \to 1} h_j(\omega) \leq 0$. Hence, $h_j(\omega)$ is negative, and as a consequence $\varphi'_j(\omega)$ is positive. Thus, we conclude that $\varphi_j(\omega)$ is increasing for all $j \in J$. □

We choose γ to be $\gamma = \max_{j \in J}(\gamma_j)$.

The equivalence of inequality (25) to constraint (11) is given by the next theorem.

Theorem 2. *Constraint $\omega \geq \omega_\gamma$, where ω_γ is the value for which in Equation (25) holds equality, is equivalent to constraint (11).*

Proof. Let ω_γ be he value for which in Equation (25) we have equality. Because function $\varphi_j(\omega)$ is increasing, then for any $\omega \geq \omega_\gamma$, Equation (25) is true. □

Therefore, once a value for γ is chosen, we determine the value for ω_γ with any root-finding numerical method. The constraint (11) in optimization problem (P2) becomes (for mobile charging station):

$$\sum_{i \in I} f_i x_{ij} \geq \mu_j \left[z_{j1} \omega_{\gamma j1} + \sum_{m=2}^{C_j} z_{jm} (\omega_{\gamma jm} - \omega_{\gamma j(m-1)}) \right], \tag{26}$$

$\forall j \in J_1$, where values $\omega_{\gamma jm}$ are determined a priori from inequality (25), those being the values for which in this equation we have equalities. The equivalent linear constraint of (11) for fixed charging station is

$$\sum_{i \in I} f_i x_{ij} \geq \mu_j z_{j1} \omega_{\gamma j1}, \quad j \in J_2 \tag{27}$$

Fixed values of $\omega_{\gamma jm}$ are computed for every possible level of probability-chosen γ, provided there are m mobile charging stations at node j.

Inequality (25) is equivalent to (11), i.e., Theorem 2 holds, and because ω can be rewritten as a function of variables x_{ij}, it yields linear deterministic constraints (26) with $\omega_{\gamma jm}$ such that (25) holds as equality.

4. Estimating Parameters

New York City, United States is one of the cities with the highest population densities, with a population of over 8 million in 2018 and an area of 784 km^2. Therefore, the service area considered in this paper is restricted to one of its districts; namely, Brooklyn, New York (Figure 2 displays the fixed stations [48]). Most fixed charging stations are located in the north of Brooklyn, with very few being located to the south of it. Most of these are placed in parking lots, those being L2 level type charging stations (95%) (only three being the fast-charging type).

In this section, we estimate some parameters of the temporary placement problem of mobile charging stations based on data from [9]. The database, which we denote by NYC, refers to New York taxi trips in 2013. For each NYC record line we have the following attributes:

- Identity number of the taxi;
- Distance (miles) and trip time (seconds);
- Customer pick-up times and arrival times at destination (day, month, year, hour);
- Pick-up and destination locations of customers (latitude and longitude).

In order to develop the EV market, a solution, even temporary, is needed. This solution, in the case of Brooklyn and similar cases, may be the mobile charging station.

Estimating the parameters of the location problem involves, firstly, building a matrix of distances and trip driving speeds between the intersections of the service area. A first step is to associate each pickup and destination location with an intersection. Using the OpenStreetMap application, the distances from each location to each intersection are calculated. The pickup and destination locations are associated with the closest intersection. For each intersection i, the road distance and the duration of the trip to intersection j, are determined with OpenStreetMap. Based on these distances we determine the driving speed from intersection i to intersection j. In this way, we obtain the driving time from intersection i to intersection j, which we denote by $D_t^M(i,j)$.

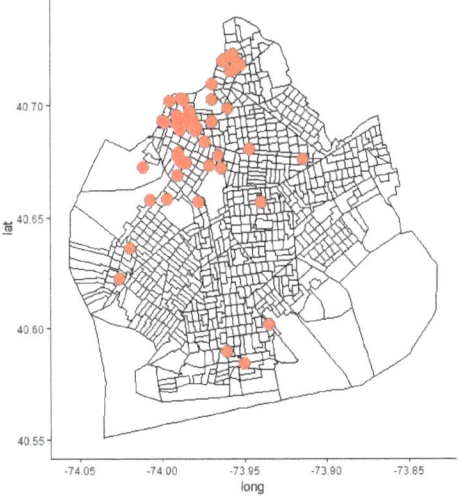

Figure 2. Charging stations in Brooklyn, New York [48].

Because energy consumption is influenced by traffic conditions, we determine how much time the taxi spends in traffic depending on the day of the week and the time of day. This is achieved with TomTom statistics [49]. These statistics tell us how much more time a car spends in traffic per day and hour, as a percentage. On Monday, there is heavy traffic between 15:00 and 16:00 and between 8:00 and 9:00, while on Saturday, there is heavy traffic in the evening between 17:00 and 20:00. Thus, we obtain the trip time from intersection i to intersection j which we denote by $D_t^T(i,j)$.

The idle trip time due to traffic conditions is given by

$$D_t^I(i,j) = D_t^T(i,j) - D_i^M(i,j) \tag{28}$$

Based on trip time, driving time and idle time, we can derive the energy consumed by the taxi for a trip from intersection i to intersection j. Similarly to [50,51], we estimate the energy consumption using a black-box approach using multiple linear regression. Hence, total energy consumption for a trip can be decomposed into moving energy consumption $E_i^M(i,j)$ and auxiliary loading energy consumption $E_t^A(i,j)$:

$$E_t^T(i,j) = E_t^M(i,j) + E_t^A(i,j) \tag{29}$$

$$E_i^M(i,j) = \eta(\alpha_1 v_t(i,j)^2 + \alpha_2 v_t(i,j) + \alpha_3)d_{ij} \tag{30}$$

$$E_t^A(i,j) = l_t D_i^T(i,j)/60 \tag{31}$$

where $v_t(i,j)$ represents the driving speed of a trip from intersection i to intersection j at time t (km/h) and $E_t^T(i,j)$ total energy consumption.

The value of the auxiliary parameter l_t is dependent on the outside temperature, the battery storage capacity of an EV being affected by it. This value can be estimated based on historical weather data from New York City. Similarly to [50], we choose l_t to be between 1 and 1.5.

The parameter η is the aggressive factor of the car driver. As the driving is more aggressive, the EV consumes more energy. In contrast, a calm driver can save from 30% to 40% of the energy consumed in the event of aggressive driving [50]. We take $\eta = 0.8$ (calm driving), $\eta = 1$ (normal driving) and $\eta = 1.2$ (aggressive driving) [50]. Based on [5,50], we have chosen the Nissan Leaf model to be the electric taxi car, for which we take $\alpha_1 = 0.1554$, $\alpha_2 = -5.4634$ and $\alpha_3 = 189.297$.

In order to estimate some parameters of the optimization problem (P2) such as the Poisson intensity f_i, a simulation approach is applied. Therefore, a taxi simulation drive is performed which is based on destination probabilities for pickup intersection i, at time t. An hourly time interval is considered for deriving these probabilities. Using the NYC database, we determine the number of clients at intersection i for moment of time t (denoted by $N_t^P(i)$) and the number of clients who took a taxi from intersection i to intersection j (denoted by $N_t^D(i,j)$). The probability of picking up a client at intersection i, at time t, having as destination intersection j is given by:

$$P_{i/j}^D = \frac{N_t^D(i,j)}{N_t^P(i)} \tag{32}$$

These probabilities are transformed into a discrete distribution for each intersection and moment of time t. Destination intersection for a taxi located at intersection i and moment of time t are chosen according to a corresponding discrete distribution constructed assuming the following:

- The taxi has enough energy to make the trip.
- If for intersection i at time t there is no probability of a destination obtained from the NYC data, then all intersections have the same probability of destination.
- Discrete distributions are constructed so that intersections with high probabilities have greater chances of random choice than others. Intersections for which we did not have a destination probability in the data were considered to have the same probability.
- The taxi picks up a customer as soon as it reaches the next intersection.
- The maximum charge capacity of an electric taxi battery is 30 kW, taking the lower and upper limits as 5% and 95%.

Based on probabilities $P_{i/j}^D$, each trip (i,j) is put into a category based on time intervals of an hour as follows: ≥ 0.5 strong chance of destination, $[0.2, 0.5)$ medium high chance of destination, $(0, 0.2)$ medium chance of destination and 0 low change of destination. The reason for not using just the probabilities $P_{i/j}^D$ is that at some point during the simulation, the program remained in only one intersection i that at a moment of time t had no probability of destination for any intersection j. For each intersection i we count the number of strong, medium-high, medium and low destination intersections. The probability of destination for a strong intersection is given by 0.5/number of strong intersections; for medium high it is 0.4/number of medium high intersections; for medium, 0.3/number of medium intersections; and for a low intersection is what is left to reach 1 over number of low intersections corresponding to pickup intersection i. In this manner, we construct discrete probability of destination for every pickup intersection i.

With these values, we can determine the intensity of charge requests f_i in a certain area within a certain time-frame.

In the analysis, different types of mobile charging stations can be considered. Depending on this, the average serving time differs from a high charge rate to a longer service time.

5. Results and Discussions

Based on simulation performed and estimated parameters, we solved the optimization problem (P2) for different days and time-frames. A number of 200 taxis were simulated starting from different locations, 100 taxis of which started the working day at 6 o'clock, and the other 100 taxis started at 7 o'clock.

Considering that taxi drivers make the decision to recharge their electric taxis when they have less than 30% battery capacity, the intersections where these decisions were made were recorded. Because the number of intersections obtained may be quite large, a hierarchical clustering approach was used to cluster intersections by considering a maximum distance of 4 km from the cluster center [52,53]. We assume that the taxi driver's style is normal ($\eta = 1$). Possible locations for mobile charging stations were randomly chosen so as to cover the entire area of the Brooklyn neighborhood in New York (see Figure 3).

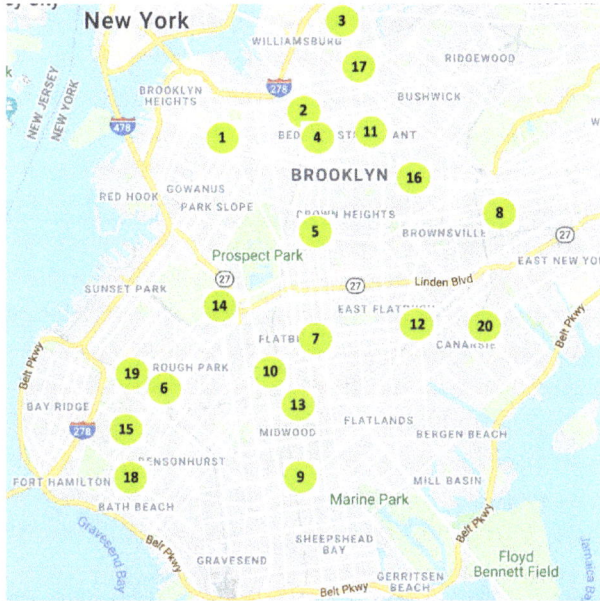

Figure 3. Possible locations for the mobile charging stations in Brooklyn, New York.

Some of the parameters of the optimization problem are as follows: $C = (1, 1, 1, \cdots)$ (meaning, we can only assign a mobile charging station in node $j \in J_1$, and in node $j \in J_2$ we only assume there is one fixed charging station); g is obtained assuming that user charging requirements follow a normal distribution with mean 9 kW and a standard deviation of 1. The reason for choosing this distribution with these values of parameters can be found in [54] by Helmus and Van Den Hoed. They presented results related to the amount of charging of EV users and fitted the data using the normal distribution. Operating cost was assumed constant for all centers j, meaning $P_j = P = 100$; the mean serving rate was $\mu_j \in [1.7 \times 10^{-6}, 1.77]$ for all $j \in J$ (unless otherwise stated) and $B_{max} = 10$.

For different values of the parameters, we have different solutions. Depending on the charger capacity of the mobile charging station, we may need more or fewer charging stations. Tables 3 and 4 shows the results obtained for Monday and Saturday, 9:00–11:00 and 12:00–15:00 time-frames for $\omega_U = 150$.

Saturday and Monday between 9 o'clock and 11 o'clock, we need two mobile charging stations with or without considering probability constraints. However, their locations differ: for situations without probability constraints (locations 12 and 15) and with probability constraints (locations 5 and 12 for constraint $b = 1$ and $\beta = 0.90$, locations 18 and 20 for constraint $a = 1, \alpha = 0.90$ or $a = 1, b = 1, \gamma = 0.90$). There may be cases wherein there is a need for more mobile charging stations just to make the service more efficient. On Monday, 12:00–15:00 time-frame, there is a need for a second mobile charging station, if the service operator decides to make sure it has at least one client with a probability of 0.9. On Saturday, within the same time-frame we require one mobile charging station in location 12 or 13.

Table 3. Results for different days for $\omega_U = 150$ and 9:00–11:00 time period.

	Without Probability Constraints	b = 1, β = 0.90	a = 1, α = 0.90	a = 1, b = 1, γ = 0.90
	Monday			
Solution	569.51	593.92	572.41	572.41
B	2	2	2	2
SoC_{max}	119	132	132	132
SoC_{min}	15	31	10	10
Location mobile charging station	12, 15	5, 12	18, 20	18, 20
Location fixed charging station	28, 55, 60, 61, 64	22, 55, 64	28, 52, 60, 61	28, 52, 60, 61
	Saturday			
Solution	729.5	760.83	733.39	733.39
B	2	2	2	2
SoC_{max}	142	149	142	142
SoC_{min}	26	26	26	26
Location mobile charging station	12, 18	5, 12	18, 20	18, 20
Location fixed charging station	26, 52, 55, 61, 64	22, 49, 55, 62, 64	26, 52, 54, 61	26, 52, 54, 61

Table 4. Results for different days for $\omega_U = 150$ and 12:00–15:00 time period.

	Without Probability Constraints	b = 1, β = 0.90	a = 1, α = 0.90	a = 1, b = 1, γ = 0.90
	Monday			
Solution	460.38	482.12	541.62	541.62
B	1	1	2	2
SoC_{max}	150	150	97	97
SoC_{min}	9	39	9	9
mean(SoC)	52	77	52	52
Location mobile charging station	9	5	15, 20	15, 20
Location fixed charging station	48, 55, 60, 61, 64	51, 55, 64	42, 50, 60, 61	42, 50, 60, 61
	Saturday			
Solution	356.57	368.31	356.98	356.98
B	1	1	1	1
SoC_{max}	142	142	142	142
SoC_{min}	21	21	25	25
mean(SoC)	55	55	73	73
Location mobile charging station	13	12	13	13
Location fixed charging station	55, 57, 59	49, 55, 62	57, 61	57, 61

Different locations are obtained for different probability constraints. As the rate of charge improves, the mean service time decreases and the service is more efficient. The temporary locations of mobile charging stations obtained in Tables 3 and 4 are in areas with no or few fixed charging stations. Most EVs are assigned to fixed charging stations when this is possible. Through computation optimization, EVS are assigned to fixed charging location in the northern part of Brooklyn (Tables 3 and 4, Figure 4) and in this area no mobile charging station is located. It is possible that if in a area there is a sufficient fixed charging infrastructure, then there is no need for a mobile charging station. In this case, the mobile charging station can only act as a car attendant service.

Figure 4. Location of fixed charging stations obtained through computation optimization, Brooklyn, New York.

6. Operational Metrics—Comparative Experiments

In the current operational mode, if we consider dynamic charging constraints, it may be that at some point on the journey to a new client, the mobile charging station may receive a new charge request [11,12], prolonging the service time for the new charge request.

In this section, we compare two operational modes of a mobile charging station considering some metrics of the service mobile charging station defined in [5]. These operational metrics quantify the performance of the operational mode of a mobile charging station. The first operational mode assumes moving the mobile station to where the EVs are located in order to recharge them, and the other one is the one introduced in this paper—the temporary location operational mode. For the first operational mode we assume a nearest-job-next (NJN) strategy, wherein the mobile station stays in the last location where it recharged an EV until a new charge request appears. Additionally, we assume that the mobile charging station has enough charger capacity (kW) to fulfill all charge requests in the service period considered.

One operational metric is the miss ratio index that measures the amount of charge requests that do not get to be satisfied within the service period. It is measured in percentage and it is defined as:

$$P_{miss} = \frac{N_{failed}}{N_{total}} \times 100 \tag{33}$$

where N_{total} represents the number of total requests received during service period, while N_{failed} represents the number of requests that failed to be satisfied during the service period.

Another metric is the response time, which is defined as the time taken from when the EV sends out its charging request to the time when this request is completed. Another metric, which applies only for the second operational mode, is the queuing time, which is defined as the time taken from when the EV arrives to the location of the mobile charging station to the time the charging request is completed. These operational metrics are used to compare the two different operational modes mentioned. For each EV, we assume a charging time of 15 min while the service time is taken to be 9–13:00 on a Monday for NYC database. For moving operational mode, the initial location is in the center of the service area of Brooklyn.

In Table 5 we displayed some empirical results obtained through simulations. Charging requests appear following a Poisson process of parameter λ, while we consider the depot to be in the center of service area. The depot is the initial location of mobile charging station in moving operational mode, and it is the temporary location in the new operational mode. Different dimensions of the service area are taken into account.

Miss ratios in both strategies are similar; however, the miss ratios of temporary location operational are always smaller or equal to those of moving operational mode. We cannot say the same about the response times of the service, which are significantly influenced by the dimensions of the service area, with the response times of moving operational mode being always greater than those for temporary location strategy (large differences can be observed). As the intensity of charge requests λ decreases along with a smaller service area (around 50 km^2), the response times of the two strategies are comparable and we may be able to choose one strategy or the other. However, for moving operational mode, we have the problem of travel distance that increases the cost of operations of a mobile charging station. The travel distance can be substantial even for a small service area.

Results obtained through the NYC database, as shown in the previous section, also suggest that the temporary location operational mode is the better one. Our comparative experiments showed a 36% miss ratio for the moving operational mode (initial location is the center of Brooklyn), while for our mode of operation the miss ratio is only 9% (temporary location is also the center of Brooklyn). This is due to the fact that as the mobile charging station stations and charges EVs, the time it should take the mobile station to reach each EV is taken over by each EV. Therefore, the service becomes more efficient, thereby allowing the mobile charging station to fulfill more charge requests.

Table 5. Operational metrics for moving and temporary location operational modes.

Service Area	Response Time	(minutes)	Miss Ratio		Queuing Time (minutes)	Traveled Distance (km)
Operational mode	Moving	Location	Moving	Location	Location	Moving
$\lambda_1 = 0.00005$						
180 km^2	179	105	0%	0%	15	864
100 km^2	91	51	20%	20%	15	578
50 km^2	37	34	0%	0%	15	192
$\lambda_2 = 0.0001$						
180 km^2	181	84	6.25%	6.25%	24	1108
100 km^2	74	54	0%	0%	17	533
50 km^2	41	33	0%	0%	15	222
$\lambda_3 = 0.0002$						
180 km^2	319	94	19%	0%	24	1276
100 km^2	106	61	23%	23%	17	776
50 km^2	60	36	18.75%	18.75%	17	509

As for the response time, the mean response time for the first mentioned operational mode is 57 min (without the EVs that could not be charged within the service period considered) or an hour and 43 min (considering all charge requests completed). Concerning our temporary location operational mode, the mean response time is 40 min, while the mean queuing time spend by an EV at the location of the mobile charging station is 21 min. Moreover, the total distance needed for the mobile charging station to fulfill all requests in the moving operational mode is 61.94 km or 1 h and 36 min of driving. Furthermore, the travel distance of a mobile charging station may vary due to traffic constraints, weather and temperature [11].

One advantage of the temporary location strategy is the reduced response time of the service. In a survey conducted in 2016 [55] the authors evaluated the different drawbacks that can occur by owning an EV. The results speak by themselves: EV's drivers are willing to make a detour to recharge, but they strongly reject waiting times. Since, for the moving operational time we have such a long period of waiting, this can be a serious issue for EV drivers. Thus, temporary location of mobile charging station may be preferable to moving the mobile charging station from one place to another. Moreover, because EV's drivers do not reject making a detour to recharge, this is an argument in favor of a temporary location strategy.

7. Conclusions

In this paper, a new operational mode for a mobile charging station which assumes temporary locations in some possible places for certain periods of time is introduced. We formulate this problem as a location-allocation problem considering queuing constraints (at most b clients and/or at least a clients).

Our contributions in this paper are summarized as follows:

- We formulated a nonlinear optimization problem and an equivalent mixed-linear optimization problem for optimal temporary location of a mobile charging station.
- We obtained new probability-queuing constraints, considering at least a clients in the queue and/or at most b clients in the queue.
- We compared two operational modes for a mobile charging station.

The key findings of our work are:

- Different temporary locations are obtained for different probability constraints and days of the week.
- The locations obtained are in areas with no or few fixed charging stations.
- The temporary location operational mode compared to moving operational mode has a smaller mean response time.
- Travel distance is an issue in moving operational mode, increasing operational costs, a problem that does not arise in the temporary location strategy.

Of course, these results have been obtained assuming a charging time of less than 30 min, which may be very realistic for the next few years. Future research will also take into account other practical constraints that may be included in the implementation of the charging station based on a fuel cell hybrid power supply [56,57] with appropriate control in current and voltage [58,59]. This paper is a basis for future directions, such as:

- The development of a new strategy in different urban areas to install predictive charging stations;
- Another study in which we intend to use installed charging stations and gather data on the number of cars, the flow of cars in certain time periods and driver behavior given by GPS data;
- A study for increasing the percentage of renewable energy that supplies charging stations; for example, a charging station with PV on the roof;
- Uncertainty analysis with fuzzy intervals;
- Extending the work by considering Markov dependence hypothesis.

Author Contributions: Conceptualization, I.B., V.P. and M.-S.R.; methodology, I.B. and V.P.; software, I.B.; data curation, I.B.; writing—Original draft preparation, I.B. and M.-S.R.; writing—review and editing, M.-S.R., I.B. and N.B.; supervision, V.P. and N.B. All authors have read and agreed to the published version of the manuscript.

Funding: This work was supported by a grant from the Romanian Ministry of Research and Innovation, CCCDI-UEFISCDI, project number PN-IIIP1-1.2-PCCDI-2017-0776/number 36PCCDI/15.03.2018, within PNCDIIII.

Acknowledgments: We thank the editors and the three reviewers for their valuable comments and suggestions that improved the paper. The reviewers gave us many interesting future directions for this work.

Conflicts of Interest: The authors declare no conflict of interest.

References

1. Global EV Outlook 2017 Two Million and Counting. Available online: https://webstore.iea.org/global-ev-outlook-2017 (accessed on 3 August 2019).
2. European Alternative Fuels. Available online: https://www.eafo.eu/countries/european-union/23640/infrastructure/electricity (accessed on 6 March 2020).
3. *Electric Vehicles in Europe*; European Environment Agency: Copenhagen, Denmark 2016.
4. IEC 61851-1 Standard. Available online: https://webstore.iec.ch/publication/33644 (accessed on 5 August 2019).
5. Huang, S.; He, L.; Gu, Y.; Wood, K.; Benjaafar, S. Design of a mobile charging service for electric vehicles in an urban environment. *IEEE Trans. Intell. Transp. Syst.* **2014**, *16*, 787–798. doi:10.1109/TITS.2014.2341695. [CrossRef]
6. Idaho National Laboratory (INL). What Is the Impact of Utility Demand Charges on a DCFC Host? INL/EXT-15-35706. June 2015. Available online: http://avt.inl.gov/pdf/EVProj/EffectOfDemandChargesOnDCFCHosts.pdf (accessed on 23 July 2019).
7. Orlando Utilities Commission (OUC). DC Fast Charging Efforts in Orland. Presentation by OUC. 9 December 2014. Available online: http://www.advancedenergy.org/portal/ncpev/blog/news/wp-content/uploads/2014/12/OUC-presentation-NCPEV-14.pdf (accessed on 15 July 2019).
8. Gong, D.; Tang, M.; Buchmeister, B.; Zhang, H. Solving Location Problem for Electric Vehicle Charging Stations-A Sharing Charging Model. *IEEE Access* **2019**, *7*, 138391–138402. doi:10.1109/ACCESS.2019.2943079. [CrossRef]
9. *Taxicab Fact Book*; NYC Taxi Limousine Commision: New York, NY, USA, 2014. Available online: https://www1.nyc.gov/site/tlc/about/request-data.page (accessed on 12 March 2019).
10. Chen, H.; Su, Z.; Hui, Y.; Hui, H. Dynamic charging optimization for mobile charging stations in Internet of Things. *IEEE Access* **2018**, *6*, 53509–53520. doi:10.1109/ACCESS.2018.2868937. [CrossRef]
11. Cui, S.; Zhao, H.; Zhang, C. Multiple types of plug-in charging facilities' location-routing problem with time windows for mobile charging vehicles. *Sustainability* **2018**, *10*, 2855. doi:10.3390/su10082855. [CrossRef]
12. Cui, S.; Zhao, H.; Chen, H.; Zhang, C. The Mobile Charging Vehicle Routing Problem with Time Windows and Recharging Services. *Comput. Intel. Neurosc.* **2018**. doi:10.1155/2018/5075916. [CrossRef] [PubMed]
13. Chen, H.; Su, Z.; Hui, Y.; Hui, H. Optimal approach to provide electric vehicles with charging service by using mobile charging stations in heterogeneous networks. In Proceedings of the 2016 IEEE 84th Vehicular Technology Conference (VTC-Fall), Montreal, QC, Canada, 18–21 September 2016.
14. Jia, L.; Zechun, H.; Yonghua, S.; Zhuowei, L. Optimal siting and sizing of electric vehicle charging stations. In Proceedings of the 2012 IEEE International Electric Vehicle Conference, Greenville, SC, USA, 4–8 March 2012.
15. Dong, J.; Liu, C.; Lin, Z. Charging infrastructure planning for promoting battery electric vehicles: An activity-based approach using multiday travel data. *Transp. Res. Part C Emerg. Technol.* **2014**, *38*, 44–55. doi:10.1016/j.trc.2013.11.001. [CrossRef]
16. Zhu, Z.H.; Gao, Z.Y.; Zheng, J.F.; Du, H.M. Charging station location problem of plug-in electric vehicles. *J. Transp. Geogr.* **2016**, *52*, 11–22. doi:10.1016/j.jtrangeo.2016.02.002. [CrossRef]
17. Csiszár, C.; Csonka, B.; Földes, D.; Wirth, E.; Lovas, T. Urban public charging station locating method for electric vehicles based on land use approach. *J. Transp. Geogr.* **2019**, *74*, 173–180. doi:10.1016/j.jtrangeo.2018.11.016. [CrossRef]
18. Xiong, Y.; Gan, J.; An, B.; Miao, C.; Bazzan, A.L. Optimal electric vehicle fast charging station placement based on game theoretical framework. *IEEE trans. Intell. Transp. Syst.* **2017**, *19*, 2493–2504. doi:10.1109/TITS.2017.2754382. [CrossRef]
19. Shukla, A.; Verma, K.; Kumar, R. Planning of Fast Charging Stations in Distribution System Coupled with Transportation Network for Capturing EV flow. In Proceedings of the 2018 8th IEEE India International Conference on Power Electronics (IICPE), Jaipur, India, 13–15 December 2018.
20. Lu, F.; Hua, G. A location-sizing model for electric vehicle charging station deployment based on queuing theory. In Proceedings of the 2015 International Conference on Logistics, Informatics and Service Sciences (LISS), Barcelona, Spain, 27–29 July 2015.

21. Said, D.; Cherkaoui, S.; Khoukhi, L. Multi-priority queuing for electric vehicles charging at public supply stations with price variation. *Wirel. Commun. Mob. Com.* **2015**, *15*, 1049–1065. doi:10.1002/wcm.2508. [CrossRef]
22. Zhu, J.; Li, Y.; Yang, J.; Li, X.; Zeng, S.; Chen, Y. Planning of electric vehicle charging station based on queuing theory. *J. Eng.* **2017**, *13*, 1867–1871. doi:10.1049/joe.2017.0655. [CrossRef]
23. Pierre, M.; Jemelin, C.; Louvet, N. Driving an electric vehicle. A sociological analysis on pioneer users. *Energy Effic.* **2011**, *4*, 511. doi:10.1007/s12053-011-9123-9. [CrossRef]
24. Hidrue, M.K.; Parsons, G.R.; Kempton, W.; Gardner, M.P. Willingness to pay for electric vehicles and their attributes. *Resour Energy Econ.* **2011**, *33*, 686–705. doi:10.1016/j.reseneeco.2011.02.002. [CrossRef]
25. Procopiou, A.T.; Quirós-Tortós, J.; Ochoa, L.F. HPC-based probabilistic analysis of LV networks with EVs: Impacts and control. *IEEE Trans. Smart Grid* **2016**, *8*, 1479–1487. doi:10.1109/TSG.2016.2604245. [CrossRef]
26. Kattmann, C.; Rudion, K.; Tenbohlen, S. Detailed power quality measurement of electric vehicle charging infrastructure. *CIRED-Open Access Proc. J.* **2017**, *2017*, 581–584. [CrossRef]
27. Hou, K.; Xu, X.; Jia, H.; Yu, X.; Jiang, T.; Zhang, K.; Shu, B. A reliability assessment approach for integrated transportation and electrical power systems incorporating electric vehicles. *IEEE Trans. Smart Grid* **2016**, *9*, 88–100. doi:10.1109/TSG.2016.2545113. [CrossRef]
28. Gjelaj, M.; Toghroljerdi, S.H.; Andersen, P.B.; Træholt, C. Optimal Infrastructure Planning for EVs Fast Charging Stations based on Prediction of User Behavior. *IET Electr. Syst. Transp.* **2019**, *10*, 1–12. [CrossRef]
29. Ding, Z.; Lu, Y.; Zhang, L.; Lee, W.J.; Chen, D. A Stochastic Resource-Planning Scheme for PHEV Charging Station Considering Energy Portfolio Optimization and Price-Responsive Demand. *IEEE Trans. Ind. Appl.* **2018**, *54*, 5590–5598. doi:10.1109/TIA.2018.2851205. [CrossRef]
30. He, F.; Wu, D.; Yin, Y.; Guan, Y. Optimal deployment of public charging stations for plug-in hybrid electric vehicles. *Transport. Res. B-Meth.* **2013**, *47*, 87–101. doi:10.1016/j.trb.2012.09.007. [CrossRef]
31. Asamer, J.; Reinthaler, M.; Ruthmair, M.; Straub, M.; Puchinger, J. Optimizing charging station locations for urban taxi providers. *Transp. Res. Part Policy Pract.* **2016**, *85*, 233–246. [CrossRef]
32. Frade, I.; Ribeiro, A.; Gonçalves, G.; Antunes, A.P. Optimal location of charging stations for electric vehicles in a neighborhood in Lisbon, Portugal. *Transp. Res. Rec.* **2011**, *2252*, 91–98. [CrossRef]
33. He, Y.; Kockelman, K.M.; Perrine, K.A. Optimal locations of US fast charging stations for long-distance trip completion by battery electric vehicles. *J. Clean. Prod.* **2019**, *214*, 452–461. [CrossRef]
34. Wang, C.; He, F.; Lin, X.; Shen, Z.J.M.; Li, M. Designing locations and capacities for charging stations to support intercity travel of electric vehicles: An expanded network approach. *Transp. Res. Part Emerg. Technol.* **2019**, *102*, 210–232. [CrossRef]
35. Davidov, S.; Pantoš, M. Planning of electric vehicle infrastructure based on charging reliability and quality of service. *Energy* **2017**, *118*, 1156–1167. [CrossRef]
36. Hu, D.; Liu, Z.W.; Chi, M. Multiple Periods Location and Capacity Optimization of Charging Stations for Electric Vehicle. In Proceedings of the 2019 China-Qatar International Workshop on Artificial Intelligence and Applications to Intelligent Manufacturing (AIAIM), Doha, Qatar, 1–4 January 2019.
37. Rahman, I.; Vasant, P.M.; Singh, B.S.M.; Abdullah-Al-Wadud, M.; Adnan, N. Review of recent trends in optimization techniques for plug-in hybrid, and electric vehicle charging infrastructures. *Renewable Sustainable Energy Rev.* **2016**, *58*, 1039–1047. doi:10.1016/j.rser.2015.12.353. [CrossRef]
38. Liu, Z.F.; Zhang, W.; Ji, X.; Li, K. Optimal planning of charging station for electric vehicle based on particle swarm optimization. In Proceedings of the IEEE PES Innovative Smart Grid Technologies, Tianjin, China, 21–24 May 2012.
39. Soares, J.; Sousa, T.; Morais, H.; Vale, Z.; Canizes, B.; Silva, A. Application-specific modified particle swarm optimization for energy resource scheduling considering vehicle-to-grid. *Appl. Soft. Comput.* **2013**, *13*, 4264–4280. doi:10.1016/j.asoc.2013.07.003. [CrossRef]
40. Awasthi, A.; Venkitusamy, K.; Padmanaban, S.; Selvamuthukumaran, R.; Blaabjerg, F.; Singh, A.K. Optimal planning of electric vehicle charging station at the distribution system using hybrid optimization algorithm. *Energy* **2017**, *133*, 70–78. doi:10.1016/j.energy.2017.05.094. [CrossRef]
41. Xu, S.; Feng, D.; Yan, Z.; Zhang, L.; Li, N.; Jing, L.; Wang, J. Ant-based swarm algorithm for charging coordination of electric vehicles. *Int. J. Distrib. Sens. Netw.* **2013**. doi:10.1155/2013/268942. [CrossRef]

42. Fazelpour, F.; Vafaeipour, M.; Rahbari, O.; Rosen, M.A. Intelligent optimization to integrate a plug-in hybrid electric vehicle smart parking lot with renewable energy resources and enhance grid characteristics. *Energy Convers. Manag.* **2014**, *77*, 250–261. doi:10.1016/j.enconman.2013.09.006. [CrossRef]
43. Yan, X; Duan, C.; Chen, X.; Duan, Z. Planning of Electric Vehicle charging station based on hierarchic genetic algorithm. In Proceedings of the 2014 IEEE Conference and Expo Transportation Electrification Asia-Pacific (ITEC Asia-Pacific), Beijing, China, 31 August–3 September 2014.
44. Marianov, V.; Serra, D. Location–allocation of multiple-server service centers with constrained queues or waiting times. *Ann. Oper. Res.* **2002**, *111*, 35–50. doi:10.1023/A:1020989316737. [CrossRef]
45. Marianov, V.; Serra, D. Probabilistic, maximal covering location—allocation models forcongested systems. *J. Reg. Sci.* **1998**, *38*, 401–424. doi:10.1111/0022-4146.00100. [CrossRef]
46. Sayarshad, H.R.; Chow, J.Y. Non-myopic relocation of idle mobility-on-demand vehicles as a dynamic location-allocation-queueing problem. *Transport. Res. E-Log.* **2017**, *106*, 60–77. doi:10.1016/j.tre.2017.08.003. [CrossRef]
47. Wolff, R.W. *Stochastic Modeling and the Theory of Queues*; Prentice Hall: Englewood Cliffs, NJ, USA, 1989; Volume 14.
48. *Electric Vehicle Charging Stations in New York*; New York Government: New York, NY, USA, 2017. Available online: https://data.ny.gov/Energy-Environment/Electric-Vehicle-Charging-Stations-in-New-York/7rrd-248n (accessed on 16 June 2019).
49. TomTom Statistics. June 2019. https://www.tomtom.com/engb/traffic-index/new-york-traffic (accessed on 16 June 2019).
50. Tseng, C.M.; Chau, S.C.K.; Liu, X. Improving viability of electric taxis by taxi service strategy optimization: A big data study of New York city. *IEEE Trans. Intell. Transp. Syst.* **2018**, *99*, 1–13. doi:10.1109/TITS.2018.2839265. [CrossRef]
51. Tseng, C.M.; Chau, C.K. Personalized prediction of vehicle energy consumption based on participatory sensing. *IEEE Trans. Intell. Transp. Syst.* **2017**, *18*, 3103–3113. doi:10.1109/TITS.2017.2672880. [CrossRef]
52. Müllner D. fastcluster: Fast Hierarchical, Agglomerative Clustering Routines for R and Python. *J. Stat. Softw.* **2013**, *53*, 1–18. [CrossRef]
53. R Core Team. *R: A Language and Environment for Statistical Computing*; R Foundation for Statistical Computing: Vienna, Austria, 2018. Available online: https://www.R-project.org/ (accessed on 12 February 2019).
54. Helmus, J.; Van Den Hoed, R. Unravelling user type characteristics: Towards a taxonomy for charging infrastructure. *World Electr. Veh. J.* **2015**, *7*, 589–604. [CrossRef]
55. Philipsen, R.; Schmidt, T.; Van Heek, J.; Ziefle, M. Fast-charging station here, please! User criteria for electric vehicle fast-charging locations. *Transp. Res. Part F Traffic Psychol. Behav.* **2016**, *40*, 119–129. doi:10.1016/j.trf.2016.04.013. [CrossRef]
56. Bizon, N. Energy efficiency for the multiport power converters architectures of series and parallel hybrid power source type used in plug-in/V2G fuel cell vehicles. *Appl. Energy* **2013**, *102*, 726–734. doi:10.1016/j.apenergy.2012.08.021. [CrossRef]
57. Bizon, N. Energy Efficiency of Multiport Power Converters used in Plug-In/V2G Fuel Cell Vehicles. *Appl. Energy* **2012**, *96*, 431–443. doi:10.1016/j.apenergy.2012.02.075. [CrossRef]
58. Bizon, N. Nonlinear control of fuel cell hybrid power sources: Part II –Current control. *Appl. Energy* **2011**, *88*, 2574–2591. doi:10.1016/j.apenergy.2011.01.044. [CrossRef]
59. Bizon, N. Nonlinear control of fuel cell hybrid power sources: Part I –Voltage control. *Appl. Energy* **2011**, *88*, 2559–2573. doi:10.1016/j.apenergy.2011.01.030. [CrossRef]

© 2020 by the authors. Licensee MDPI, Basel, Switzerland. This article is an open access article distributed under the terms and conditions of the Creative Commons Attribution (CC BY) license (http://creativecommons.org/licenses/by/4.0/).

www.ingramcontent.com/pod-product-compliance
Lightning Source LLC
LaVergne TN
LVHW070456100526
838202LV00014B/1733

MDPI
St. Alban-Anlage 66
4052 Basel
Switzerland
Tel. +41 61 683 77 34
Fax +41 61 302 89 18
www.mdpi.com

Mathematics Editorial Office
E-mail: mathematics@mdpi.com
www.mdpi.com/journal/mathematics